Topics in Soliton Theory and Exactly Solvable Nonlinear Equations

Topics in Soliton Theory and Exactly Solvable Nonlinear Equations

edited by **M. Ablowitz**
(Clarkson Univ., Potsdam NY)

B. Fuchssteiner
(Univ. of Paderborn)

M. Kruskal
(Princeton Univ.)

Proceedings of the Conference on Nonlinear Evolution Equations, Solitons and the Inverse Scattering Transform

Oberwolfach, Germany
July 27–August 2, 1986

World Scientific

Published by

World Scientific Publishing Co. Pte. Ltd.
5 Toh Tuck Link, Singapore 596224
USA office: 27 Warren Street, Suite 401-402, Hackensack, NJ 07601
UK office: 57 Shelton Street, Covent Garden, London WC2H 9HE

British Library Cataloguing-in-Publication Data
A catalogue record for this book is available from the British Library.

TOPICS IN SOLITON THEORY AND EXACTLY SOLVABLE NONLINEAR EQUATIONS

ISBN-13 978-9971-5-0253-9
ISBN-10 9971-5-0253-4

PREFACE

The conference on Nonlinear Evolution Equations, Solitons and the Inverse Scattering Transform took place in Mathematische Forschungsinstitut in Oberwolfach, a small village in a beautiful rural area of southern Germany. It was the first conference at this charming place which was solely dedicated to this area of mathematics and physics. The conference took place during the period July 27 — August 2, 1986.

The conference was organized by Mark J. Ablowitz, Benno Fuchssteiner and Martin Kruskal, all of whom served as editors of these proceedings.

The participants came from a variety of European countries (France, Italy, the Netherlands and Germany), from North America, Japan and China. It was regretted that none of the invited researchers of the Soviet Union were able to take part in the meeting.

One of the aims of the organizers was to bring together some of the most active researchers in this rapidly moving field which now has entered its third decade. This subject continues to be very popular amongst physicists and mathematicians alike. The conference brought together thirty mathematicians and physicists interested in various aspects of soliton theory and their applications.

The lecture program contained both expository and technical talks and was supplemented, in accordance with the Oberwolfach tradition, by numerous informal discussions.

The program consisted of 29 lectures. The subjects covered by these lectures include Soliton Theory and the Inverse Scattering Transform 1+1 and 2+1 dimensions, Topics in Pure Inverse Scattering Theory, Symmetry and Nonlinear Equations, Theory of Recursion Operators in 1+1 and 2+1 Dimensions, Integrable Dynamical Systems, Novel Exactly Solvable Nonlinear Equations, Geometrical Aspects, Direct Linearizations, Painlevé Analysis, Computer Algebra Algorithms and Various Applications.

These proceedings contain essentially the same topics. This volume does not exactly reflect the program of the conference since the authors were encouraged to expand and update their contributions and to include additional details as well as results not presented at the meeting. Some authors even took the liberty to completely change the subject of their contribution.

Some of the most recent and significant developments in the field are contained in these proceedings. We certainly hope that the range of applications and new mathematical and physical insights will be of interest to a wide range of researchers. The interested reader will discover a wealth of information in this book which, we believe, accurately reflects the state of the art.

We are particularly pleased to thank the Mathematische Forschungsinstitut for providing the unique atmosphere responsible for an intense and successful meeting.

We are indebted to the authors for the hard work they invested in their contribution, and we would like to express our appreciation to World Scientific Publishing for putting this volume in print.

TABLE OF CONTENTS

SYMMETRIES AND BI-HAMILTONIAN STRUCTURES OF 2+1 DIMENSIONAL SYSTEMS

P.M.Santini[+] and A.S.Fokas[°]

[+]Dipartimento di Fisica, Università di Roma I, 00185 Roma.
I.N.F.N. Sezione di Roma.

[°]Department of Mathematics,Clarkson University,Potsdam,N.Y.13676,USA.

ABSTRACT

Thetheory associated with the recursion operators of classes of integrable nonlinear evolution equations in 2+1 dimensions is summarized. In particular the notions of symmetries, gradients of conserved quantities, strong and hereditary symmetries, Hamiltonian operators are generalized to equations in multidimensions. Applications to physically relevant equations like the Kadomtsev-Petviashvili equation are illustrated. Integro-differential evolution equations like the Benjamin-Ono equation are shown to be also described by this generalized theory.

1. INTRODUCTION

In recent years a great progress in the understanding of the algebraic-geometrical structure of integrable nonlinear evolution equations in 1+1 dimensions has been achieved. A central rôle in the theory is played by an integro-differential operator Φ, the recursion operator. Φ possesses an algebraic-geometrical property, called hereditary [1] or Nijenhuis [2] property, and then it generates commuting symmetries. Moreover: i) its adjoint Φ^* maps gradients of conserved quantities into gradients of conserved quantities; ii) Φ admits a symplectic-cosymplectic factorization and then it generates constants of motion in involution [3]; iii) Φ times the first Hamiltonian $\Theta^{(1)}$ is the second Hamiltonian $\Theta^{(2)}$ [2], then the associated evolution equations have a bi-Hamiltonian nature, a fundamental property underlying integrability [4].

The general theory of recursion operators and their connection to a bi-Hamiltonian formulation in 1+1 dimensions has been developed by Magri [2], Gel'fand and Dorfman [5] and Fokas and Fuchssteiner [3].

An analogous theory for integrable evolution equations in 2+1 dimensions has not been developed prior to our work. In particular no example of multidimensional recursion operator was known (apart from very special cases, like the 2+1 dimensional Burgers equation, that can be linearized through a generalized Cole-Hopf transformation [6]). Various and interesting efforts to obtain recursion operators in 2+1 dimensions were made for instance in [7,8], and essentially showed that recursion operators of a certain form do not exist. It is interesting to notice that a similar situation was known for integro-differential equations in 1+1 dimensions [9], like the Benjamin-Ono equation

$$q_t = Hq_{xx} + 2qq_x, \quad Hf(x) \doteq \pi^{-1} \fint_{-\infty}^{\infty} dx'(x'-x)^{-1} f(x'). \quad (1.1)$$

In this case, in order to bypass the absence of the recursion operator, an alternative approach, the so-called master-symmetries approach, was introduced to generate commuting symmetries [9]. Such an approach has been later successfully used in 2+1 [10] and in 1+1 [12] dimensions, as well as for finite dimensional systems [11].

After the discovery[13] of the recursion operator Φ_{12} and of the two Hamiltonian operators $\Theta_{12}^{(1)}$ and $\Theta_{12}^{(2)}$ associated with the Kadomtsev-Petviashvili (KP) equation

$$q_t = q_{xxx} + 6qq_x + 3\alpha^2 D^{-1} q_{yy}, \tag{1.2}$$

we have developed the theory associated with these operators[14,15] and applied it to two explicit examples: the classes of evolution equations associated with (1.2) and with the Davey-Stewardson equation

$$iq_t + \frac{1}{2}(q_{xx} + \alpha^2 q_{yy}) = q(u - |q|^2),\ u_{xx} - \alpha^2 u_{yy} = 2|q|^2_{xx}. \tag{1.3}$$

In particular, in Ref.14, i) we present a systematic derivation of the recursion operator Φ_{12} from the underlying spectral problem, ii) we show that Φ_{12} generates (what we call) extended symmetries $\sigma_{12}^{(m)}$ and its adjoint generates extended gradients $\gamma_{12}^{(m)}$ of conserved quantities, iii) we show that $\sigma_{12}^{(m)}(q_1,q_2) = 0$ are Bäcklund Transformations (BT) and $\sigma_{11}^{(m)} \doteq \sigma_{12}^{(m)}\big|_{y_2 = y_1}$ and $\gamma_{11}^{(m)} \doteq \gamma_{12}^{(m)}\big|_{y_2 = y_1}$ are, respectively, commuting symmetries and gradients of conserved quantities in involution for the associated class of equations. In order to deal with the "extended" objects of this theory, i) we introduce a novel (operator) directional derivative, ii) we deal with a Lie algebra of operators, as opposed to a Lie algebra of functions. In Ref.15, exploiting the richness of this operator Lie algebra, i) we generate, via Φ_{12}, time-dependent symmetries, ii) we use an isomorphism between Lie and Poisson brackets to show that all these symmetries correspond to extended gradients, and hence give rise to conserved quantities, iii) we show that the already known master-symmetries are connected to these symmetries, and, since they correspond to gradients, they cannot be used to generate Φ_{12}; iv) we find a non-gradient master-symmetry that can be used to generate Φ_{12}, v) we finally motivate and verify some of the aspects of this generalized theory by establishing that equations in 2+1 dimensions are <u>exact reductions</u> of certain nonlocal evolution equations, whose algebraic properties are rather straightforward.

In this note we essentially present a summary of the results contained in [13,14,15], referring to these papers for proofs and details. We additionally exhibit the recursion operator of the Benjamin-Ono (BO) class, showing that integro-differential equations in 1+1 dimensions fit into this generalized theory [16].

2. ALGEBRAIC PROPERTIES IN 2+1 DIMENSIONS

We consider evolution equations of the form

$$q_t = K(q),$$

where q is an element of some space S of functions on the plane vanishing rapidly for $|x|, |y| \to \infty$, and K is some differentiable map on this space (for convenience independent of x,y,t). We use the KP equation (1.2) as an illustrative example.

2.1 Representations Of Integrable Equations In 2+1 Dimensions

Integrable equations in 2+1 dimensions belong to some hierarchy (generated by a recursion operator Φ_{12}); <u>fundamental</u> in our theory is to write these equations in the form

$$q_{1_t} = \beta_n \int_{\mathbb{R}} dy_2 \delta_{12} \Phi_{12}^n \mathcal{R}_{12}^0 \cdot 1 \doteq \beta_n \int_{\mathbb{R}} dy_2 \delta_{12} K_{12}^{(n)} \doteq K_{11}^{(n)}, \tag{2.1}$$

where $\delta_{12} = \delta(y_1 - y_2)$ denotes the Dirac delta function, $q_i = q(x, y_i, t)$, $i = 1,2$, $K_{12}^{(n)} = K^{(n)}(q_1, q_2) \in \tilde{S}$ and Φ_{12}, $\mathcal{R}_{12}^0$ are operator valued functions on $\tilde{S}$. Through this paper m and n are non-negative integers.

For example the recursion operator associated with the KP class is

$$\Phi_{12} = D^2 + q_{12}^+ + Dq_{12}^+ D^{-1} + q_{12}^- D^{-1} q_{12}^- D^{-1}, \quad q_{12}^{\pm} = q_1 \pm q_2 + \alpha(D_1 \mp D_2), \tag{2.2}$$

where $D = \partial/\partial_x$, $D^{-1} = \int_{-\infty}^{x} dx'$ and $D_i = \partial/\partial_{y_i}$, $i = 1,2$. This operator generates two classes of evolution equations corresponding to two different starting operators $\mathcal{R}_{12}^0$, given by

$$\mathcal{N}_{12} = q_{12}^-, \quad \mathcal{M}_{12} = Dq_{12}^+ + q_{12}^- D^{-1} q_{12}^-. \tag{2.3}$$

Then the KP equation is obtained from (2.1) for n = 1, $\beta_1 = 1/2$ and $\mathcal{R}_{12}^0 = \mathcal{M}_{12}$, while the first two non trivial equations of the second class ($\mathcal{R}_{12}^0 = \mathcal{N}_{12}$)

are

$$q_t = \alpha q_y, \quad n = 1, \ \beta_1 = 1/4, \tag{2.4a}$$

$$q_t = \alpha(q_{xxy} + 4qq_y + 2q_x D^{-1} q_y + \alpha^2 D^{-2} q_{yyy}), \ n=2, \ \beta_2 = 1/8. \tag{2.4b}$$

The recursion operator Φ_{12} enjoys a simple commutator relation with $h_{12} = h(y_1 - y_2)$:

$$[\Phi_{12}, h_{12}] = -\beta h'_{12}, \quad h'_{12} \doteq \partial h_{12}/\partial y_1, \tag{2.5}$$

which implies that $\delta_{12} K_{12}^{(n)} \doteq \delta_{12} \Phi_{12}^n \hat{R}_{12}^0 \cdot 1 = \sum_{\ell=0}^{n} \beta^\ell \binom{n}{\ell} \Phi_{12}^{n-\ell} \delta_{12} \hat{R}_{12}^0 \cdot 1$, where $\delta_{12}^\ell \doteq \partial^\ell \delta_{12}/\partial y_1^\ell$. Moreover the commutation relation of $\hat{R}_{12}^0$ with h_{12} implies that $\delta_{12} K_{12}^{(n)}$ can be written in the following <u>alternative form</u>

$$\delta_{12} K_{12}^{(n)} = \sum_{\ell=0}^{n} b_{n,\ell} \Phi_{12}^{n-\ell} \hat{R}_{12}^0 \cdot \delta_{12}^\ell \ . \tag{2.6}$$

For the two classes of equations associated with the KP equation we have that

$$\beta = -4\alpha, \quad [\hat{N}_{12}, h_{12}] = 0, \ [\hat{M}_{12}, h_{12}] = -\tilde{\beta} D h'_{12}, \ \tilde{\beta} \doteq \beta/2,$$

$$b_{n,\ell} \doteq \begin{cases} \beta^\ell \binom{n}{\ell}, \ \text{if } \hat{R}_{12}^0 = \hat{N}_{12}, \\ \sum_{s=0}^{\ell} \beta^{\ell-s} \tilde{\beta}^s \binom{n-s}{\ell-s}, \ \text{if } \hat{R}_{12}^0 = \hat{M}_{12}. \end{cases} \tag{2.7}$$

2.2 A New Directional Derivative

In ref.s 14,15 we present a systematic approach to derive recursion operators and classes of integrable 2+1 dimensional equations in the form (2.1) from the underlying spectral problems. This derivation is based on the use of integral representations of operators depending on q and ∂/∂_y. In the KP case, for example, the basic operator

$$\hat{q} = q + \partial/\partial_y \tag{2.8}$$

appearing in the underlying spectral problem

$$w_{xx} + \hat{q} w = 0, \tag{2.9}$$

is represented by

$$\hat{q}_1 f_{12} = (q_1 + \alpha D_1) f_{12} \doteq \int_{\mathbb{R}} dy_3 q_{13} f_{32}, \qquad (2.10a)$$

where $g_{ij} \doteq g(x, y_i, y_j, t)$, $i,j = 1,2,3$.

The above mapping between an operator and its kernel induces a mapping between derivatives:

$$\hat{q}_{1_d} [g_{12}] f_{12} = \int_{\mathbb{R}} dy_3 g_{13} f_{32}, \qquad (2.10b)$$

where $\hat{q}_{1_d} [g_{12}]$ denotes the directional d-derivative of the operator valued function $\hat{q}_1$ in the direction g_{12}. Using an appropriate bilinear form (see (4.3a)), it is possible to obtain the d-derivative of the adjoint $\hat{q}_1^* = q_2 - \alpha D_2$ of $\hat{q}_1$ and, consequently, of the basic operators $q_{12}^{\pm} = \hat{q}_1 \pm \hat{q}_1^*$:

$$q_{12_d}^{\pm} [g_{12}] f_{12} = \int_{\mathbb{R}} dy_3 (g_{13} f_{32} \pm f_{13} g_{32}). \qquad (2.11)$$

Since Φ_{12} and $\mathcal{R}_{12}^0$ are polynomials in $q_{12}^{\pm}$, their directional derivatives $\Phi_{12_d} [g_{12}]$ and $\mathcal{R}_{12_d}^0 [g_{12}]$ are well defined.

The <u>d-derivative</u> is a <u>novel aspect</u> of the theory in 2+1 dimensions; its connection with the usual Fréchet derivative (hereafter indicated by the subscript f) is given by the following projective formula

$$K_{12_d} [\delta_{12} g_{12}] = K_{12_f} [g_{12}] \doteq K_{12_{q_1}} [g_{11}] + K_{12_{q_2}} [g_{22}], \qquad (2.12)$$

where K_{12} is an arbitrary function in $\tilde{S}$ and $K_{12_{q_i}}$ denotes the Fréchet derivative of K_{12} with respect to q_i, i.e.

$$K_{12_{q_i}} [g_{ii}] = \frac{\partial}{\partial \varepsilon} K_{12}(q_i + \varepsilon g_{ii}, q_j) \big|_{\varepsilon = 0}, \quad i,j=1,2;\ i \neq j. \qquad (2.13)$$

Operator valued functions on $\tilde{S}$ for which d-derivatives are defined are called admissible.

2.3 The Lie Algebra Of The Starting Symmetries

The starting symmetries K^0_{12} are written as $\hat{K}^0_{12}\cdot 1$, where $\hat{K}^0_{12}$ are admissible operators. Crucial aspect of this theory is that the operators $\hat{K}^0_{12}$, acting on suitable functions H_{12} belonging to the Ker of the first Hamiltonian operator $\Theta^{(1)}_{12}$ (i.e. such that $\Theta^{(1)}_{12}H_{12}=0$), form a Lie algebra. Then, with respect to the 1+1 dimensional case characterized by a Lie algebra of functions, equations in 2+1 dimensions present a richer algebraic structure, characterized by a Lie algebra of operators.

For the equations associated with the KP equation, H_{12} is an arbitrary function independent of x, i.e.

$$H_{12}= H(y_1,y_2), \tag{2.14}$$

and the Lie algebra of the starting operators $\hat{N}_{12}$ and $\hat{M}_{12}$ is given by

$$[\hat{N}_{12}H^{(1)}_{12}, \hat{N}_{12}H^{(2)}_{12}]_d = -\hat{N}_{12}H^{(3)}_{12}, \quad [\hat{N}_{12}H^{(1)}_{12}, \hat{M}_{12}H^{(2)}_{12}]_d = -\hat{M}_{12}H^{(3)}_{12},$$

$$[\hat{M}_{12}H^{(1)}_{12}, \hat{M}_{12}H^{(2)}_{12}]_d = -\Phi_{12}\hat{N}_{12}H^{(3)}_{12}, \tag{2.15}$$

in terms of the Lie brackets $[,]_d$, $[\,,\,]_I$, defined by

$$[K^{(1)}_{12}, K^{(2)}_{12}]_d \doteq K^{(1)}_{12_d}[K^{(2)}_{12}] - K^{(2)}_{12_d}[K^{(1)}_{12}], \tag{2.16a}$$

$$H^{(3)}_{12} \doteq [H^{(1)}_{12}, H^{(2)}_{12}]_I \doteq \int_{\mathbb{R}} dy_3(H^{(1)}_{13}H^{(2)}_{32} - H^{(2)}_{13}H^{(1)}_{32}). \tag{2.16b}$$

3. SYMMETRIES AND HEREDITARY SYMMETRIES

3.1 The Notion Of Extended Symmetries, Connection To Symmetries And Bäcklund Transformations

The recursion operator Φ_{12} generates a sequence of functions $\sigma^{(m)}_{12} = \sigma^{(m)}(q_1,q_2) \in \tilde{S}$ defined by

$$\sigma^{(m)}_{12} \doteq \Phi^m_{12}\hat{K}^0_{12}\cdot 1. \tag{3.1}$$

These functions live in the extended space $\tilde{S}$ and, in order to give them a characterization and establish their connection to the integrable evolution equation

$$q_{1_t} = \int_{\mathbb{R}} dy_2\delta_{12}K_{12} = K_{11}, \tag{3.2}$$

the following definition is introduced.

Definition 3.1

The function $\sigma_{12} \in \tilde{S}$ is called extended symmetry of equation (3.2) iff

$$\sigma_{12\,f}[K] = (\delta_{12}K_{12})_d[\sigma_{12}]. \tag{3.3}$$

Remark 3.1

i) Definition (3.3) makes sense only if $(\delta_{12}K_{12})_d$ exists; for equations $q_{1_t} = \int_{\mathbb{R}} dy_2\, \delta_{12}K_{12}^{(n)} = K_{11}^{(n)}$ generated by Φ_{12}, $(\delta_{12}K_{12})_d$ is well defined and reads

$$(\delta_{12}K_{12}^{(n)})_d = (\delta_{12}\Phi_{12}^n\, \mathring{K}_{12}\cdot 1)_d = \sum_{\ell=0}^{n} b_{n,\ell}\,(\Phi_{12}^{n-\ell}\mathring{K}_{12}\cdot\delta_{12}^{\ell})_d, \tag{3.4}$$

where equation (2.6) is clearly used to write $\delta_{12}K_{12}^{(n)}$ in admissible form. If, for example, $\mathring{K}_{12} = \hat{M}_{12}$ and n=0 and 1, we have

$$(\delta_{12}K_{12}^{(0)})_d = 2D, \tag{3.5a}$$

$$(\delta_{12}K_{12}^{(1)})_d = 2(D^3+6D(q_1+q_2)-3\alpha(D^{-1}(q_{1_{y_1}}-q_{2_{y_2}}))+6\alpha(q_1-q_2)D^{-1}(D_1+D_2)+$$
$$+6\alpha^2 D^{-1}(D_1+D_2)^2). \tag{3.5b}$$

ii) The projective property (2.12) implies that σ_{12} is an extended symmetry of (3.2) iff σ_{12} commutes with $\delta_{12}K_{12}$, namely iff

$$[\sigma_{12}, \delta_{12}K_{12}]_d = 0. \tag{3.6}$$

iii) In the above definition we assume that σ_{12} does not explicitely depend on t, otherwise $\sigma_{12\,f}[K]$ should be replaced by $\partial\sigma_{12}/\partial t + \sigma_{12f}[K]$.

The usefulness of extended symmetries follows from the fact that they give rise to symmetries and Bäklund transformations. Precisely we have the following

Theorem 3.1

If σ_{12} is an extended symmetry of equation (3.2), then

i) $\sigma_{11} = \sigma_{12}|_{y_2=y_1}$ is a symmetry of equation (3.2), namely

$$\sigma_{11_f}[K_{11}] = K_{11_f}[\sigma_{11}]; \tag{3.7a}$$

ii) the equation

$$\sigma_{12} = \sigma(q_1,q_2) = 0 \tag{3.7b}$$

is a Bäcklund transformation for (3.2) where, of course, q_1 and q_2 are

are now viewed as two different solutions of (3.2).

3.2 Strong And Hereditary Symmetries

The introduction of the d-derivative allows a natural generalization of the notions of strong and hereditary symmetries in 2+1 dimensions; precisely we have the following

Definition 3.2

i) The admissible operator valued function Φ_{12} is a strong symmetry for A_{12} iff

$$\Phi_{12_d}[A_{12}] + [\Phi_{12}, A_{12_d}] = 0. \tag{3.8}$$

ii) The admissible operator valued function Φ_{12} is called hereditary symmetry iff

$$\Phi_{12_d}[\Phi_{12} f_{12}]\, g_{12} - \Phi_{12}\, \Phi_{12_d}[f_{12}]\, g_{12} \text{ is symmetric w.r.t. } f_{12}, g_{12}. \tag{3.9}$$

Remark 3.2

i) Equation (3.8) makes sense if the d-derivative of A_{12} exists; if A_{12} is generated by an admissible operator $\hat{A}_{12}$ on H_{12} ($A_{12} = \hat{A}_{12}\, H_{12}$), then A_{12} is admissible and $A_{12_d}[\cdot] = \hat{A}_{12_d}[\cdot]H_{12}$. Since a linear combination of admissible functions is admissible, the equation (2.6) implies that $\delta_{12} K_{12}^{(n)}$ is an admissible function.

ii) If $A_{12} = \delta_{12} K_{12}$, then equation (3.8) reads

$$\Phi_{12_f}[K] + [\Phi_{12}, (\delta_{12} K_{12})_d] = 0, \tag{3.10}$$

and we say that Φ_{12} is a strong symmetry of the evolution equation (3.2). In this case Φ_{12} maps extended symmetries of (3.2) to extended symmetries of (3.2).

As for equations in 1+1, hereditary operators generate infinitely many commuting symmetries; we have precisely the following

Theorem 3.2

If the hereditary operator Φ_{12} is a strong symmetry of the starting symmetry $\hat{K}_{12}^{o} H_{12}$, and if the starting operator $\hat{K}_{12}^{o}$ satisfies the following condition

$$\left[\mathcal{R}^{\circ}_{12} H^{(1)}_{12},\ \mathcal{R}^{\circ}_{12} H^{(2)}_{12} \right]_d = 0 \qquad \text{for} \qquad \left[H^{(1)}_{12},\ H^{(2)}_{12} \right]_I = 0, \tag{3.11}$$

then $\sigma^{(m)}_{12} \doteq \Phi^m_{12} \mathcal{R}^{\circ}_{12} \cdot 1$ are extended symmetries of every evolution equation (2.1).

Corollary 3.1

If Φ_{12} generates two classes of evolution equations corresponding to two starting operators $\mathcal{R}^{\circ}_{12}$, given by $\mathcal{N}_{12}$ and $\mathcal{M}_{12}$, and if $\mathcal{N}_{12}$ and $\mathcal{M}_{12}$ satisfy a Lie algebra of the type (2.15,16), then $\Phi_{12}{}^m \mathcal{N}_{12} \cdot 1$ and $\Phi_{12}{}^m \mathcal{M}_{12} \cdot 1$ are extended symmetries of both classes of evolution equations (2.1).

It turns out that the recursion operator (2.2) is hereditary and is a strong symmetry of the starting symmetries (2.3), then it follows that $\sigma^{(m)}_{12} = \Phi_{12}{}^m \mathcal{R}^{\circ}_{12} \cdot 1$ (for $\mathcal{R}^{\circ}_{12} = \mathcal{N}_{12}$ and/or $\mathcal{M}_{12}$) are extended symmetries of each equation of the KP classes, and Theorem 3.1 implies that $\sigma^{(m)}_{11}$ are symmetries and $\sigma^{(m)}_{12} = \sigma^{(m)}(q_1, q_2) = 0$ are BT of each member of the two hierarchies.

3.3 Isospectral Problems Yield Hereditary Symmetries

The previous section illustrates the importance of hereditary symmetries. For equations in 1+1 and in 2+1 dimensions isospectral problems yield hereditary operators:

Proposition 3.1

Let

$$\frac{dV}{dx} = U(\hat{q}, \lambda)V \tag{3.12}$$

be an isospectral two dimensional problem, where $\hat{q}$ is an operator depending on q(x,y) and $\partial/\partial y$, and λ is an eigenvalue. Assume that $(G_\lambda)_{12}$, the extended gradient of λ (see (4.2)), satisfies

$$\Psi_{12} (G_\lambda)_{12} = \mu(\lambda) (G_\lambda)_{12}. \tag{3.13}$$

Then if $\Phi_{12} = \Psi^{*}_{12}$ has a complete set of eigenfunctions, it is hereditary.

For example the isospectral problem

$$v_{xx}+\hat{q}v = \lambda v \tag{3.14}$$

associated with the KP classes implies that

$$(G_\lambda)_{12} = v_1 v_2^+, \tag{3.15}$$

where v^+ is a solution of $v_{xx}^+ + (q - \alpha\partial/\partial_y)v^+ = \lambda v^+$. Since Φ_{12}, defined by (2.2), satisfies

$$\Phi_{12}^* v_1 v_2^+ = 4\lambda v_1 v_2^+, \tag{3.16}$$

it follows that Φ_{12} is hereditary.

4. HAMILTONIAN FORMALISM

4.1 Bilinear Forms, Gradients Of Conserved Quantities

Integrable evolution equations in 2+1 dimensions posses infinitely many constants of motion of the form

$$I = \operatorname{tr} \int_{\mathbb{R}^2} dxdy_1 \rho_{11} = \operatorname{tr} \int_{\mathbb{R}^3} dxdy_1 dy_2 \delta_{12} \rho_{12} , \tag{4.1}$$

where $\rho_{12} = \rho(q_1,q_2)$ (the trace operation is obviously omitted if q is a scalar).

As in 1+1 dimensions, it is more convenient to deal with the gradients of conserved quantities; in this case the double representation (4.1) of a functional I allows the introduction of the <u>extended gradient</u> $\mathrm{grad}_{12}I$ and of the gradient grad I of I, defined by

$$I_d[g_{12}] = \operatorname{tr} \int_{\mathbb{R}^3} dxdy_1 dy_2 \delta_{12} \rho_{12_d}[g_{12}] \doteq < \mathrm{grad}_{12}I,\ g_{12}> , \tag{4.2a}$$

$$I_f[g_{11}] = \operatorname{tr} \int_{\mathbb{R}^2} dxdy_1 \rho_{11_f}[g_{11}] \doteq (\ \mathrm{grad}\ I,\ g_{11}\ ), \tag{4.2b}$$

where

$$< g_{12},\ f_{12} > \doteq \operatorname{tr} \int_{\mathbb{R}^3} dxdy_1 dy_2 g_{21} f_{12}, \tag{4.3a}$$

$$(\ g_{11},\ f_{11}\ ) \doteq \operatorname{tr} \int_{\mathbb{R}^2} dxdy_1 g_{11} f_{11}, \tag{4.3b}$$

are the proper <u>symmetric bilinear forms</u> coupling arbitrary elements $g_{12} \in \tilde{S}^*$, $f_{12} \in \tilde{S}$ and $g_{11} \in S^*$, $f_{11} \in S$ respectively ($\tilde{S}^*$ and S^* are obviously the duals of $\tilde{S}$ and S).

If L_{12}^* and L^+ denote the <u>adjoints</u> of the operator valued functions L_{12} and L on $\tilde{S}$ and S with respect to the bilinear forms (4.3a) and (4.3b) respectively, namely if

$$< L^{*}_{12}g_{12}, f_{12} > \doteq <g_{12}, L_{12}f_{12}> , \tag{4.4a}$$

$$(L^{+}g_{11}, f_{11}) \doteq (g_{11}, Lf_{11}), \tag{4.4b}$$

then one can prove the following

Proposition 4.1

i) γ_{12} and γ_{11} are extended gradient and gradient functions respectively, iff $\gamma_{12d}^{*} = \gamma_{12d}$ and $\gamma_{11f}^{+} = \gamma_{11f}$.

ii) If γ_{12} is an extended gradient, then γ_{11} is a gradient corresponding to the same potential, namely if $\gamma_{12} = \text{grad}_{12}I$, then $\gamma_{11} = \text{grad}\, I$.

If a functional I is conserved with respect to the evolution equation (3.2), this corresponds to the mathematical notion of conserved covariant.

Definition 4.1

The function γ_{12} is an extended conserved covariant of (3.2) iff

$$\gamma_{12_f}[K] + (\delta_{12}K_{12})^{*}_{d}[\gamma_{12}] = 0. \tag{4.5}$$

Then we have the following

Proposition 4.2

i) If γ_{12} is an extended conserved covariant of (3.2), γ_{11} is a conserved covariant of (3.2), namely

$$\gamma_{11_f}[K_{11}] + K_{11_f}^{*}[\gamma_{11}] = 0. \tag{4.6}$$

ii) If the functional I is a conserved quantity of (3.2), then $\gamma_{12} = \text{grad}_{12}I$ is an extended conserved covariant of (3.2). Conversely, if γ_{12} is an extended conserved covariant of (3.2) and it is the extended gradient of a functional I, then I is a constant of motion for (3.2).

As an illustration of the notions presented in this section, we have that the functionals I_j defined by

$$I_j = \frac{1}{2(2j+3)} \int_{\mathbb{R}^3} dx dy_1 dy_2 \delta_{12}\, \gamma_{12}^{(j+1)} , \; j=0,1, \tag{4.7}$$

where

$$\gamma_{12}^{(j)} = D^{-1}\Phi_{12}^{\,j}\, \hat{R}^{o}_{12}\cdot 1 , \; \hat{R}^{o}_{12} = \hat{N}_{12} \text{ and } \hat{M}_{12}, \tag{4.8}$$

are constants of motion of the KP hierarchies, corresponding to the extended gradients of conserved quantities

$$\gamma_{12}^{(j)} = \mathrm{grad}_{12} I_j = D^{-1}\, \Phi_{12}^{j}\, \mathring{K}_{12}\cdot 1 \ , \quad j=o,1. \tag{4.9}$$

Moreover $q_{12}^{\pm\,*} = \pm\, q_{12}^{\pm}$, then

$$\Phi_{12}^{*} = D^2 + D^{-1} q_{12}^{+} D + q_{12}^{+} + D^{-1} q_{12}^{-} D^{-1} q_{12}^{-}. \tag{4.10}$$

4.2 Bi-Hamiltonian Structures

The existence of a recursion operator for 2+1 dimensional systems allows a characterization of their bi-Hamiltonian nature. The "extended nature" of the operators we are dealing with, leads to a definition of Hamiltonian system in an "extended sense".

Definition 4.2

An equation (3.2) is of a Hamiltonian form (or is a Hamiltonian system) if it can be written as

$$q_{1_t} = \int_{\mathbb{R}} dy_2\ \delta_{12}\, \Theta_{12}\, \gamma_{12} \quad , \tag{4.11}$$

where γ_{12} is an extended gradient function of the form $\gamma_{12} = \hat{\gamma}_{12}\cdot 1$ and Θ_{12} is a Hamiltonian operator, i.e.

i) $$\Theta_{12}^{*} = -\,\Theta_{12} \ , \tag{4.12a}$$

ii) Θ_{12} satisfies the Jacobi identity w.r.t. the bracket

$$\{a_{12}, b_{12}, c_{12}\} \doteq \langle a_{12},\ \Theta_{12}\, d\left[\Theta_{12} b_{12}\right] c_{12} \rangle \ . \tag{4.12b}$$

The associated Poisson bracket of two functionals $I^{(1)}$ and $I^{(2)}$ is given by

$$\{\, I^{(1)},\ I^{(2)}\} \doteq \langle \mathrm{grad}_{12} I^{(1)},\ \Theta_{12}\ \ \mathrm{grad}_{12} I^{(2)} \rangle \ . \tag{4.13}$$

$\Theta_{12}^{(1)} = D$ and $\Theta_{12}^{(2)} = \Phi_{12} D$, where Φ_{12} is defined in (2.2), are examples of Hamiltonian operators. Then, since $\gamma_{12}^{(0)} = D^{-1}\hat{M}_{12}\cdot 1$ and $\gamma_{12}^{(1)} = D^{-1}\Phi_{12}\hat{M}_{12}\cdot 1$ are extended gradients, the KP equation has a double Hamiltonian structure (is a bi-Hamiltonian system):

$$q_{1_t} = \int_{\mathbb{R}} dy_2 \delta_{12}\ \Theta_{12}^{(1)}\ \gamma_{12}^{(1)} = \int_{\mathbb{R}} dy_2 \delta_{12}\, \Theta_{12}^{(2)} \gamma_{12}^{(0)} \ . \tag{4.14}$$

As in 1+1 dimensions, the existence of a compatible pair of Hamiltonian operators plays a fundamental rôle in the theory, emphasized by the following Theorem

Theorem 4.1

Let $\Theta^{(1)}_{12} + \nu\Theta^{(2)}_{12}$ be a Hamiltonian operator for all constant values of ν. Assume that $\Theta^{(1)}_{12}$ is invertible. Define $\Phi_{12} \doteq \Theta^{(2)}_{12}(\Theta^{(1)}_{12})^{-1}$; $\hat{\gamma}^{\circ}_{12} \doteq (\Theta^{(1)}_{12})^{-1}\hat{K}^{\circ}_{12}$. Assume that the operator Φ_{12} is a strong symmetry for the starting symmetry $\hat{K}^{\circ}_{12}H_{12}$ that satisfies (3.11); further assume that $\hat{\gamma}^{\circ}_{12}H_{12}$ is an extended gradient function. Then

i) Equations (2.1) are bi-Hamiltonian systems.

ii) Φ_{12} is hereditary.

iii) $K^{(m)}_{12} = \Phi^{m}_{12}\hat{K}_{12}\cdot 1$ and $\gamma^{(m)}_{12} = (\Phi^{*}_{12})^{m}\hat{\gamma}^{\circ}_{12}\cdot 1$ are extended symmetries and extended gradients of conserved quantities respectively of equations (2.1).

iv) $K^{(m)}_{11}$ and $\gamma^{(m)}_{11}$ are symmetries and gradients of conserved quantities in involution for equations (2.1), namely

$$K^{(m)}_{11_f}[K^{(n)}_{11}] = K^{(n)}_{11_f}[K^{(m)}_{11}] , \tag{4.15a}$$

$$\{ I^{(m)}, I^{(n)}\} = \langle \delta_{12}\gamma^{(m)}_{12}, \Theta_{12}\gamma^{(n)}_{12} \rangle = 0, \quad \Theta_{12} = \Theta^{(1)}_{12} \text{ or } \Theta^{(2)}_{12}, \tag{4.15b}$$

$$\gamma^{(j)}_{12} \doteq \mathrm{grad}_{12} I^{(j)} . \tag{4.15c}$$

v) $K^{(m)}_{12} = K^{(m)}(q_1,q_2) = 0$ are auto-BT for equations (2.1).

Remark 4.1

If $\Phi_{12} = \Theta^{(2)}_{12}(\Theta^{(1)}_{12})^{-1}$, where $\Theta^{(1)}_{12}$ and $\Theta^{(2)}_{12}$ are skew-symmetric, then $\Phi_{12}\Theta^{(1)}_{12} = \Theta^{(1)}_{12}\Phi^{*}_{12}$ (the operators Φ_{12} and $\Theta^{(1)}_{12}$ are well coupled).

The hypothesis of Theorem 4.1 are satisfied by the Hamiltonian operators $\Theta^{(1)}_{12}$, $\Theta^{(2)}_{12}$ and by the starting operators $\hat{N}_{12}$ and $\hat{M}_{12}$ of the KP classes; then they enjoy the propetries i)-v).

5. OTHER ASPECTS OF THE THEORY

5.1 Time Dependent Symmetries, Connection To Master-Symmetries

In order to investigate the properties of time independent commuting symmetries of the integrable evolution equations (2.1), one uses only special choices of H_{12} (given by $H_{12}= 1$ and δ^{ℓ}_{12}) for which the Lie algebra of the starting symmetries is abelian. More general choices of H_{12} make the Lie algebra of the starting symmetries non-abelian and give rise to time dependent extended symmetries.

A time dependent symmetry

$$\sigma_{12} = \sum_{j=0}^{r} t^j \, \Sigma^{(j)}_{12} \tag{5.1}$$

of equation (3.2) must satisfy the following equations

$$\Sigma^{(j)}_{12} = -\frac{1}{j}\left[\Sigma^{(j-1)}_{12}, \delta_{12} K_{12}\right]_d, \quad j=1,\ldots,r, \tag{5.2a}$$

$$\left[\Sigma^{(r)}_{12}, \delta_{12} K_{12}\right]_d = 0. \tag{5.2b}$$

This implies that constructing a symmetry of order r in time is equivalent to finding a function $\Sigma^{(0)}_{12}$ with the property that its $r+1^{st}$ Lie commutator with $\delta_{12} K_{12}$ is zero. Considering the KP class and using the structure of the Lie algebra (2.15,16) and equation (2.6), it is possible to show that t-dependent symmetries of order r are generated through equations (5.2) starting with

$$\Sigma^{(0)}_{12} = \tau^{(m,r)}_{12} \doteq \Phi^{m}_{12} \hat{R}^{\circ}_{12} H^{(r)}_{12}, \tag{5.3}$$

where $\hat{R}^{\circ}_{12} = \hat{N}_{12}$ and/or $\hat{M}_{12}$, and $H^{(r)}_{12}$ is defined by

$$H^{(r)}_{12} \doteq (y_1+y_2)^r, \tag{5.4}$$

or, more generally, by any homogeneous polynomial of degree r in y_1 and y_2. For example σ_{12} given by (5.1), with

$$\Sigma^{(0)}_{12} = \Phi^{m}_{12} \hat{M}^{\circ}_{12} H^{(r)}_{12}, \tag{5.5a}$$

$$\Sigma^{(j)}_{12} \doteq \sum_{s_1,\ldots,s_j=0}^{p_n} \nu(r,j,s)\, \Phi_{12}^{(m+jn-\sum_{\ell=1}^{j}(2s_\ell+1))} \hat{M}_{12} H_{12}^{(r-\sum_{\ell=1}^{j}(2s_\ell+1))}, \tag{5.5b}$$

$$\nu(r,j,s) \doteq \frac{(-2)^j}{j!}\left(\prod_{\mu=1}^{j} \theta\Big(r-\sum_{\ell=1}^{\mu}(2s_\ell+1)\Big)\right)\left(\sum_{\ell=1}^{j} b_{n,2s_\ell+1}\right)\frac{r!}{\left(r-\sum_{\ell=1}^{j}(2s_\ell+1)\right)!}, \tag{5.5c}$$

$$\theta(a) \doteq \begin{cases} 1, & a \geq 0, \\ 0, & a < 0, \end{cases} \quad p_n \doteq \begin{cases} (n-1)/2, & n \text{ odd} \\ (n-2)/2, & n \text{ even} \end{cases} \quad b_{n,\ell} \doteq (-4\alpha)^{\ell} \binom{n}{\ell}, \tag{5.5d}$$

is a time dependent extended symmetry of order r of equation $(2.1)_n$, corresponding to $\mathring{K}_{12} = \mathring{N}_{12}$. The discovery of the existence of a hereditary operator Φ_{12}, together with the structure of the Lie algebra of the starting symmetries, allows a simple and elegant characterization of the 2+1 dimensional master-symmetries. Here we briefly remark that

$$\tau_{11}^{(m,r)} \doteq \int_{\mathbb{R}} dy_2\, \delta_{12}\, \tau_{12}^{(m,r)} \tag{5.6}$$

are the so-called <u>master-symmetries of degree r of KP</u> [12].

5.2 Gradient And Non-Gradient Master-Symmetries

Using an isomorphism between Lie and Poisson brackets and the Lie algebra of the starting symmetries $\mathring{K}_{12}H_{12}$, it is easy to prove that $(\Theta_{12}^{(1)})^{-1}\Phi_{12}^{m}\mathring{K}_{12}H_{12}$ are extended gradients. This implies that the extended symmetries associated with them give rise to conserved quantities. For example, the t-dependent symmetry

$$\sigma_{12} = \mathring{M}_{12}^{(m)}(y_1+y_2) + t12\alpha\, \mathring{N}_{12}^{(m+1)}\cdot 1 \tag{5.7}$$

of the KP equation $q_t = 2(q_{xxx} + 6qq_x + 3\alpha^2 D^{-1} q_{yy})$ corresponds to the extended gradient $D^{-1}\sigma_{12}$, then it gives rise to the t-dependent conserved quantity

$$I = \int_{\mathbb{R}^2} dx dy_1 \Big(\frac{1}{2(2m+3)} \big(D^{-1}\mathring{M}_{12}^{(m+1)}(y_1+y_2)\big)_{y_2=y_1} +$$

$$+ \frac{3\alpha t}{m+2} \big(D^{-1}\mathring{N}_{12}^{(m+2)}\cdot 1\big)_{y_2=y_1} \Big). \tag{5.8}$$

Since the master-symmetries $\tau_{12}^{(m,r)}$ are related to these gradients, they cannot be used to generate Φ_{12}. Nevertheless <u>non-gradient master-symmetries</u> of 2+1 dimensional equations exist, for example $T_{12} \doteq \Phi_{12}^{2}\delta_{12}$ is a non-gradient master-symmetry of the KP classes, satisfying the following equation

$$[\Phi_{12}^{n}\mathring{K}_{12}\cdot 1, T_{12}]_d = b_n \Phi_{12}^{n+1}\mathring{K}_{12}\cdot 1, \tag{5.9}$$

where $b_n = 4n$ and $2(2n+1)$, if $\mathcal{R}^{0}_{12} = \hat{N}_{12}$ and $\hat{M}_{12}$ respectively.

Since $(\Theta^{(1)}_{12})^{-1} T_{12}$ is not a gradient function, T_{12} can be used to generate Φ_{12}, exactly as in 1+1 dimensions, through the formula

$$8\Phi_{12} = T_{12_d} + \Theta^{(1)}_{12} T_{12_d} (\Theta^{(1)}_{12})^{-1}. \tag{5.10}$$

The existence of the T_{12} master-symmetry finally implies a simple derivation of the equation

$$I_n = \frac{1}{b_{n+1}} \int_{\mathbb{R}^2} dxdy_1 \gamma^{(n+1)}_{11}, \tag{5.11}$$

where I_n is the potential of $\gamma^{(n)}_{11}$ ($\gamma^{(n)}_{11}$ = grad I_n), and $\gamma^{(j)}_{12} = (\Phi^{*}_{12})^{j}\, \Theta^{-1}_{12} \mathcal{R}^{0}_{12} \cdot 1$.

6. THE BENJAMIN-ONO CLASS

The remarkable connections (algebraic and analytical) between the KP and the BO equations are also confirmed by the fact that, although the BO equation is a 1+1 dimensional system, its recursion operator lives in an extended space. In fact it is possible to show that the BO class can be represented in the following way[16])

$$q_{1_t} = \beta_n \int_{\mathbb{R}} dx_2 \delta_{12} \Phi^{n}_{12} q^{-}_{12} \cdot 1, \tag{6.1}$$

where $\delta_{12} = \delta(x_1 - x_2)$. The hereditary operator Φ_{12} is defined by

$$\Phi_{12} = q^{+}_{12} - iq^{-}_{12} H, \qquad q^{\pm}_{12} = q_1 \pm q_2 + i(D_1 \mp D_2),$$

$$q_i = q(x_i, t), \; D_i = \partial/\partial x_i, \; i=1,2, \tag{6.2}$$

and the operator H, whose action is defined on functions $f_{12} = f(x_1, x_2)$ of the type

$$f_{12} = a_1 + b_2 + c_{12}, \; H_1 c_{12} = H_2 c_{12}, \tag{6.3}$$

is given by

$$Hf_{12} = H_1 a_1 + H_2 b_2 + H_1 c_{12}, \tag{6.4}$$

where H_i, i=1,2, is the Hilbert transform with respect to the variable

x_i :

$$H_i g_{ij} = \pi^{-1} \int_R dx_i'(x_i'-x_i)^{-1} g(x_i',x_j), \qquad i \neq j . \tag{6.5}$$

It is possible to show that all the extended symmetries $\sigma_{12}^{(j)}$ generated by Φ_{12} :

$$\sigma_{12}^{(j)} \doteq \Phi_{12}^{\ j} q_{12}^{-} \cdot 1 \tag{6.6}$$

are functions of the type (6.3), then formula (6.1) is well defined. The BO equation (1.1) corresponds to n=2 and $\beta_2=(4i)^{-1}$. One can show that the algebraic properties of the BO class (6.1) are described by the generalized theory summarized in this paper.

REFERENCES

1) B.Fuchssteiner, Nonlinear Anal. TMA 3, 849 (1979).

2) F.Magri, J.Math.Phys. 19, 1156 (1978); Nonlinear Evolution Equations and Dynamical Systems, edited by M.Boiti, F.Pempinelli and G.Soliani, Lecture Notes in Physics, Vol.120 (Springer, New York, 1980), p.233.

3) A.S.Fokas and B.Fuchssteiner, Lett. Nuovo Cimento 28, 299 (1980); B.Fuchssteiner and A.S.Fokas, Physica 4D, 47 (1981).

4) I.M.Gel'fand, private communication.

5) I.M.Gel'fand and I.Y.Dorfman, Funct. Anal. Appl. 13, (1979); 14, 71 (1980).

6) M.Salerno,"On the phase manifold geometry of the two-dimensional Burgers equations",Preprint CNS, Los Alamos National Laboratory, 1985.

7) V.E.Zakharov and B.G.Konopelchenko, Comm. Math. Phys. 94,483 (1984).

8) B.G.Konopelchenko and V.G.Dubrovsky, Physica 16D, 79 (1985).

9) A.S.Fokas and B.Fuchssteiner, Phys. Lett. A86, 341 (1981).

10) W.Oevel and B.Fuchssteiner, Phys. Lett. A88, 323 (1982). H.H.Chen, Y.C.Lee and J.E.Lin, Physica 9D, 439 (1983).

11) W.Oevel,"Master-symmetries for finite dimensional integrable systems: the Calogero-Moser system", Preprint, Paderborn University, (1986).

12) B.Fuchssteiner, Progr. Theoret. Phys. 70, 150 (1983). I.Ya.Dorfman,"Deformations of Hamiltonian structures and integrable systems", Preprint. W.Oevel,"Ageometrical approach to integrable systems admitting scaling symmetries", University of Paderborn, 1986.

13) A.S.Fokas and P.M.Santini, Stud.Appl.Math. 75, 179 (1986).

14) P.M.Santini and A.S.Fokas,"Recursion operators and bi-Hamiltonian structures in multidimensions.I", Preprint INSN, Clarkson University, 1986; Comm. Math.Phys. (submitted to).

15) A.S.Fokas and P.M.Santini," Recursion operators and bi-Hamiltonian structures in multidimensions.II", Preprint INSN, Clarkson University 1986; Comm. Math. Phys. (submitted to).

16) P.M.Santini and A.S.Fokas, Recursion operator and bi-Hamiltonian structure of the Benjamin-Ono equation", Preprint INSN, Clarkson University, 1986.

EXACTLY SOLVABLE MULTIDIMENSIONAL NONLINEAR EQUATIONS AND INVERSE SCATTERING

Mark J. Ablowitz

Clarkson University
Department of Mathematics and Computer Science
Potsdam, New York 13676 U.S.A.

ABSTRACT

A review of recent results associated with exactly solvable multidimensional nonlinear systems and related questions of direct and inverse scattering is given.

In this lecture a review of some recent results associated with exactly solvable multidimensional nonlinear systems will be given. The motivation for much of this work has come via what is commonly referred to as the Inverse Scattering Transform (I.S.T.; as a reference see, for example, [1]). IST is a method to solve certain nonlinear equations by associating them with appropriate compatible linear equations, one of which is identified as a scattering problem and the others(s) serves to fix the "time evolution" of the scattering data.

In one spatial dimension the prototype problem is the (KdV) equation

$$u_t + 6uu_x + u_{xxx} = 0. \tag{1}$$

The KdV equation is compatible with

$$v_{xx} + u(x,t)v = \lambda v \tag{2}$$

$$v_t = (\gamma+u_x)v - (4\lambda+2u)v_x \tag{3}$$

i.e. $v_{xxt} = v_{txx}$ implies (1). Equation (2) is the Schrödinger scattering problem, λ the eigenvalue (γ= const. in (3)). The solution of (1) on the line: $-\infty<x<\infty$ for initial values $u(x,t=0)$ vanishing sufficiently rapidly at infinity is obtained by studying the associated direct and inverse scattering problem of (2) and using (3) to fix the time evolution of the scattering data. It turns out that the inverse problem amounts to solving a matrix Riemann-Hilbert boundary value problem (RHBVP) whose jump discontinuity depends explicitly on the scattering data. Calling $\lambda=-k^2, v(x,k)=\mu(x,k)e^{-ikx}$ the RHBVP takes the following form,

$$(\mu_+-\mu_-)(x,t,k) = \mu_-(x,t,\alpha(k))\ V(x,t,k) \text{ on } \Sigma$$
$$\mu_\pm\to 1, \quad |k|\to\infty \tag{4}$$

where

$V(x,t,k) - r(k,t)\ e^{2ikx}$, $\alpha(k) = -k$, $\Sigma=\{k:k\in\mathbb{R}\}$, and $\mu_\pm$ are the boundary values as $\mathrm{Im}k\to 0\pm$ of meromorphic functions in the upper (+) lower(-) half plane. (4) may be converted into a linear integral equation by taking the minus projections and the potential is reconstructed via

$$u(x,t) = -\frac{1}{\pi}\frac{\partial}{\partial x}\int_C \mu(k,x,t,-k)\ V(x,t,k)dk \qquad (5)$$

where the contour is taken above all poles of $r(k,t)$; of which there is at most a finite number, $k_j = i\kappa_j$, $\kappa_j>0$ $j = 1,---N$. The scattering data: the reflection coefficient, $r(k,t)$ evolves simply in time

$$r(k,t) = r(k,0)\ e^{8ik^2t} \qquad (6)$$

The above scheme may be extended so as to solve a surprisingly large number of interesting nonlinear evolution equations. There are two scattering problems of particular interest in one dimension:

(i) Scalar scattering problems:

$$\frac{d^n v}{dx^n} + \sum_{j=2}^{n} u_j(x)\frac{d^{n-j}v}{dx^{n-j}} = \lambda v,$$

$$v(x,k),\quad u_j \in \mathbb{C}$$

(ii) First order systems - generalized AKNS

$$\frac{dv}{dx} = i\ k\ J\ v + q\ v$$

$$v(x,k),q(x) \in \mathbb{C}^{N\times N},\ J = \text{diag}\ (J^1,---J^n)$$

$$J^i \neq J^j, i \neq j$$

$$q^{ii} = 0.$$

Via an appropriate transformation the inverse problem associate with (i), (ii) can be expressed as a matrix RHBVP of the form (4). The potentials u_j,q can be shown to satisfy nonlinear evlution equations by appending to (i), (ii) suitable linear time evolution equations. One then finds that the scattering data $V(x,t,k)$ evolves simply in time. Well known solvable nonlinear equations include the Boussinesq, modified KdV, sine-Gordon, nonlinear Schrodinger, and three wave interaction equations. The reader may wish to consult for example[2a-e] for a detailed discussion of some of this material.

It is most significant that these concepts can be generalized to 2 spatial plus one time dimension. Here the prototype euqation is the Kadomtsev-Petviashvili (K-P) equation:

$$(u_t + 6uu_x + u_{xxx})_x = -3\sigma^2 u_{yy} \tag{7}$$

which is the compatibility equation between the following linear problems:

$$v_y + v_{xx} + u(x,y,t)v = 0 \tag{8}$$

$$v_t + 4v_{xxx} + 6uv_x + 3(u_x - \sigma \int_{-\infty}^{x} u_y dx')v + \gamma v = 0 \tag{9}$$

(γ = const.). We shall consider the question of solving (7) for $u(x,y,0)$ decaying sufficiently rapidly in the plane $r^2 = x^2 + y^2 \to \infty$. Physically speaking, both cases $\sigma^2 = -1$ (KPI) $\sigma^2 = +1$ (KPII) are of interest. Whereas KPI can be related to a RHBVP of a certain type (nonlocal; see ref.[3]) KPII turns out to require new ideas. Letting

$$v = \mu(x,y,k)e^{ikx + k^2 y/\sigma}$$

$\sigma = \sigma_R + i\sigma_I$, $\sigma_R \neq 0$. Then there exist functions μ bounded for all x,y satisfying $\mu \to 1$ as $|k| \to \infty$. However such a function turns out to be nowhere analytic in k, rather it depends nontrivially on both the real and imaginery parts of $k = (k_R + ik_I)$. $\mu = \mu(x,y,k_R,k_I)$.

In fact μ satisfies a generalization of a RHBVP - namely a $\bar{\partial}$ (DBAR) problem where μ satisfies,

$$\frac{\partial\mu}{\partial\bar{k}} = \mu(x,y,\xi_0,k_I)\ V(x,y,k_R,k_I) \tag{10}$$

where $\frac{\partial}{\partial\bar{k}} = \frac{1}{2}(\frac{\partial}{\partial k_R} + i\frac{\partial}{\partial k_I})$ and V has the structure

$$V(x,y,k_R,k_I) = \frac{\mathrm{sgn}(k_0)e^{i\beta(x,y,k_R,k_I,\xi_0)}}{\pi|\sigma R|} T(k_R,k_I)$$

$$\beta(x,y,k_R,k_I,\xi_0) = (x + 2y\frac{k_I}{\bar{\sigma}_R})(\xi_0 - k_R) = -2(x + 2y\frac{k_I}{\sigma_R})k_0$$

$$\xi_0 = -k_R - \frac{2\sigma_I}{\sigma_R}k_I,\quad k_0 = k_R + \frac{\sigma_I}{\sigma_R}k_I \tag{11}$$

(11) may be converted into a linear integral equation by employing the generalized Cauchy formula. $T(k_R,k_I)$ is viewed as the ("nonphysical" data, i.e. inverse scattering data: i.e. inverse data) and the potential is reconstructed via

$$u(x,y) = \frac{2i}{\pi}\frac{\partial}{\partial x}\iint \mu(x,y,\xi_0,k_I)V(x,y,k_R,k_I)dk_R dk_I . \tag{12}$$

For K-P the evolution of the data obeys ($\gamma = 4ik^3$ in (9))

$$\frac{\partial T}{\partial t} = (8ik_0)(6kk_0 - 4k_0^2 - 3k^2)T \tag{13}$$

where $k_0 = k_R + \frac{\sigma_I k_I}{\sigma_R}$, $k = k_R + ik_I$.

Similar ideas apply to higher order scalar problems

(iii) $$\sigma\frac{\partial v}{\partial y} + \frac{\partial^n v}{\partial x^n} + \sum_{j=2}^{n} u_j(x)\frac{\partial^{n-j} v}{\partial x^{n-j}} = 0$$

where: v, $u_j \in \mathbb{C}$ and to first order systems

(iv) $$\sigma\frac{\partial v}{\partial y} + J\frac{\partial v}{\partial x} + q(x,y)v = 0$$

where: $v, q \in \mathbb{C}^{N\times N}$, $J=\text{diag}(J^1,\dots,J^N)$, $J^i \neq J^j$, $i \neq j$ with $q^{ii} = 0$. Interested readers many consult reference[4a,b] for associated details.

The notion of $\bar{\partial}$ extends to higher dimensional scattering and inverse scattering problems. However as we shall mention, despite the fact that the inverse scattering problem is essentially tractable there does not appear to be any local nonlnear evolution equations in dimensions greater than 2 + 1 associated with multidimensional generalizations of (iii) or (iv).

Our prototype scattering problem will be

$$\sigma v_y + \Delta v + u(x,y)v = 0$$

$$\Delta = \sum_{\ell=1}^{n}\frac{\partial^2}{\partial x_\ell^2}, \qquad x \in \mathbf{R}^n, \quad y \in \mathbf{R} . \tag{14}$$

Letting

$$v = \mu(x,y,k)e^{ik\cdot x + k^2 y/\sigma}$$

$$k = k_R + ik_I, \quad k \in \mathbb{C}^n$$

$$k\cdot x = \sum_1^n k_j j, \quad \sigma = \sigma_R + i\sigma_I .$$

Then there exist functions μ bounded for all x, y satisfying $\mu\to 1$, as $|k_j| \to \infty$, $j = 1,\dots,n$. When $\sigma_R \neq 0$ μ turns out to be nonanalytic in each of the variables k, i.e. $\mu = \mu(x,y,k_{R_1},\dots k_{R_n}, k_{I_1},\dots,k_{I_n})$ and satisfies a $\bar{\partial}$ problem linear in μ, in each of the variables k_j; i.e. μ satisfies an equation of the form,

$$\frac{\partial\mu}{\partial\bar{k}_j} = \tilde{T}_j(\mu); \qquad j = 1,\dots,n \tag{15}$$

where $\tilde{T}_j(\mu)$ is an appropriate linear integral operator which depends only on one scalar scattering function T : $\tilde{T}_j = \tilde{T}_j[T]$, $T=T(k_R,\tilde{k}_I,\xi)$ ξ being (n-1) integration parameters in the nonlocal operator $\tilde{T}_j$. The inverse problem is redundant, i.e. we are given $T(k_R,k_I,\xi)$(3n-1) parameters) and we must reconstruct a local potential $u(x,y)$(n+1 parameters). A serious issue is how to characterize admissible inverse data T, i.e. data that really arises from a local potential (small generic changes in $T(k_R,k_I,\xi)$ cannot be expected to arise from a local potential $u(x,y)$). Insight into this question is obtained by requiring $\partial^2\mu/\partial\bar{k}_i\partial\bar{k}_j = \partial^2\mu/\partial\bar{k}_j\partial\bar{k}_i$ $(i \neq j)$. The form of this constraint is given by

$$\mathcal{L}_{ij}(T) = \tilde{N}_{ij}[T] \tag{16}$$

where $\mathcal{L}_{ij}$ is a linear operator and $\tilde{N}_{ij}$ a nonlinear (quadratic) nonlocal operator. Details can be found in[5a,b]. Equation (16) can be integrated and this integrated version may be used to reconstruct $u(x,y)$ as well as give a characterization for admissible scattering data: $T(k_R,k_I,\xi)$. However (16) also indicates why simple local nonlinear evolution equationshave not been associated with equation (8). Namely in the previous lower dimensional (2+1 and 1+1) problems the time evolution of the scattering data obeyed a particularly simple equation, (e.g. $\frac{\partial T}{\partial t} = \omega(k_R,k_I)T$. However in this case such a simple flow will not be maintained - due to the nonlinear constraint (16).

These ideas can be generalized to first order systems:

(v) $$\frac{\partial v}{\partial y} + \sigma\sum_{j=1}^{n} J_j \frac{\partial v}{\partial x_j} = qv$$

$v,\ q \in \mathbb{C}^{N\times N}$, $J_j = \mathrm{diag}(J_j^1,\dots,J_j^N)$

$J_j^k \neq J_j^\ell$, $\quad k \neq \ell$.

with similar results obtained[6a,b]. Again the scattering data satisfies a nonlinear constraint. In general, there is no compatible local nonlinear evolution equation associated with (v). However when certain restrictions are put on J_j then the constraint equation becomes linear and the so-called N wave interaction equations are compatible with the system (v). Nachman and Ablowitz[6a] showed that at most, the system would be 3+1 dimensional, and Fokas[6b] showed that indeed the system is reducible to 2+1 dimensions by a transformation of independent variables (characteristic variables).

Beals and Coifman have given an alternative but similar formulation[7a,b] in the scalar case.

There is an n-dimensional problem which also fits within the framework of IST: The so-called generalized wave and generalized sine-Gordon euqation (GWE and GSGE). These equations arise in the context of differential geometry and serve to extend the classical results of Bäcklund for the sine-Gordon equation to n-dimensions[8]. The n-dimensional Bäcklund tranformation is given by:

$$dX + XA^tX = A - XB, \tag{17}$$

where

$$dX = \sum_{j=1}^{n} \frac{\partial X}{\partial x_j}\, dx_j,$$

$$A_{ij} = \beta_i(z) a_{ij} dx_j,$$

$$B_{ij} = \frac{1}{a_{1i}} \frac{\partial a_{1j}}{\partial x_i}\, dx_j - \frac{1}{a_{1j}} \frac{\partial a_{1i}}{\partial x_j} dx_i, \quad 1 \le i, j \le n, \tag{18}$$

and $a = \{a_{ij}\} \in \mathbf{R}^{nxn}$. Equations (17-18) reduce to the Bäcklund transformation for the generalized sine-Gordon equation (GSGE) when

$$\beta_i(z) = (z^2 + (2\delta_{i1} - 1))/2z, \tag{19}$$

and for the generalized wave equation (GWE) when

$$\beta_i(z) = -(1-z^2)/2z \equiv \lambda(z). \tag{20}$$

The compatibility condition required for the existence of solutions to these Bäcklund transformations results in a system of second-order partial differential equations for an orthogonal n x n matrix $a = \{a_{ij}\}$ in (17) which is a function of n independent variables $a = a(x_1, x_2, \ldots, x_n)$. The equation has the form

$$\frac{\partial}{\partial x_i}\left(\frac{1}{a_{1i}}\frac{\partial a_{1j}}{\partial x_i}\right) + \frac{\partial}{\partial x_j}\left(\frac{1}{a_{1k}}\frac{\partial a_{1i}}{\partial x_j}\right)$$

$$+ \sum_{k \neq i,j} \frac{1}{a_{1k}^2}\frac{\partial a_{1i}}{\partial x_k}\frac{\partial a_{1j}}{\partial x_k} = \varepsilon a_{1i}a_{1j}, \quad i \neq j,$$

$$\frac{\partial}{\partial x_k}\left(\frac{1}{a_{1j}}\frac{\partial a_{1i}}{\partial x_j}\right) \qquad \frac{1}{a_{1k}a_{1j}}\frac{\partial a_{1i}}{\partial x_k}\frac{\partial a_{1k}}{\partial x_j}, \quad i,\ j,\ k \ \text{distinct},$$

$$\frac{\partial a_{jk}}{\partial x_k} = \frac{\partial a_{ji}}{a_{1i}}\frac{\partial a_{1k}}{\partial x_i}, \qquad i \neq k, \tag{21}$$

where $\varepsilon = 1$ for the GSGE and $\varepsilon = 0$ for the GWE.

We observe that when $n = 2$ and $\varepsilon = 1$ (GSGE), the orthogonal matrix $a = \{a_{ij}\}$ given by

$$a = \begin{pmatrix} \cos\frac{1}{2}u & \sin\frac{1}{2}u \\ -\sin\frac{1}{2}u & \cos\frac{1}{2}u \end{pmatrix} \tag{22}$$

for the function $u = u(x,t)$ reduces the GSGE to the classical sine-Gordon equation ($\kappa = -1$),

$$u_{tt} - u_{xx} - \kappa \sin u = 0. \tag{23}$$

On the other hand when $n = 2$ and $\kappa = 0$, then with (22) the GWE reduces to the wave equation (23). When $n \geq 3$ the generalization of the wave equations discussed here is nonlinear.

The Bäcklund transformations (17) described above are in fact matrix Riccati equations. Linearizations of such a system can be performed in a striaghtforward manner (see for example[9]). Introducing the transformation

$$X = UV^{-1}, \tag{24}$$

where U, V and n x n matrix functions of $x_1, \ldots, x_n$, the following linear system is deduced:

$$\begin{pmatrix} dU \\ dV \end{pmatrix} = \begin{pmatrix} 0 & A \\ A^t & B \end{pmatrix} \begin{pmatrix} U \\ V \end{pmatrix} \tag{25}$$

with the components of A, B given by (18). Compatibility ensures that the orthogonal matrix $a = \{a_{ij}\}$ satisfies the GSGE with (19) and GWE with (20). Alternatively, if we call

$$\begin{pmatrix} U \\ V \end{pmatrix} = \psi ,$$

the following linear system of 2n o.d.e.'s are obtained:

$$\frac{\partial \psi}{\partial x_j} = \lambda \tilde{A}_j \psi + C_j \psi, \tag{26}$$

where $\tilde{A}_j$, C_j are 2n x 2n matrices with the block structure

$$\tilde{A}_j = \begin{pmatrix} 0 & \tilde{a}_j \\ \tilde{a}_j^t & 0 \end{pmatrix}, \quad C_j = \begin{pmatrix} 0 & 0 \\ 0 & \gamma_j \end{pmatrix}. \tag{27}$$

Here $\tilde{a}_j$, $\tilde{\gamma}_j$ are n x n matrices having the following structure:

$$\tilde{a}_j = (\frac{\delta}{\lambda} - 1)e_1 a_j + a_j,$$
$$a_j = a e_j \tag{28}$$

where $e_j = \{e_j\}_{ik}$ is the unit matrix

$$\{e_j\}_{ik} = \begin{cases} 1 & i = k = j, \\ 0 & \text{otherwise}, \end{cases} \tag{29}$$

and in component form λ_j takes the form

$$(\lambda_j)_{k\ell} = (1-\delta_{kj})\frac{1}{a_{\ell k}} \frac{\partial a_{\ell j}}{\partial x_k} \delta_{\ell j} - (1-\delta_{\ell j})\frac{1}{a_{1\ell}} \frac{\partial a_{1j}}{\partial x_\ell} \quad {}_{kj}. \tag{30}$$

In (28) a is the orthogonal matrix $\mathbb{R}^n \to SO(n)$ associated with the GWE when $\delta = \lambda$ and with the GSGE when $\delta = \frac{1}{2}(z + 1/z)$, $\lambda = \frac{1}{2}(z - 1/z)$, and γ_j is the matrix (30): $\mathbb{R}_n \to M_n(\mathbb{R})$, $\gamma_j + \gamma_j^t = 0$. Equations (21) arise as the compatibility condition associated with (26). More explicitly, for the GWE the scattering problem takes the form $[\psi = \psi(x,\lambda)]$

$$\frac{\partial \psi}{\partial x_j} = \lambda A_j \psi + C_j \psi \tag{31}$$

with

$$A_j = \begin{pmatrix} 0 & a_j \\ a_j^t & 0 \end{pmatrix}, \tag{32}$$

and C_j given by (27,30).

For the GSGE the scattering problem for $\psi = \psi(x,z)$ takes the form

$$\frac{\partial}{\partial x_j} = \delta(z) \begin{pmatrix} 0 & e_1 a_j \\ a_j^t e_1 & 0 \end{pmatrix} \psi$$

$$+ \lambda(z) \begin{pmatrix} 0 & (I-e_1)a_j \\ a_j^t(I-e_1) & 0 \end{pmatrix} \psi + C_j \psi, \tag{33}$$

$\delta(z)$, $\lambda(z)$, C_j given above, or equivalently

$$\frac{\partial}{\partial x_j} = \frac{z}{2} A_j \psi + \frac{z}{2} B_j \psi + C_j \psi, \tag{34}$$

where

$$B_j = \begin{pmatrix} 0 & u a_j \\ a_j^t u & 0 \end{pmatrix}, \quad u = \mathrm{diag}(+1, -1, \ldots, -1). \tag{35}$$

In[8] it is shown how these linear problems may be viewed as a direct and inverse scattering problem for the GWE and GSGE. Namely the direct and inverse problem may be solved for matrix potentials, depending in the orthogonal matrix a, tending to the identity sufficiently fast on certain "generic" directions. It should be noted that solving the n-dimensional GWE and GSGE reduces to the study of the scattering and inverse scattering associated with a coupled system of n one-dimensional o.d.e.'s. This is in marked contrast to other attempts described earlier to isolate solvable (local) multidimensional nonlinear evolution equation which are compatibility conditions of two Lax-type operators, e.g.,

$$L\psi = \lambda\psi \tag{36}$$

$$\psi_t = M\psi \tag{37}$$

where L is a partial differential operator with the variable t entering only parametrically. Although as we have seen nonlinear evolution equations in three independent variables can be associated with such Lax pairs (e.g. the K-P, Davey-Stewartson, three wave interaction equations, etc.) little progress via this route has been made in more than three dimensions. As discussed earlier one has to overcome a serious constraint inherent in the scattering/inverse scattering theory for higher dimensional partial differential operators in order to be able to isolate associated solvable nonlinear equations, i.e. the scattering data generally satisfies a nonlinear equation (e.g. (16)). The analysis associated with the GWE and GSGE avoids these difficulties since the GWE and GSGE problems are simply a compatible set of nonlinear one-dimensional o.d.e.'s. The results in [8] demonstrate that the initial value problem is posed with given data along lines and not on (n-1) dimensional manifolds.

Similar ideas apply to certain n-dimensional extensions of the so-called anti-self-dual Yang-Mills equations (SDYM). In[9] it is shown that these multi-dimensional nonlinear equations are associated with compatible two-dimensional linear systems. Broad classes of solutions may be calculated by the $\bar{\partial}$ method. Since the overall compatible linear systems are coupled two-dimensional equations, the scattering data does not satisfy the nonlinear constraint discussed earlier.

Finally we remark that there is a class of nonlocal equations which can be reduced to exactly solvable equations. In the context of multidimensional nonlinear equations perhaps the most interesting example is

$$(u_t + u_{xxx} + 2(uH_z u)_x)_x = -3\sigma^2 u_{yy}, \tag{38}$$

where

$$(H_z u)(x,y,z,t) = \frac{1}{\pi} \fint_{-\infty}^{\infty} \frac{u(x,y,\xi,t)}{\xi - z}\, d\xi \tag{39}$$

and $\fint_{-\infty}^{\infty}$ denotes the Cauchy principal value integral. (38) is reduced to the K-P euqation

$$(w_t + w_{xxx} - i(w^2)_x)_x = -3\sigma^2 w_{yy} \tag{40}$$

via the transformation

$$w = u + iH_z u. \tag{41}$$

Details and other examples are given in [16].

ACKNOWLEDGEMENTS

I am most pleased to acknowledge the many crucial contributions of my colleagues: A.S. Fokas, D. Bar Yaacov, A.I. Nachman, R. Beals and K. Tenenblat. This work was supported in part by the National Science Foundation under grant numer DMS-8202117, the Office of Naval Research under grant number N00014-76-C-0867, and the Air Force Office of Scientific Research under grant number AFOSR-84-0005.

REFERENCES

[1] Ablowitz, M.J. and Segur, H., "Solitons and the Inverse Scattering Transform", SIAM Appl. Math., Phila., PA, 4 (1981).

[2a] Ablowitz, M.J. and Fokas, A.S., "Comments on the Inverse Scattering Tranform and Related Nonlinear Evolution Equations", Lect. Notes in Phys., 189, Proc. CIFMO School and Workshop held at Oaxtepec, Mexico (K.B. Wolf, ed.), Springer, 1982.

[2b] Beals, R. and Coifman, Commun. Pure Appl. Math., 1984, pp. 39-90.

[2c] Beals, R., The Inverse Problems for Ordinary Differential Operators on the Line; to appear, Amer. J. of Math.

[2d] Shabat, A.B., Func. Annal and Appl. 9, (1975) 75; Diff. eq. XV (1979) 1824.

[2e] Caudrey, P., Physica 6D (1982) 51.

[3a] Fokas, A.S. and Ablowitz, M.J., Stud. Appl. Math. 69, 211 (1983).

[3b] Manakov, S.V., Phys. 3D, 420 (1981).

[4a] Ablowitz, M.J., BarYaacov, D., and Fokas, A.S., Stud. Appl. Math. 69, 135, (1983).

[4b] Fokas, A.S. and Ablowitz, M.J., The Inverse Scattering Transform for Multidimensional (2+1) Problems, Lect. Notes in Phys., 189, Proc. CIFMO School and Workshop held at Oaxtepec, Mexico, 1982, K.B. Wolf, ed.

[5a] Nachman, A.I. and Ablowitz, M.J., Stud. Appl. Math. 71, 243-250 (1984).

[5b] Ablowitz, M.J. and Nachman, A.I., Physica 18D, 223 (1986).

[6a] Nachman, A.I. and Ablowitz, M.J., Stud. Appl. Math. 71 251-262 (1984).

[6b] Fokas, A.S., Phys. Rev. Lett., 57, No. 2, 159 (1986).

[7a] Beals, R., and Coifman, R.R., Multidimensional Inverse Scattering on Nonlinear PDE, Proc. Symposium Pure Math, 43, 45 (1985).

[7b] Beals, R. and Coifman, R.R., Physica 18D, 242, (1986).

[8] Ablowitz, M.J., Beals, R., and Tenenblat, K., Stud. Appl. Math., 74, 177-203 (1986).

[9] Ablowitz, M.J., Costa, D.G., and Tenenblat, K., Solutions of Multidimensional Extensions of the Anti-Self-Dual Yang-Mills Equations, preprint, INS #66, 1986.

[10] Ablowitz, M.J., Fokas, A.S., and Kruskal, M.D., Note on Solutions to a Class of Nonlinear Singular Integro-Differential Equations, preprint, INS #64, 1986.

ON THE INVERSE SCATTERING TRANSFORM FOR THE n-DIMENSIONAL SCHRÖDINGER OPERATOR

Richard B. Lavine and Adrian I. Nachman
Department of Mathematics, University of Rochester

ABSTRACT

An open problem at the heart of multidimensional potential inverse scattering has been to determine whether the Faddeev-Lippman-Schwinger equation admits homogeneous solutions for real values of the appropriate parameters. We show that any reasonable potential with at least one bound state will give rise to such real exceptional points. We also include a brief review of the $\bar{\partial}$ method which leads to this integral equation.

1. INTRODUCTION

We present some recent analytical results concerning the inverse scattering problem for the time-independent Schrödinger equation in $\mathbb{R}^n$, $n \geq 3$:

$$-\Delta\phi(x,k) + V(x)\phi(x,k) = k^2\phi(x,k). \tag{1}$$

To fix the notation, let G_+ be the usual outgoing Green's function for the unperturbed equation, given by

$$G_+(x - y,k) = -\frac{1}{(2\pi)^n}\int \frac{e^{i(x-y)\cdot\xi}}{\xi^2 - k^2 - i0}\,d\xi \; . \tag{2}$$

Let $\phi_+(x,k)$ be the family of solutions of (1) obtained by solving the Lippman-Schwinger equation:

$$\phi_+(x,k) = \exp(ix\cdot k) + \int G_+(x-y),k)V(y)\phi_+(y,k)dy . \qquad (3)$$

The scattering amplitude is defined as

$$A(\xi,k) = \int e^{-ix\cdot\xi} V(x)\phi_+(x,k)dx , \qquad (4)$$

and can be read off "on shell" (that is, for those vectors ξ in $\mathbb{R}^n$ with $|\xi| = |k|$) from the large $|x|$ asymptotics of ϕ_+ :

$$\phi_+(x,k) = e^{ix\cdot k} + c\frac{|k|^{\frac{n-3}{2}}}{|x|^{\frac{n-1}{2}}} e^{i|k||x|}A\left(\frac{x}{|x|}|k|, k\right) + o\left(|x|^{\frac{-n+1}{2}}\right). \qquad (5)$$

One would like to characterize among all functions $A(\xi,k)$ (given for all ξ,k in $\mathbb{R}^n$ with $|\xi|=|k|$) those which correspond to local (real valued, short range) potentials V . In the investigation of this problem one is led to study another family of solutions to the Schrödinger equation (1) (corresponding to a different choice of the Green's function in (3)) and the main difficulty turns out to be the solvability of the integral equation thus obtained. Our results are that this integral equation will be uniquely solvable if the potential is small or the energy is large, but it will have homogeneous solutions (even for some real vectors k) if the potential admits bound states. The precise statements and a physical intepretation are given in section 3; the main ingredients of the proofs are sketched in section 4. Section 2 is a brief and informal review of the motivating background.

2. A BRIEF REVIEW OF THE $\bar{\partial}$ METHOD IN QUANTUM INVERSE PROBLEMS

It was shown in [8] that a way to understand the nonlinear, overdetermined inverse scattering problem in $\mathbb{R}^n$ is to relate it to the linear overdetermined system of Cauchy-Riemann equations in $\mathbb{C}^n$ (the "$\bar{\partial}$ problem"). In [8] this was achieved by first working with the operator $(-\varepsilon+ i)\frac{\partial}{\partial t} + \Delta - V(t,x)$; although the behaviour for $\varepsilon \neq 0$ is qualitatively that of the heat equation, a scattering transform T_ε can be defined by requiring it to measure the departure from analyticity of a certain family of solutions. The compatibility conditions for the $\bar{\partial}$ problem then translate into explicit conditions for T_ε .

The straightforward limit of T_ε as $\varepsilon \to 0$ turned out to be too singular to be useful; a more careful limit (parametrized by directions $\gamma \in \mathbb{R}^n$ to resolve the singularity) yields a characterization and a reconstruction procedure for the time-dependent Schrödinger equation. We recall below (in notation somewhat different from that in [8]) the (formal) results thus obtained for the time-independent operator $-\Delta + V(x)$.

For any $\gamma \in \mathbb{R}^n \setminus 0$ (and $k \in \mathbb{R}^n$), define a Green's function for the Helmholtz equation by

$$G(x - y,k,\gamma) = -\frac{1}{(2\pi)^n} \int \frac{e^{i(x-y)\cdot\xi}}{\xi^2 - k^2 + i0\gamma\cdot(\xi-k)} d\xi \qquad (6)$$

(the symbol in the denominator is to be understood as $+i0$ for $\gamma\cdot\xi > \gamma\cdot k$, $-i0$ when $\gamma\cdot\xi < \gamma\cdot k$). Let $\phi(x,k,\gamma)$ be the family of solutions of (1) obtained by solving the integral equation

$$\phi(x,k,\gamma) = \exp(ix\cdot k) + \int G(x-y,k,\gamma)V(y)\phi(y,k,\gamma)dy \; . \qquad (7)$$

Define the "scattering transform" $T(\xi,k,\gamma)$ of V by

$$T(\xi,k,\gamma) = \int e^{-ix\cdot\xi} V(x)\phi(x,k,\gamma)dx \; . \qquad (8)$$

Note the analogy to (2), (3), (4).

The family of solutions ϕ defined by (6) and (7) and the corresponding T are precisely the objects which one is led to consider if one starts from the dissipative case $\varepsilon \neq 0$ and follows the route indicated above. These objects also turn out to be identical to the ones introduced by Faddeev [2,3] as natural extensions of the Gelfand-Levitan solutions in one dimension.

The explicit compatibility conditions obtained for T_ε from $\bar\partial$ give the following characterization equation for T (in the case when (7) admits no homogeneous solutions for any k in $\mathbb{R}^n$ - this issue will be discussed in section 3): for any ξ,k with $\xi^2 = k^2$, $\gamma\cdot\xi = \gamma\cdot k$ and every w perpendicular to both ξ and k but not parallel to γ

$$\hat{V}(\xi-k) = T(\xi,k,\gamma) - \iint w\cdot N[T](sw+\xi,sw+k,uw+\gamma)\frac{du\,ds}{s+i0u} \quad , \qquad (9)$$

where the nonlinear vector valued expression $N[T]$ is defined as

$$N[T](\xi,k,\gamma) = \frac{1}{(2\pi)^n} \int (\eta-k)T(\xi,\eta,\gamma)T(\eta,k,\gamma)\delta(\eta^2-k^2)\delta(\gamma\cdot(\eta-k))d\eta. \quad (10)$$

How does this help in deciding whether a candidate A for a scattering amplitude is admissible? By expressing the solutions of (7) in terms of the generalized eigenfunctions constructed from (3) one finds:

$$T(\xi,k,\gamma) = A(\xi,k) + \frac{i}{(2\pi)^{n-1}} \int_{\gamma\cdot\eta>\gamma\cdot k} \delta(\eta^2 - k^2)A(\xi,\eta)T(\eta,k,\gamma)d\eta . \quad (11)$$

This equation was known to Faddeev and Newton, who independently used it in a Gelfand-Levitan inversion procedure [4],[9]. Given A (on shell) one can view (11) as an integral equation to find T (on shell). Then (9) gives not only an explicit test for the admissibility of T (the right side should depend on the variable ξ-k only and have the appropriate decay) but also a simple way to (re)construct V from T purely by quadratures.

An explicit characterization and reconstruction based on the $\bar\partial$ problem was independently discovered by Beals and Coifman; working directly with the operator $-\Delta + V(x)$ on the variety $k^2 = 0$ in $\mathbb{C}^n$, they were led to introduce a scattering transform t different from the one above (t and T are related by analytic continuation). For small potentials (and $n = 3$) they have proved that the compatibility condition dictated by $\bar\partial$ is necessary and sufficient to characterize t . There seems to be, however, no useful way of constructing t from knowledge of the on-shell amplitude A.

One way to better understand the characterization equation (9) is to observe that it is equivalent to the following two conditions:

i) $w\cdot\nabla_\gamma T(\xi,k,\gamma) = -2\pi i\, w\cdot N[T](\xi,k,\gamma)$; (12)

if we write $k = s\gamma + k_\perp$ and $\xi = s\gamma + \xi_\perp$ with $k_\perp\cdot\gamma = 0$ and $\xi_\perp\cdot\gamma=0$ then

ii) the function $s \mapsto T(s\gamma + \xi_\perp, s\gamma + k_\perp,\gamma) - \hat{V}(\xi_\perp - k_\perp)$ extends to a Hardy function in the upper-half complex s-plane.

Condition ii) is essentially Faddeev's characterization. Since any solution T of (11) satisfies i) (whether or not A is admissible), he did not include i) as part of his characterization (of A).

When studying i) and ii) it is not necessary to retrace and justify every step of the limiting procedure of [8]. One can start from (6), (7) and (8). The major benefit of the $\bar{\partial}$ approach (aside from having provided a systematic route to the Faddeev characterization as well as to the only inversion method known for the time-dependent Schrödinger problem) may be the insight that there is a scattering transform T (related to A via a linear integral equation) which can be explicitly characterized. Although an explicit characterization of A itself is still not available, for several of the problems where it would be needed (reconstruction from partial data, reconstruction from noisy data) one may be able to use that of T instead. For example, given an arbitrary $\tilde{T}$, one may be able to construct (using the explicit conditions 12i) and ii)) an admissible T closest to $\tilde{T}$. Then, given a noisy (possibly unadmissible) $\tilde{A}$, solving (11) would lead to the corresponding noisy $\tilde{T}$ hence to a closest admissible T; the latter would yield a most reasonable candidate for a potential associated to $\tilde{A}$.

3. EXISTENCE AND NONEXISTENCE OF EXCEPTIONAL POINTS

The usefulness of the scattering transform T hinges on the solvability of the integral equation (11). The question of existence of homogeneous solutions to this equation had remained open since Faddeev's paper [2]; its importance has been emphasized by R. Newton [9], [10], [11]. It turns out that the inverse equation (11) has homogeneous solutions if and only if the integral equation (7) for the direct problem does. The decay at $|x| \to \infty$ expected of such a solution (for reasonable V) is that of G, namely $O\left(\frac{1}{|x|^{\frac{n-1}{2}}}\right)$; this motivates us to work in the weighted L^2 space

$$L^2_{-\delta}(\mathbb{R}^n) = \{f : \|f\|^2_{-\delta} =: \int (1 + x^2)^{-\delta} |f(x)|^2 \, dx < \infty\} \text{ with } \delta > \tfrac{1}{2} . \qquad (13)$$

A value of k for which (7) admits a nontrivial homogeneous solution in $L^2_{-\delta}$ will be called an exceptional point. We will use the notation $k = s\gamma + k_\perp$ (with $k_\perp \cdot \gamma = 0$). To establish the characterization equation for T one also needs to know about the possibility of exceptional points with $\operatorname{Im} s > 0$ (the integral equation used to construct the analytic continuation of ϕ will be described in section 4).

In the discussion below it will be useful to refer to the scattering matrix $S(\kappa) = I + A(\kappa)$ with $A(\kappa)$ the operator defined on $L^2(S^{n-1}, d\omega)$ (the square integrable functions on the unit sphere) by

$$A(\kappa)f(\alpha) = -\frac{i\kappa^{n-2}}{2(2\pi)^{n-1}} \int_{S^{n-1}} A(\kappa\alpha,\kappa\beta)f(\beta)d\omega(\beta) \quad , \quad |\alpha| = 1 . \tag{14}$$

We are now ready to list some results on nonexistence of exceptional points.

Theorem 1. If $c_n \int |V(x)|^{n/2} dx < 1$ then there are no exceptional points (an explicit constant c_n can be given).

Theorem 2. Suppose $|V(x)| \le C(1 + x^2)^{-\frac{1}{2} - \varepsilon}$. There are no exceptional points $k = s\gamma + k_\perp$, $\operatorname{Im} s \ge 0$ with $|k| \ge d_n(\varepsilon)C$ (a constant d_n can be given explicitly).

Remark. Here and in the other theorems in this section we can also treat potentials with local singularities no worse than $|x - x_0|^{-2+\varepsilon}$. For the next result, note that since $S(\kappa)$ is unitary, the spectrum of $A(\kappa)$ consists of points λ in the complex plane lying on the circle $|\lambda + 1| = 1$.

Proposition. If the real part of $\lambda > -1$ for all λ in the spectrum of $A(\kappa)$ then there are no real exceptional points at energy $\kappa = \sqrt{s^2 + k_\perp^2}$.

Corollary. If the operator norm $\|A(\kappa)\| < \sqrt{2}$ then there are no real exceptional points of energy κ .

Remark. On the large energy inversion:

It is known (and also follows from our Main Lemma in section 4) that $A(\kappa) \to 0$ in operator norm as $\kappa \to \infty$; so given A , one can (in principle) find M such that there will be no real exceptional points with $|k| \geq M$. For values of $\kappa \geq M$ one can thus solve for T in the integral equation (11) which takes place on the energy shell $|\xi| = |k| = \kappa$. By taking $|\xi_\perp| = |k_\perp| \geq M$ one can insure that only the values of T at energy $\geq M$ enter into (9). Therefore, one should be able to check for compatibility of those values of the data $A(\xi,k)$ with $|\xi| = |k| \geq M$ and then (re)construct V from knowledge of those values. In contrast to the Born inversion, this large energy procedure includes a characterization of the data lying outside an energy shell and does not rely exclusively on the limiting values of $A(\xi,k)$ for $|k| \to \infty$.

We now turn to our result on existence of exceptional points. The assumptions below allow (for any dimension $n \geq 3$) potentials with local singularities no worse than $|x - x_0|^{-2+\varepsilon}$ and which decay at infinity slightly faster than $|x|^{-2}$.

<u>Theorem 3.</u> Let $n \geq 3$. Suppose $V = V_1 + V_2$ with $|V_1(x)| \leq C(1+x^2)^{-1-\varepsilon}$ and $(1+x^2)^{\frac{1}{2}+\varepsilon} V_2(x) \in L^{n/2+\varepsilon}$. If V has any bound state, then for every γ there will be at least one real exceptional point.

The proof actually shows that if V admits a complex exceptional point then it will have real ones as well.

A preliminary version of Theorem 3 was announced in [7] .

We conclude this section with a physical interpretation of the existence of real exceptional points. Since the Faddeev Green's function is a "mixture" of both incoming and outgoing solutions, it was considered difficult to relate a homogeneous solution of the Faddeev-Lippman-Schwinger equation (7) to objects with clear physical meaning. Yet, looking at the integral equation (11), it is easy to express the existence of a real exceptional point as a simple property of the scattering matrix S .

Define a projection operator on $L^2(S^{n-1}, d\omega)$ by

$$Pf(\alpha) = f(\alpha) \text{ if } \gamma\cdot\alpha > \gamma\cdot k/|k| , \; Pf(\alpha) = 0 \text{ otherwise.} \tag{15}$$

Then (11) admits a homogeneous solution at energy $\kappa = |k|$ if $f = -A(\kappa)Pf$ for some square integrable f on the unit sphere with $Pf \not\equiv 0$. Writing $g = Pf$ we have $PS(\kappa)Pg = 0$. Thus, Theorem 3 shows that for any γ there is a cone Γ in $\mathbb{R}^n$ with axis γ and (roughly speaking) a state which in its momentum representation has support in Γ, with the property that the corresponding scattered state is supported outside Γ. It is as if no forward scattering takes place in this case (!). We know, however, that there always is forward scattering: the phenomenon just described will not occur for cones Γ with small openings (it is not hard to show that if the angle between γ and k is small, k cannot be a real exceptional point). The intuitive explanation we are left with is that it is interference which for this special choice of initial state causes nothing to be seen in the "forward directions."

Although the existence of real exceptional points appears to be a major stumbling block for the methods based on Faddeev's Green's function which generalize the Gel'fand-Levitan [4], [9] or Marchenko [10] procedures, for the scheme proposed in [8] the difficulties may be surmountable: the characterization equation (written as (9) in the case without poles) will now have some additional terms corresponding to the poles of T at the exceptional points; yet one may continue to use it to recover $\hat{V}$, without having to solve any additional (difficult) factorization problems.

4. OUTLINE OF THE PROOFS OF THE MAIN RESULTS

To make the integral operator in (7) compact we multiply every term of the equation, as is customary, by $|V(x)|^{1/2}$. Writing $k = s\gamma + k_\perp$ (with γ fixed from now on, and of unit length) we also multiply through by $\exp(-is\gamma\cdot x)$ to obtain an operator with nice analytic continuation to $\mathrm{Im}\, s > 0$. This yields the operator $L(s,k_\perp)$:

$$L(s,k_\perp)f(x) = |V|^{1/2}(x) \int \tilde{G}(x-y,s,k_\perp)\, V^{1/2}(y)f(y)dy, \qquad (16)$$

with

$$\tilde{G}(x,s,k_\perp) = -\frac{1}{(2\pi)^n}\int \frac{e^{ix\cdot\xi}}{\xi^2+2s\gamma\cdot\xi-k_\perp^2}\,d\xi \quad \text{for} \quad \mathrm{Im}\, s>0 \ . \tag{17}$$

It can be shown that, for potentials satisfying the assumptions of Theorem 3, L is in the Schatten class I_q, for some q depending on n. In this notation, an exceptional point is a pair $(s,k_\perp)$ for which $L(s,k_\perp)$ has the eigenvalue 1.

The brunt of the proof of Theorem 3 is to show that an appropriate modified Fredholm determinant $d(s,k_\perp) = \det_q(I-L(s,k_\perp))$ has the following properties (let $\mathbb{R}^n_\perp$ denote the set of $k_\perp \in \mathbb{R}^n$ with $k_\perp \cdot \gamma = 0$):

i) $d(s,k_\perp)$ is continuous on $\{\mathrm{Im}\, s \geq 0\}\times\mathbb{R}^n_\perp \setminus (0,0)$. (18)

ii) for any $k_\perp$ in $\mathbb{R}^n_\perp$, $d(s,k_\perp)$ is an analytic function of s in the upper-half complex plane.

iii) $d(s,k_\perp)\to 1$ as $\sqrt{|s|^2 + |k_\perp|^2}\to\infty$ uniformly in $\{\mathrm{Im}\, s\geq 0\}\times \mathbb{R}^n_\perp$.

Once properties i), ii), iii) have been established, it is easy to complete the proof: pick M such that $\sqrt{|s|^2+|k_\perp|^2}\geq M$ implies $|d(s,k_\perp)-1|<1/2$. Let $\Omega_\delta = \{s: |s|<M, \mathrm{Im}\, s\geq\delta\}$. One shows that for any $\delta>0$, $d(s,k_\perp)$ has a zero with $\mathrm{Im}\, s=\delta$ (for some $k_\perp$); for if not, Rouché's theorem would imply that the number of zeroes of d inside Ω_δ doesn't change with $k_\perp$. Yet, for $k_\perp = 0$ this is precisely the number of (non-zero) bound states, while for $|k_\perp|\geq M$ there are no zeroes in Ω_δ.

The proof of the properties (18) rests on the following estimate.

<u>Main Lemma</u>. For all $s,k_\perp$ with $\mathrm{Im}\, s\geq 0$ and $\sqrt{|s|^2 + |k_\perp|^2} \geq \kappa_0$,

$$\|\tilde{G}(\cdot,s,k_\perp)*f\|_{-\delta} \leq \frac{c}{\sqrt{|s|^2+|k_\perp|^2}}\,\|f\|_\delta \tag{19}$$

if $\delta>\frac{1}{2}$ and $n\geq 3$; an explicit $c = c(\kappa_0,\delta,n)$ can be given.

When $k_\perp = 0$ we recover the "limiting absorption principle" for the resolvent, well known in scattering theory. For the case $s = i|k_\perp|$, a similar result has been proved by Sylvester and Uhlmann [12] in a different context. The main difficulty here is to obtain a bound which remains uniform for $k_\perp$ or Ims or Res near zero. Our proof consists of a one dimensional calculation together with certain localizations in momentum space, which change, depending on the relative sizes of $k_\perp$, Res and Ims. The details will be given in [6].

The above lemma shows that L is a bounded operator on L^2 if V is as in Theorem 2. To prove compactness of L and also allow for singular potentials, one exploits the smoothing properties of $\tilde{G}$.

Theorem 2 is an immediate consequence of (19).

To prove Theorem 1, one needs an estimate which does not worsen as $k = s\gamma + k_\perp$ approaches zero. We show the following

Lemma 2. For all $k \in \mathbb{R}^n$

$$\| |V|^{1/2} G(\cdot,k,\gamma) * (V^{1/2} f)\|_{L^2} \le c_n' \|f\|_{L^2} \|V\|_{L^{n/2}} \quad . \tag{20}$$

The argument uses complex interpolation in a manner related to the harmonic analysis proofs of the sharp restriction theorems for the Fourier transform [13] . From Lemma 2 we know that potentials with small $L^{n/2}$ norm cannot have real exceptional points; the proof of Theorem 3 then shows that there cannot be any complex ones either.

After returning from the meeting in Oberwolfach, where the above results were presented, it was brought to our attention that G. M. Henkih and R. Novikov have independently solved the exceptional points problem (for potentials V which are C^∞ and rapidly decreasing at ∞). The precise statements and proof are to appear in [5], along with several other very important advances on the inverse problem.

Acknowledgement

This research was supported in part by the National Science Foundation under grant No. DMS-8602033.

REFERENCES

1. Beals, R. and Coifman, R.R., "Multidimensional Inverse Scattering and Nonlinear PDE," Proc. Symp. Pure Math. 43, American Math. Soc., Providence, 45-70 (1985).
2. Faddeev, L.D., "Growing Solutions of the Schrödinger Equation," Dokl. Akad. Nauk SSSR, 165, 514-517 (1965) (transl. Sov. Phys. Dokl. 10, 1033).
3. Faddeev, L.D., "Factorization of the S-matrix for the Multidimensional Schrödinger operator," Dokl. Akad. Nauk SSSR, 167, 69-72 (1966) (transl. Sov. Phys. Dokl. 11, 209-211.
4. Faddeev, L.D., "Inverse Problem of Quantum Scattering Theory,II," J. Soviet Math. 5, 334-396 (1976).
5. Henkin, G.M. and Novikov, R., "$\bar{\partial}$-Equation in the Multidimensional Inverse Scattering Problem," to appear in Uspekhi Mat. Nauk.
6. Lavine, R.B. and Nachman, A.I., "Exceptional Points in Multidimensional Inverse Scattering, in preparation.
7. Nachman, A.I., "Multidimensional Inverse Scattering for Schrödinger Operators," talk given at the International Conferench on Differential Equations and Mathematical Physics, Birmingham, Alabama (March 1986).
8. Nachman, A.I. and Ablowitz, M.J., "A Multidimensional Inverse Scattering Method," Studies in Appl. Math 71, 243-250 (1984).
9. Newton, R.G., "The Gel'fand-Levitan Method in the Inverse Scattering Problem," in Scattering Theory in Mathematical Phys., (ed. J.A. Lavita and J.P. Marchand), Reidel, Dordrecht, 1974.
10. Newton, R.G., "A Fadeev-Marchenko Method for Inverse Scattering in Three Dimensions," Inverse Problems 1, 127-132 (1985).
11. Newton, R.G., "Some Open Questions in Multidimensional Inverse Problems," plenary lecture at the International Conference on Differential Equations and Mathematical Physics, Birmingham, Alabama (March 1986).
12. Sylvester, J. and Uhlmann, G., "A Global Uniquenesss Theorem for an Inverse Boundary Value Problem," to appear in the Annals. of Math.

13. Tomas, P.A., "Restriction Theorems for the Fourier Transform," Proc. Symp. Pure Math., 35, Part I, 111-114 (1979).

ASPECTS OF INTEGRABILITY IN ONE AND SEVERAL DIMENSIONS

A. S. Fokas
Department of Mathematics and Computer Science and
Institute for Nonlinear Studies
Clarkson University
Potsdam, New York 13676

INTRODUCTION

In this paper I summarize some results obtained in the period June 1985 - July 1986. It is my pleasure to acknowledge collaboration with the following colleagues: M.J. Ablowitz, P. Clarkson, M. Kruskal, R.A. Leo, L. Martina, U. Mugan, V. Papageorgiou, P.M. Santini, and G. Soliani.

The results on Inverse Scattering in multidimensions and on the algebraic properties of equations in 2+1 (i.e. two spatial and one temporal) dimensions should be of particular interest: With respect to algebraic properties of equations in 2+1 we note that the question of finding the recursion operator and the bi-Hamiltonian formulation of these equations has remained open for a rather long time. It was even doubted in the literature if the relevant results in 1+1 could be extended to 2+1. P.M. Santini and the author have recently shown that equations in 2+1 solvable via the Inverse Scattering Transform are bi-Hamiltonian systems. They have also given explicitly the recursion and bi-Hamiltonian operators for large classes of equations in 2+1, including the Kadomtsev-Petviashvili (a two dimensional analogue of the Korteweg-deVries) and the Davey-Stewartson (a two dimensional analogue of the nonlinear Schrödinger) equations.

In this paper I emphasize the basic ideas and results. Furthermore an attempt is made to put these results into perspective. Details can be found in the sited papers.

1. AN INVERSE PROBLEM FOR N x N AKNS IN MULTIDIMENSIONS

This problem has been studied in[1], where I considered the inverse problem associated with the following system of N first-order equations in n+1 dimensions:

$$\Psi_{x_0} + \sigma \sum_{\ell=1}^{n} J_\ell \Psi_{x_\ell} = q\Psi,$$

$$\sigma = \sigma_R + i\sigma_I, \ \sigma_I \neq 0, \ n > 1, \tag{1.1}$$

where $q(x_0,x)$ is an N x N matrix-valued off-diagonal function in $\mathbb{R}^{n+1}$, decaying suitably fast for large x_0,x, and the J_ℓ are constant real diagonal N x N matrices (we denote the diagonal entries of J_ℓ by $J_\ell^1,\ldots,J_\ell^N$). Alternatively, using the transformation

$$\Psi(x_0,x,k) = \mu(x_0,x,k)\exp[i \sum_{\ell=1}^{n} k_\ell(x_\ell - \sigma x_0 J_\ell)], \quad k \in \mathbb{C}^n, \tag{1.2}$$

I considered

$$\mu_{x_0} + \sigma \sum_{\ell=1}^{n} (J_\ell \mu_{x_\ell} + ik_\ell[J_\ell,\mu]) = q\mu. \tag{1.3}$$

I assume that $n \leq N$, otherwise the entries of the J_ℓ matrices will be linearly related and one can always reduce n by a change of coordinates. An inverse problem in this case is defined as follows: Given appropriate inverse data T, where T is an N x N matrix-valued off-diagonal function of suitable inverse parameters, reconstruct the potential q.

There is a twofold motivation for considering such an inverse problem.

(a) If $\sigma = -1$ then the above reduces to the formulation of a physically important inverse scattering problem: Given the scattering amplitude $S(\lambda,k)$, λ, $k \in \mathbb{R}^n$, which is a function of the scattering parameters λ, k, reconstruct q.

(b) In recent years a deep connection has been discovered between inverse scattering of linear eigenvalue problems in one spatial dimension and the initial value problem of certain nonlinear evolution equations in 1+1 (i.e., one spatial and one temporal dimension). Recently a similar connection has been used to extend the above results to nonlinear evolution equations in 2+1 (i.e., two spatial and one temporal dimension)[2-6]. In particular the inverse scattering of (1.3) with $\sigma = -1$ and n=1 has been used to linearize the N-wave interaction equations in 2+1 (see Ref.[4]), the Davey-Stewartson (DS) I (see Ref.[4]) (a 2+1 analog of the nonlinear Schrödinger), and the modified Kadomtsev-Petviashvili (MKP) I (see Ref.[4]) (a 2+1 analog of the modified KdV) equations. Furthermore the inverse problem of (1.3) with $\sigma = i$ and n=1 has been used to linearize[5] DSII and MKPII. However, in spite of the above success in 2+1, no physically interesting equation is known to be related to (1.3) for $n > 1$ and $\sigma_I \neq 0$ [the N-wave interaction equations in n+1 spatial and one temporal dimension[7] are related to (1.3) but with $\sigma_I = 0$].

The novelty associated with inverse problems in greater than two spatial dimensions ($n > 1$) stems from the fact that while the potential $q(x_0,x)$ depends on $n + 1$ variables, the inverse data $T(k_R,k_I,m_2,\ldots,m_n)$, $k_R \in \mathbf{R}^n$, $k_I \in \mathbf{R}^n$, $m_\ell \in \mathbf{R}$, depends on $3n - 1$ variables. This has important implications:

(a) The inverse data must be appropriately constrained. This "characterization" of the inverse data is conceptually analogous to the characterization of the inverse scattering data in the multidimensional Schrödinger equation[8-11].

(b) The existence of "redundant" scattering parameters in the inverse scattering of the Schrödinger equation is used to reconstruct the potential in closed form in terms of the scattering amplitude function. This is the well-known Born approximation[12]. Can one use the redundancy of the inverse parameters here to also reconstruct q in closed form?

In the above paper, I do the following.

(a) Following A. Nachman and M.J. Ablowitz, I derive an equation that characterizes inverse data:

$$\hat{T}^{ij}(w_0,w) \doteq T^{ij}(w_0,w,\chi) - \frac{1}{\pi}\int_{\mathbf{R}^2} \frac{d\chi'_{p_R}\, d\chi'_{p_I}\, N^{ij}_{1p}[T](w_0,w,\chi^{p'})}{\chi_p - \chi'_p}, \quad (1.4)$$

where $w_0 \in \mathbf{R}$, $w \in \mathbf{R}^n$, $\chi \in \mathbb{C}^{n-1}$ are related to k, m and N is a quadratic function of T. That is, $T^{ij}(k,m)$ is appropriate inverse data iff the right-hand side of (1.4) is independent of χ. Hence, Eq. (1.4) serves as both characterizing T^{ij} and defining $\hat{T}^{ij}$.

(b) I reduce the general problem of reconstructing an N x N potential q in n + 1 dimensions to one of reconstructing a 2 x 2 potential with entries q^{ij}, q^{ji} in two dimensions. The inverse data needed for this reconstruction is precisely $\hat{T}^{ij}$, $\hat{T}^{ji}$. This reduction makes crucial use of the existence of redundant scattering parameters. In this sense it is the analog of the Born approximation. However, the crucial difference is that while in the inverse scattering of the multidimensional Schrödinger equation one can reconstruct the potential in closed form, here one can only reduce the general problem to one for 2 x 2 matrices in two dimensions. This reduced problem was solved in[4]].

Partial results about the case $\sigma = i$ were given in[13]]. Equation (1.3) was also considered in[14]] where, although the characterization problem was solved (an equation very similar to (1.4), the so-called "T equation", was first obtained in[14]]) the problem of an effective reconstructon of q was left open.

The basic steps are as follows:

1.1 Bounded Eigenfunctions.

The function $\mu(x_0,x,k)$ defined below, solves equation (1.3), is bounded for all complex values of k and tends to I for large k:

$$\mu^{ij}(x_0,x,k) = \delta^{ij}+\mathrm{sgn}\frac{(\sigma_I J_1^i)}{2\pi i}\int_{\mathbf{R}^2} d\xi_0 d\xi_1 \frac{\exp[i\beta^{ij}(x_0-\xi_0,x_1-\xi_1,k)]}{(x_1-\xi_1) - \sigma J_1^i(x_0-\xi_0)}$$

$$\times (q\mu)^{ij}(\xi_0,\xi_1,x_2-(x_1-\xi_1)J_2^i/J_1^i,\ldots,x_n-(x_1-\xi_1)J_n^i/J_1^i,k),\ k \in \mathbb{C}^n, \quad (1.5)$$

where β^{ij} is defined by

$$\beta^{ij}(x_0,x_1,k) \doteq \sum_{\ell=1}^{n} \frac{J_\ell^i-J_\ell^j}{\sigma_I}\left[x_0|\sigma|^2 k_{\ell_I} - \frac{x_1(\sigma k_\ell)_I}{J_1^i}\right],\ k_\ell = k_{\ell_R}+ik_{\ell_I}. \quad (1.6)$$

Equivalently μ_{ij} satisfies

$$\mu^{ij}(x_0,x,k) = \delta^{ij}+\frac{\mathrm{sgn}(\sigma_I J_1^i)}{2\pi i}\int_{\mathbf{R}^{n+1}} d\xi_0 d\xi [c_{n-1}\int_{\mathbf{R}^{n-1}} dm^2 e^{i\alpha^i(x-\xi,m)}]$$

$$\times \frac{\exp[i\beta^{ij}(x_0-\xi_0,x_1-\xi_1,k)](q\mu)^{ij}(\xi_0,\xi,k)}{x_1-\xi_1-\sigma J_1^i(x_0-\xi_0)}, \quad (1.7)$$

where

$$dm^2 \doteq dm_2 \ldots dm_n, \quad \alpha^i(x,m) \doteq \sum_{\ell=2}^{n} m_\ell (x_\ell - x_1 \frac{J_\ell^i}{J_1^i}), \quad c_n \doteq \frac{1}{(2\pi)^n}. \tag{1.8}$$

1.2 Departure from Holomorphicity.

Let μ^{ij} be defined by eq. (1.5). Then

$$\frac{\partial \mu}{\partial \bar{k}_p}(x_0, x, k) = \sum_{i,j} \gamma^i (J_p^i - J_p^j) \exp[i\beta^{ij}(x_0, x_1, k)]$$

$$\times\, c_{n-1} \int_{\mathbf{R}^{n-1}} dm^2 \exp[i\alpha^i(x,m)] T^{ij}(k,m) \mu(x_0, x, \lambda^{ij}(k,m)) E_{ij}, \tag{1.9}$$

where $\beta^{ij}(x_0, x_1, k)$, $\alpha^i(x,m)$ are defined by (1.6), (1.8) respectively; E_{ij} is an N x N matrix with zeros in all its entries except the ij^{th}, which equals 1; and λ^{ij} and T^{ij} are given by

$$\lambda_1^{ij}(k,m) \doteq (k_{1_R}^{ij} - \sum_{\ell=2}^{n} m_\ell \frac{J_\ell^i}{J_1^i}, k_{1_I}), \quad \lambda_r^{ij}(k,m) = (k_{r_R} + m_r, k_{r_I}); \; r = 2, \ldots n.$$

$$\gamma^i \doteq \bar{\sigma}/4\pi i |J_1^i \sigma_I|,$$

$$T^{ij}(k,m) \doteq \int_{\mathbf{R}^{n+1}} d\xi_0 d\xi \exp[-i\beta^{ij}(\xi_0, \xi_1, k) - i\alpha^i(\xi, m)](q\mu)^{ij}(\xi_0, \xi, k). \tag{1.10}$$

1.3 Characterization of T.

(a) Assume that $\partial\mu/\partial\bar{k}_p$ is given by Eq. (1.9) and the $T^{ij}(k,m)$ is given by (1.10). Then

$$L_{rp}^{ij} T^{ij}(k,m) = -\sum_{\ell=1}^{n} c_{n-1} \int_{\mathbf{R}^{n-1}} dM^2 T^{i\ell}(\lambda^{\ell j}(k,M), m-M) T^{\ell j}(k,m)$$

$$\times\, [(J_p^\ell - J_p^j)(J_r^i - J_r^\ell) - (J_r^\ell - J_r^j)(J_p^i - J_p^\ell)] \doteq N_{rp}^{ij}[T](k,m), \tag{1.11}$$

where

$$L_{rp}^{ij} \doteq (J_p^i - J_p^j)\frac{\partial}{\partial \bar{k}_r} - (J_r^i - J_r^j)\frac{\partial}{\partial \bar{k}_p}. \tag{1.12}$$

(b) Assume that $\partial\mu/\partial\bar{k}_p$ is given by Eq. (1.9) and that $\partial^2\mu/\partial\bar{k}_r\partial\bar{k}_p$ is symmetric with respect to r,p. Then $T^{ij}(k,m)$ solves (1.11).

Following A. Nachman and M.J. Ablowitz I introduce appropriate Born variables. Then equation (1.11) can be integrated. Furthermore, we can compute the limit of T^{ij} in the new coordinates as $|\chi_p| \to \infty$ (see below):

Let $w_0^{ij}, w_1^{ij}, w_\ell$, $\ell = 2,\ldots,n \in \mathbf{R}^1$ and $\chi_\ell \in \mathbb{C}^1$, $\ell = 2,\ldots,n$, be defined by

$$w_0^{ij} \doteq \sum_{r=1}^{n} \frac{J_r^i - J_r^j}{\sigma_I} |\sigma|^2 k_{r_I}, \quad w_1^{ij} \doteq -\sum_{r=1}^{n} \frac{J_r^i - J_r^j}{\sigma_I J_1^i} (\sigma k_r)_I - \sum_{r=2}^{n} m_r \frac{J_r^i}{J_1^i},$$

$$w_\ell \doteq m_\ell, \quad \chi_\ell^{ij} \doteq \frac{k_\ell}{J_1^j - J_1^i}, \quad \ell = 2,\ldots,n. \tag{1.13}$$

Assume that

$$(J_1^r - J_1^j)(J_p^i - J_p^j) \neq (J_1^i - J_1^j)(J_p^r - J_p^j), \text{ for all distinct } i,j,r \text{ and } p \neq 1. \tag{1.14}$$

For convenience of writing we usually suppress the superscripts, i,j in w_0, w_1, χ. Let k denote $k_1,\ldots,k_n$, m denote $m_2,\ldots,m_n$, χ denote $\chi_2,\ldots,\chi_n$, w denote $w_1,\ldots,w_n$. Then we have the following.

(a) The inverse of the transformation $k,m \to w_0,w,\chi$ is given by

$$k_\ell = \chi_\ell (J_1^j - J_1^i), m_\ell = w_\ell, \ell = 2,\ldots,n, \quad k_1 = -\sum_{r=2}^{n} (J_r^j - J_r^i)\chi_r + \frac{(\bar\sigma/|\sigma|^2) w_0 + \Sigma_{r=1}^n w_r J_r^i}{J_1^j - J_1^i}. \tag{1.15}$$

(b) In the new coordinates, Eq. (1.11) with r=1 becomes

$$\frac{\partial T^{ij}}{\partial \bar\chi_p}(w_0,w,\chi) = N_{1p}^{ij}[T](w_0,w,\chi), \quad p = 2,\ldots,n. \tag{1.16}$$

(c) In the new coordinates

$$T^{ij}(w_0,w,\chi) = \int_{\mathbf{R}^{n+1}} d\xi_0 d\xi \exp[-i(w_0\xi_0 + w\xi)](q\mu)^{ij}(\xi_0, \xi, w_0, w, \chi),$$

$$\text{where } w\xi = \sum_{r=1}^{n} w_r \xi_r. \tag{1.17}$$

(d) Let

$$\mu_i^{\ell j} \doteq \mu^{\ell j}(x_0, x, w_0^{ij}, w^{ij}, \chi^{ij}), \quad \hat\mu_i^{\ell j} = \lim_{|\chi_p| \to \infty} \mu_i^{\ell j}. \tag{1.18}$$

Then the $\hat\mu_i^{\ell j}$ satisfy

$$\hat{\mu}_i^{ij}(x_0,x,w_0,w) = \operatorname{sgn}\frac{(\sigma_I J_1^i)}{2\pi i} c_{n-1}\int_{\mathbf{R}^{2n}} \times$$

$$\times \frac{dx_0' dx' dm^2 \exp[i\{(x_0-x_0')w_0+w_0+(x-x')w\}]}{x_1-x_1' - \sigma J_1^i(x_0-x_0')} q^{ij}(x_0',x')\hat{\mu}_i^{jj}(x_0',x',w_0,w),$$

$$\hat{\mu}_i^{jj}(x_0,x,w_0,w) = 1 + \frac{\operatorname{sgn}(\sigma_I J_1^j)}{2\pi i} c_{n-1}\int_{\mathbf{R}^{2n}} \times$$

$$\times \frac{dx_0' dx' dm^2 q^{ji}(x_0',x')\hat{\mu}_i^{ij}(x_0',x',w_0,w)}{x_1-x_1' - \sigma J_1^j(x_0-x_0')}, \quad \hat{\mu}_i^{\ell j}=0, \text{ for all } \ell, \ell\neq i, \ell\neq j. \tag{1.19}$$

(e) $$\lim_{|\chi_p|\to\infty} T^{ij}(w_0,w,\chi) = \int_{\mathbf{R}^{n+1}} d\xi_0 d\xi \exp[-i(w_0\xi_0+w\xi)] \times$$

$$\times q^{ij}(\xi_0,\xi)\hat{\mu}_i^{jj}(\xi_0,\xi,w_0,w) \doteq \hat{T}^{ij}(w_0,w). \tag{1.20}$$

(f) The basic characterization equation is given by

$$\hat{T}^{ij}(w_0,w)=T^{ij}(w_0,w,\chi) - \frac{1}{\pi}\int_{\mathbf{R}^2} \frac{d\chi_{p_R}' d\chi_{p_I}' N_{1p}^{ij}[T](w_0,w,\chi^{p'})}{\chi_p - \chi_p'}. \tag{1.21}$$

where $\chi^{p'}$ denotes $\chi_2,\ldots,\chi_{p-1},\chi_p',\chi_{p+1},\ldots,\chi_n$.

1.4 Reconstruction of q.

It follows from the above that as $|\chi_p|\to\infty$, the μ^{ij}'s decouple. Furthermore, the $\hat{\mu}_i^{ij}$, $\hat{\mu}_i^{jj}$ satisfy a system of two equations depending on q^{ij}, q^{ji}. It turns out that: a) By introducing appropriate spatial variables ξ, the $\hat{\mu}_i^{ij}$, $\hat{\mu}_i^{jj}$ satisfy equations in two spatial equations. b) The invers data needed to reconstruct $\hat{\mu}_i^{ij}$, μ_i^{jj} (and hence q^{ij}, q^{ji}) can be obtained from $\hat{T}^{ij}$:

Let

$$\alpha_r \doteq \frac{J_2^j J_r^i - J_r^j J_2^i}{J_1^i J_2^j - J_1^j J_2^i}, \quad \beta_r = \frac{J_1^i J_r^j - J_1^j J_r^i}{J_1^i J_2^j - J_1^j J_2^i}, \quad r = 1,\ldots,n, \tag{1.22}$$

where for convenience of writing we have suppressed the dependence of α_r, β_r on i,j. Let $\xi_0 \in \mathbf{R}$, $\xi \in \mathbf{R}^n$,

$$x_0 = \xi_0, \quad x_1 = \xi_1, \quad x_2 = \xi_2,$$
$$x_\ell = \xi_\ell + \alpha_\ell \xi_1 + \beta_\ell \xi_2, \quad \ell = 3,\dots,n. \tag{1.23}$$

Then we have the following.

(a) The system (1.19) becomes

$$\hat{\mu}_i^{ij}(\xi_0,\xi,\hat{k}) = \operatorname{sgn}\frac{\sigma_I J_1^i}{2\pi i}\int_{\mathbf{R}^2} d\xi_0' d\xi_1' [\xi_1 - \xi_1' - \sigma J_1^i(\xi_0 - \xi_0')]^{-1}$$

$$\times \exp[i\hat{\beta}^{ij}(\xi_0-\xi_0', \xi_1 - \xi_1', \hat{k})] q^{ij}\hat{\mu}_i^{jj}(\xi_0', \xi_1', \xi_2$$
$$- (\xi_1-\xi_1')J_2^i/J_1^i, \xi_3,\dots,\xi_n, k),$$

$$\hat{\mu}_i^{jj}(\xi_0,\xi,\hat{k}) = 1 + \operatorname{sgn}\frac{\sigma_I J_1^j}{2\pi i}\int_{\mathbf{R}^2} d\xi_0' d\xi_1' [\xi_1 - \xi_1' - \sigma J_1^j(\xi_0 - \xi_0')]^{-1}$$

$$\times q^{ji}\hat{\mu}_i^{ij}(\xi_0', \xi_1', \xi_2 - (\xi_1-\xi_1')J_2^j/J_1^j, \xi_3,\dots,\xi_n,\hat{k}), \tag{1.24}$$

where

$$\hat{k} \doteq \sum_{r=1}^{n}(k_r\alpha_r + \frac{J_2^j - J_2^i}{J_1^j - J_1^i} k_r\beta_r), \quad \hat{\beta}^{ij}(x_0,x_1,\hat{k})$$

$$\hat{\beta}^{ij}(x_0,x_1,\hat{k}) \doteq \frac{J_1^i - J_1^j}{\sigma_I}[x_0|\sigma|^2 k_I - x_1\frac{(\sigma k)_I}{J_1^i}]. \tag{1.25}$$

(b) $\hat{T}^{ij}$ in the new coordinates becomes

$$\hat{T}^{ij}(\hat{k},\hat{m}) = \int_{\mathbf{R}^{n+1}} d\xi_0' d\xi' \exp[-i\hat{\beta}^{ij}(\xi_0', \xi_1', \hat{k}) +$$

$$+ \hat{m}_2(\xi_2' - \xi_1'\frac{J_2^i}{J_1^i}) + \sum_{r=3}^{n}\hat{m}_r\xi_r']q^{ij}\hat{\mu}_i^{jj}(\xi_0', \xi', \hat{k}), \tag{1.26}$$

where

$$\hat{m}_2 \doteq m_2 + \sum_{r=3}^{n} m_r\beta_r, \quad \hat{m}_\ell = m_\ell, \quad \ell = 3,\dots,n. \tag{1.27}$$

(c) The inverse data associated with (1.24) and the analogous problem for $\hat{\mu}_j^{ji}, \hat{\mu}_j^{ii}$ are given by $\tilde{T}^{ij}$, $\tilde{T}^{ji}$. Let

$$\tilde{T}^{ij}(k,\xi_2-\xi_1 J_2^i/J_1^i,\xi_3,\ldots,\xi_n) \doteq c_{n-1}\int_{\mathbf{R}^{n-1}} d\hat{m}\, \exp[i\hat{m}_2(\xi_2-\xi_1\frac{J_2^i}{J_1^i}) + i\sum_{r=3}^{n}\hat{m}_r\xi_r]\hat{T}^{ij}(\hat{k},\hat{m}). \tag{1.28}$$

Then

$$\tilde{T}^{ij}(\hat{k},\xi_2-\xi_1 J_2^i/J_1^i,\xi_3,\ldots,\xi_n) \doteq \int_{\mathbf{R}^2} d\xi_0' d\xi_1' \exp[-i\hat{\beta}^{ij}(\xi_0',\xi_1',k)] \times (q^{ij}\hat{\mu}_i^{jj})(\xi_0',\xi_1',\xi_2-(\xi_1-\xi_1')J_2^i/J_1^i,\xi_3,\ldots,\xi_n,\hat{k}). \tag{1.29}$$

2. INVERSE SCATTERING FOR THE HYPERBOLIC N x N AKNS IN MULTIDIMENSIONS

This problem has been studied in[15], where I considered equations (1.1) and (1.3) with $\sigma = -1$. This system appears to be physically more interesting than (1.1)-(1.3): (a) Since the system is hyperbolic one may consider the physically important question of inverse scattering (IS); i.e., given a scattering amplitude function $S(\lambda,k)$ find the potential $q(x_0,x)$. (b) A special case of the above system, namely if the J_ℓ's are constrained by

$$\frac{J_p^\ell-J_p^j}{J_r^\ell-J_r^j} = \frac{J_p^i-J_p^j}{J_r^i-J_r^j}, \quad p,r=1,\ldots,n, \quad i,j,\ell=1,\ldots,N, \tag{2.1}$$

is associated with the N-wave interaction in n+1 spatial and one temporal dimensions[7,16].

With respect to (a), (b) above the following results are obtained:

(a) I first define $S(\lambda,k)$, λ, $k \in \mathbf{R}^n$, in terms of an eigenfunction $\mu_L(x_0,x,k)$. $S(\lambda,k)$ motivates the introduction of the Born variables $w_0 \in \mathbf{R}^1$, $w \in \mathbf{R}^n$, $\chi \in \mathbf{R}^{n-1}$. I then define $T(\lambda,k)$ in terms of an eigenfunction $\mu(x_0,x,k)$. The crucial new step is that μ is required to have analyticity in one of the χ's, say χ_2, as opposed to one of the k's. Let

$$\hat{T}(w_0,w) := \lim_{|\chi_2|\to\infty} T(w_0,w,\chi);$$

then T^{ij} satisfies

$$\frac{1}{2}\hat{T}^{ij}(w_0,w) = T^{ij}(w_0,w,\chi)-(P_{\chi_2}T^{ij})(w_0,w,\chi), \tag{2.2}$$

where P_{χ_2} denotes a (+) or (-) projection in the variable χ_2, i.e.,

$$(P^{\pm}_{\chi_2}f)(\chi_2) := \int_{\mathbf{R}^1} d\chi_2' f(\chi_2')/2\pi i(\chi_2' - \chi_2 \mp i0).$$

The sign of certain parameters $\varepsilon_2^{\ell ji}$, where

$$\varepsilon_r^{\ell ji} := (J_1^j - J_1^i)(J_r^\ell - J_r^j) - (J_r^j - J_r^i)(J_1^\ell - J_1^j), \tag{2.3}$$

determines whether the (+) or the (-) projection is needed. Equation (2.2) defines $\hat{T}^{ij}$, which actually depends on q^{ij}, and

$$\hat{\mu}_i^{jj} := \lim_{|\chi_2|\to\infty} \mu_i^{jj}(x_0,x,w_0,w,\chi).$$

The question whether (2.2) is sufficient for the characterization of T^{ij} remains open. With a proper coordinate transformation, $\hat{\mu}_i^{ij}$, $\hat{\mu}_i^{jj}$, $\hat{\mu}_j^{ji}$, $\hat{\mu}_j^{ii}$ define an IS problem for a 2 x 2 potential with entries q^{ij}, q^{ji}, in two spatial dimensions. The inverse data needed for this problem are simply related to $\hat{T}^{ij}$, $\hat{T}^{ji}$. The solution of such an IS problem was given in[4]. It is interesting that there exists a simple relationship between T and S. Actually if N = 3 then T = S. This is quite remarkable since for the first time a closed-form expression can be obtained between S and T. Furthermore, μ_L satisfies a Volterra as opposed to a Fredholm equation; hence one immediately excludes the possibility of bound states.

With a proper coordinate transformation, the N-wave interaction equations in n + 1 spatial dimensions can always be reduced to two spatial dimensions. Thus a genuine three-spatial-dimensional nonlinear evolution equation, related to an IS problem, remains to be found.

Let $\mu_L(x_0,x,k)$, $x_0 \in \mathbf{R}^1$, $x,k \in \mathbf{R}^n$ be the solution of (1.3) which also solves

$$\mu_L^{\ell j}(x_0,x,k) = \delta^{\ell j} + \int_{-\infty}^{x_0} dx_0' \exp[ik(J - J^j)(x_0-x_0')]$$

$$\times (q\mu_L)^{\ell j}(x_0', x + J^\ell(x_0 - x_0'), k), \tag{2.4}$$

where

$$kJ^\ell := \sum_{r=1}^{n} k_r J_r^\ell, \quad mx = \sum_{r=1}^{n} m_r x_r, \quad c_n := (2\pi)^{-n},$$

and $x + J^\ell x_0$ denotes $x_1 + J_1^\ell x_0, \ldots, x_n + J_n^\ell x_0$. Let $\Phi(x_0, x)$ be the general solution of (2.1) with $\sigma = -1$ tending to $f(x + x_0 J)$ or $g(x + x_0 J)$ as $x_0 \to -\infty$ or $x_0 \to \infty$, respectively. The scatteirng operator is defined by $g = \tilde{S}f$, and $\tilde{S}$ is uniquely defined in terms of $S(\lambda,k)$, where

$$S^{ij}(\lambda,k) := c_n \int_{R^{n+1}} dx_0 dx \exp[i(k-\lambda)x + i(kJ^j - \lambda J^i)x_0]$$

$$\times (q\mu_L)^{ij}(x_0, x, k), \quad i \neq j, \tag{2.5}$$

and

$$kx := \sum_{\ell=1}^{n} k_\ell x_\ell .$$

The Born variables are defined by

$$w_0 := kJ^j - \lambda J^i, \; w_\ell := k_\ell - \lambda_\ell, \; \chi_\ell := \frac{k_\ell}{J_1^j - J_1^i}, \; \ell = 2, \ldots, n, \tag{2.6}$$

where, for convenience of writing, we suppress the dependence of w_0, χ on i,j.

$T^{ij}(\lambda,k)$ satisfies an equation similar to S^{ij} where $\mu_L^{\ell j}$ is replaced by $\mu_i^{\ell j}$. The eigenfunction $\mu_i^{\ell j}$ satisfies an equation similar to (2.4) where $\int_{-\infty}^{x_0}$ is replaced by $\int_{\ell j i}$. This integral is either $\int_{-\infty}^{x_0}$ or $-\int_{x_0}^{\infty}$ according to the following requirements: (1). If $\varepsilon_2^{\ell ji} = 0$ then choose $\int_{-\infty}^{x_0}$. (2) If $\varepsilon_2^{\ell ji} \neq 0$ then $(x_0 - x_0')\ \varepsilon^{\ell ji}$ must have the same sign for all ℓ (i,j are fixed). (3) If (2) can be satisfied with either $\int_{-\infty}^{x_0}$ or $-\int_{-\infty}^{x_0}$, choose the first integral. To illustrate the above, consider N = 4. Since $\varepsilon_2^{\ell ji} = \varepsilon_2^{\ell ji} = 0$, there exist only two nonzero $\varepsilon^{\ell ji}$, if they are of the same sign choose both integrals to be $\int_{-\infty}^{x_0}$, if they are of opposite signs then one integral is $\int_{-\infty}^{x_0}$ and the other is $-\int_{x_0}^{-\infty}$. In the Born variables, $T_{ij}(w_0, w, \chi)$ depends on

$\mu_i^{\ell j}(x_0,x,w_0,w,\chi)$, which satisfy

$$\mu_i^{\ell j}(x_0,x,w_0,w,\chi) = \delta^{\ell j} + \int_{\ell j i} dx_0' \exp\{i[w_0 - wJ^i + \sum_{r=2}^{n} \varepsilon_r^{\ell j i}\chi_r]$$

$$\times (x_0-x_0')\}(q\mu_i)^{\ell j}(x_0', x + J^\ell(x_0-x_0'), w_0, w, \chi). \tag{2.7}$$

Equation (2.7) implies the following: (1) $\mu_i^{\ell j}$ and hence T_{ij} have analyticity properties with respect to χ_2: $T^{ij}(\mu_i^{\ell j})$ is a (+) or (-) function with respect to χ_2 (i.e., it is holomorphic in the upper or lower χ_2 half-planes, respectively) according to whether $(x_0-x_0')\,\varepsilon_2^{\ell ji}$ is >0, or <0. (2) As $|\chi_2| \to \infty$, $\mu_i^{\ell j} \to \hat{\mu}_i^{\ell j}(w_0,x,w_0,w)$, where $\hat{\mu}_i^{\ell j}$ satisfy a reduced system: $\hat{\mu}_i^{\ell j} = 0$ for all $\ell \neq 1,j$, and $\hat{\mu}_i^{ij}$, $\hat{\mu}_i^{jj}$ satisfy a system of two integral equations of the Volterra type (see below). Hence as $|\chi_2| \to \infty$,

$$T^{ij} \to \hat{T}^{ij}(w_0,w) := c_n \int_{R^{n+1}} dx_0 dx \exp[i(w_0x_0 + wx)]$$

$$\times \; q^{ij}(x_0,x)\hat{\mu}_i^{jj}(x_0,x,w_0,w).$$

Since T^{ij} is a (+) or (-) function of χ_2 tending to $\hat{T}^{ij}$ as $|\chi_2| \to \infty$, its (-) or (+) projection must satisfy Eq. (2.2). [We define $p^+(1) = \frac{1}{2}$, $p^-(1) = -\frac{1}{2}$.]

Given T^{ij}, T^{ji} one can compute $\hat{T}^{ij}$, $\hat{T}^{ji}$, which actually can be used to reconstruct the 2 x 2 matrix potential with entries q^{ij}, q^{ji}: We consider the reduced system. The crucial fact is that it corresponds to N = 2, and hence

$$J_r^\ell = \alpha_r J_1^\ell + \beta_r J_2^\ell,\ \ell = i \text{ or } j,\ \alpha_r := \frac{J_2^j J_r^i - J_r^j J_2^i}{J_1^i J_2^j - J_1^j J_2^i},\ \hat{k} := \frac{k(J^j - J^i)}{J_1^j - J_1^i}; \tag{2.8}$$

β_r is defined as α_r with $2 \leftrightarrow 1$, $j \leftrightarrow i$, and again for convenience of writing we suppress the ij dependence of α_r, β_r. Since one can introduce a single $\hat{k}$, it follows that the reduced system must be transformable to a system in two dimensions. This is indeed the case. Let

$$x_0 = \xi_0,\ x_1 = \xi_1,\ x_2 = \xi_2 - \gamma^{ij}\xi_0 + \delta^{ij}\xi_1,\ x_\ell = \xi_\ell + \alpha_\ell \xi_1 + \beta_\ell \times$$

$$\times\ (\xi_2-\gamma^{ij}\xi_0 + \delta^{ij}\xi_1), \quad \ell = 3,\ldots,n, \tag{2.9}$$

where $\gamma^{ij} := (J_1^iJ_2^j - J_1^jJ_2^i)/(J_1^i - J_1^j)$, $\delta^{ij} := (J_2^i - J_2^j)/(J_1^i - J_1^j)$.
Then the reduced system satisfies

$$\hat{\mu}_i^{\ell j}(\xi_0,\xi,\hat{k}) = \delta^{\ell j} + c_n\int_{-\infty}^{x_0} d\xi_0'\int_{\mathbf{R}^{2n}} d\xi' dm\, \exp\{i[\hat{k}(J_1^\ell-J_1^j) + m_1J_1^\ell]\times$$
$$\times(\xi_0-\xi_0')+im_1J_1^\ell(\xi_1-\xi_1') + i\sum_{r=2}^{n} m_r(\xi_r-\xi_r')\}(q\hat{\mu}_i)^{\ell j}(\xi_0',\xi',\hat{k}),\ \ell=i,j, \tag{2.10}$$

where $(q\hat{\mu}_i)^{ij} = q^{ij}\hat{\mu}_i^{jj}$, $(q\hat{\mu}_i)^{jj} = q^{ji}\hat{\mu}_L^{ij}$. Hence $\hat{\mu}_i^{\ell j}$ satisfy two integral equations in the variables ξ_0, ξ_1:

$$\hat{\mu}_i^{\ell j}(\xi_0,\xi_1,\hat{k}) = \delta^{\ell j}+c_1\int_{-\infty}^{x_0} d\xi_0'\int_{\mathbf{R}^2} d\xi_1' dm_1 \exp\{i[\hat{k}(J_1^\ell-J_1^j) + m_1J_1^\ell \times$$
$$\times[(\xi_0-\xi_0') + im_1J_1^\ell(\xi_1-\xi_1')\}(q\hat{\mu}_i)^{\ell j}(\xi_0',\xi_1',\xi_2,\ldots,\xi_n,k). \tag{2.11}$$

The inverse data associated with (2.11) can be obtained from $\hat{T}^{ij}$: Let

$$\hat{w}_\ell=w_\ell,\ \ell = 3,\ldots,n,\quad \hat{w}_2=w_2 + \sum_{r=3}^{n} w_r\beta_r,\ \hat{\lambda} = \sum_{r=1}^{n}(\alpha_r+\beta_r)\lambda_r. \tag{2.12}$$

Then, since

$$w_0x_0 + wx = (\hat{k}J_1^j - \hat{\lambda}J_1^i)\xi_0 + (\hat{k}-\hat{\lambda})\xi_1 + \sum_{r=2}^{n}\hat{w}_r\xi_r,$$

$\hat{T}^{ij}$ in the new coordinates becomes

$$\hat{T}^{ij}(\hat{k},\hat{\lambda},\hat{w}) = c_n\int_{\mathbf{R}^{n+1}} d\xi_0 d\xi \exp[i(\hat{k}J_1^j - \hat{\lambda}J_1^i)\xi_0 + i(\hat{k}-\hat{\lambda})\xi_1 + i\sum_{r=2}^{n}\times$$
$$\times\hat{w}_r\xi_r](q^{ij}\hat{\mu}_i^{jj})(\xi_0',\xi',\hat{k}).$$

Hence, when we take the Fourier transform of $\hat{T}^{ij}$ with respect to $\hat{w}$, it follows that

$$\tilde{T}^{ij}(\hat{k},\hat{\lambda},\xi_2,\ldots,\xi_n) := \int_{\mathbf{R}^{n-1}} d\hat{w}\ \hat{T}^{ij}(\hat{k},\hat{\lambda},\hat{w})\exp(-i\hat{w}\xi) =$$
$$= c_n\int_{\mathbf{R}^2} d\xi_0 d\xi_1 \exp[i(\hat{k}J_1^j-\hat{\lambda}J_1^i)\xi_0 + i(\hat{k}-\hat{\lambda})\xi_1]q^{ij}\hat{\mu}_i^{jj}(\xi_0',\xi_1',\xi_2,\ldots,\xi_n,\hat{k}). \tag{2.13}$$

The inverse data $\tilde{T}^{ij}$ and its analog $\tilde{T}^{ji}$ are precisely what is needed to solve the inverse problem for q^{ij}, q^{ji} associated with Eq. (2.11) and its analogs for $\hat{\mu}_j^{ij}$, $\hat{\mu}_j^{ii}$, see[4].

S^{ij}, T^{ij} are defined in terms of $\mu_L^{\ell j}$, respectively. It is possible to find a simple relationship between $\mu_L^{\ell j}$, $\mu_i^{\ell j}$ which then yields a simple relationship between S^{ij}, T^{ij}. Actually if $N = 3$ then $T^{ij} = S^{ij}$.

It was shown in[14] that the N-wave interactions can always be reduced to three spatial dimensions. It is shown in[15] that they can actually be reduced to two spatial dimensions[17].

3 RECURSION OPERATORS AND BI-HAMILTONIAN FORMULATION OF EQUATIONS IN 2+1

This is joint work with P.M. Santini[18-20]. Since it is summarized in these proceedings in a separate contribution, I shall limit myself to a few remarks.

Since the discovery of an exact approach to nonlinear evolution equations in 1+1 (i.e., one spatial and one temporal dimension)[21], two interrelated aspects have recieved much attention in the literature:

(i) The development of a method of solving suitable initial-value problems. For initial data decaying at infinity such a method is the inverse scattering transform (IST)[22]. This method crucially utilizes the existence of an associated isospectral linear eigenvalue problem.

(ii) The investigation of the "algebraic" properties of the given equation. A fundamental role with respect to the algebraic properties is played by an integrodifferential operator, given various names in the literature: squared eigenfunction operator[23], recursion operator[24], strong symmetry[25], hereditary symmetry[25], Kahler operator[26], regular operator[27]. This operator has the following properties:

(a) It generates the associated hierarchy.

(b) It generates infinitely many symmetries (in particular, if it has a certain property which Fuchssteiner calls hereditary, it generates a set of commuting symmetries).

(c) Its adjoint generates gradients of conserved quantities (in particular, it generates a set of involutionary constants of the motion).

(d) Its multiplication by one Hamiltonian structure generates (under certain conditions[28]) a second Hamiltonian.

(e) The eigenfunctions of its adjoint are quadratic products of the eigenfunctions of the associated isospectral problem and form a complete set[29].

It should be noted that given the isospectral eigenvalue problem, there exists an algorithmic technique for obtaining the recursion operator (see for example[30]). This is, from a unification point of view, quite satisfactory, since both the method of solution (IST) and the algebraic properties (recursion operator) stem from the same entity (isospectral eigenvalue problem). For the Korteweg-deVries (KdV) equation $q_t = q_{xxx} - 6qq_x$, $q = q(x,t)$, the recursion operator Φ is $D^2 - 4q - 2q_x D^{-1}$, where $D \doteq \partial_x$, $(D^{-1}f)(x) \doteq \int_{-\infty}^{x} f(\xi)d\xi$. If Δ is the adjoint of Φ, then Δ satisfies $\Delta\psi^2 = 4\lambda\psi^2$, where ψ solves $\psi_{xx} - (q+\lambda)\psi = 0$.

The above two aspects have been thoroughly investigated for a number of physically important equations in 1+1. Each of these equations has physically significant two-spatial-dimensional analogues. For example, the KdV is generalized to the Kadomtsev-Petviashvili (KP) equation, the modified KdV to the modified KP, the non-linear Schrödinger to the Davey-Stewartson, etc. Furthermore, these equations are also related to isospectral eigenvalue problems which are appropriate generalizations of the corresponding one-dimensional ones. It is therefore natural to investigate aspects (i), (ii) above for two-spatial-dimensional (2+1) exactly solvable equations.

The extension of the IST to equations in 2+1 has been recently established in[2-6], (see also[31]). However, the problem of finding recursion operators in 2+1 has remained open; actually even the existence of such operators has been doubted in the literature. In this respect note:

1. The IST of the Benjamin-Ono equation has all the features of an equation in 2+1[32]. It is thus not surprising that its recursion operator has not been found. One of the authors (A.S.F.) and B. Fuchssteiner, after failing to find the recursion operator of the Benjamin-

Ono equation, introduced an alternative approach[33] for generating symmetries. This approach uses a certain function, called τ; it was subsequently applied to a number of equations, including the KP[39]. However, for equations in 2+1: (a) The relationship between τ and the eigenvalue problem has not been established. (b) There does not exist an algorithmic way of finding τ. (c) It is not known if τ can be used to obtain the second Hamiltonian. (d) τ is not hereditary.

2. The bi-Hamiltonian nature of equations in 1+1 has been emphasized as the fundamental property underlying integrability[35]. However, the bi-Hamiltonian nature of all equations in 2+1 as well as of the Benjamin-Ono has remained open. The existence of a recursion operator would directly imply the second Hamiltonian, since all these equations have one known Hamiltonian.

3. A number of important results pertinent to the algebraic properties of equations in 2+1 have obtained in the Soviet Union[36]. In particular Zakharov and Konopelchenko, in a very interesting paper[37], claimed that recursion operators are purely one-dimensional phenomena (i.e., they do not exist in more than one dimension). A careful reading of their work reveals that indeed recursion operators of a certain form do not exist in more than one dimension.

4. Several authors have noticed that mastersymmetries also exist for equations in 1+1; let us call such a mastersymmetry T. Actually, T comes from a nongradient function and can be used to generate Φ. However, τ comes from a gradient function and fails to generate a recursion operator.

The extension of the inverse scattering in 2+1 necessitated the introduction of a new idea, the use of $\bar{\partial}$ (DBAR). The extension of the theory of recursion operators and bi-Hamiltonian structures to equations in 2+1 necessitated the introduction of distributions, or more precisely the introduction of integral representations of certain differential operators.

We note that the proper analogue of T is not τ but a function denoted in[20] by T_{12}. This function also generates recursion operators in analogy with T.

4. THE KADOMTSEV-PETVIASHVILI EQUATION PERIODIC IN ONE SPATIAL DIMENSION AND DECAYING IN THE OTHER

This is joint work with V. Papageorgiou[38]; more details of this work can be found in[39]. We have considered four different problems: KPI periodic in x, KPI periodic in y, KPII periodic in x, KPII periodic in y. Here I will only summarize some results for KPI periodic in x of period 2ℓ.

4.1. Analytic Eigenfunctions.

The x-part of the Lax pair is given by

$$i\mu_y + \mu_{xx} + 2ik\mu_x = -u\mu. \tag{4.1}$$

Consider these solutions of (4.1) which also solve

$$\mu^{\pm}(x,y,k) = 1 + \frac{i}{2\ell}\int_{-\infty}^{y} d\eta \int_{-\ell}^{\ell} d\xi \sum_{m\in k\mp r\mathbb{Z}^+} \theta(m-k,x-\xi,y-\eta)u(\xi,\eta)\mu^{\pm}(\xi,\eta,k)$$

$$-\frac{i}{2\ell}\int_{y}^{\infty} d\eta \int_{-\ell}^{\ell} d\xi \sum_{m\in k\pm\tau\mathbb{Z}_0^+} \theta(m-k,x-\xi,y-\eta)u(\xi,\eta)\mu^{\pm}(\xi,\eta,k), \tag{4.2$^\pm$}$$

where

$$\tau = \pi/\ell, \quad \theta(m,x,y) = e^{imx-im^2y}. \tag{4.3}$$

These eigenfunctions are ⊕ and ⊖ with respect to the complex k-plane, i.e. they are the boundary values of functions analytic in the upper or lower halves of the k-complex plane.

A simple calculation shows

$$\mu^{+}(x,y,k) - \mu^{-}(x,y,k) = \sum_{m\in k+\tau\mathbb{Z}} T(k,m)N(x,y,m,k), \tag{4.4}$$

where

$$T(k,m) \doteq \frac{i}{2\ell}\,\mathrm{sgn}(k-m)\int_{-\infty}^{\infty} d\eta \int_{-\ell}^{\ell} d\xi\, u(\xi,\eta)\mu^{+}(\xi,\eta,k)\theta(k-m,\xi,\eta), \tag{4.5}$$

$\lambda-k$, $m-k$ are integer multiples of τ and N solves the same equation as $\mu^{-}(x,y,k)$ but with the forcing replaced by $\theta(\lambda-k,x,y)$.

4.2. A Symmetry Condition.

N, μ are related by (for simplicity of writing we suppress x,y)

$$N(\lambda,k) = \mu^-(\lambda)\theta(\lambda-k,x,y) + \begin{cases} \sum\limits_{m\varepsilon[k+\tau,\lambda]} F(\lambda,m)\mu^-(m)\theta(m-k,x,y) & \text{if } \lambda>k \\ -\sum\limits_{m\varepsilon[\lambda+\tau,k]} F(\lambda,m)\mu^-(m)\theta(m-k,x,y) & \text{if } \lambda<k, \end{cases} \tag{4.6}$$

where $[k+\tau,\lambda] = \{k+\tau,\ k+2\tau,\ldots,k+n\tau=\lambda\}$, and

$$F(\lambda,m) = \frac{i}{2\ell}\int_{-\infty}^{\infty} d\eta \int_{-\ell}^{\ell} d\xi N(\xi,\eta,\lambda,m)u(\xi,\eta). \tag{4.7}$$

4.3. Scattering Equation.

Using (4.4) and (4.6) it follows that

$$\mu^+(k)-\mu^-(k) = \sum_{m\varepsilon k+\tau\mathbb{Z}} f(k,m)\theta(m-k,x,y)\mu^-(m);\ f(k,m)=\text{sgn}(k-m)F(k,m). \tag{4.8}$$

Equation (4.8) can be viewed as a Riemann-Hilbert (RH) problem with a shift for $\mu^-(k)$. The inverse problem consists of finding μ^- in terms of the scattering data $f(k,m)$. The time evolution of $f(k,m)$ is given by

$$f(k,m,t) = f(k,m,0)e^{4i(m^3-k^3)t} \tag{4.9}$$

and $f(k,m,0)$ can be obtained in terms of initial data using (4.7).

Remarks

i) Localized solutions, periodic in x and decaying in y correspond to homogeneous solutions of $(4.2)^{\pm}$. Such solutions have been obtained in[40].

ii) The above formalism also follows by proper discretization of the results of[2].

iii) It is interesting that a symmetry condition of the type first introduced in[2] is necessary not only for KPI but also for KPII.

iv) Some of the above results were first obtained by P. Caudrey[41] viewing KP as a singular limit $N \to \infty$ of a matrix N x N one-dimensional problem.

5. THE INITIAL VALUE PROBLEM OF CERTAIN PAINLEVÉ EQUATIONS

This is joint work with U. Mugan and M.J. Ablowitz[42]; more details of this work can be found in[43].

The mathematical and physical significance of the six Painlevé transcendents, PI-PVI[44], has been well established: Mathematically, these are the only equations of the form $q_{tt} = F(q_t, q, t)$, where F is rational in q_t, algebraic in q and locally analytic in t which have the Painlevé property (i.e. their solutions are free from movable critical points). Physically, are closely related[45] to physically significant solvable PDE's and have appeared in several physical applications, see for example[46-50].

Central in the integrability of PDE's in 1+1 and 2+1 is their relation to isospectral eigenvalue problems. Similarly, central to the integrability of the Painlevé equations is their relation to isomonodromic problems (see Sato et al[51], Ueno[52], Flaschka and Newell[53], and Jimbo et al[54]).

We have systematically considered the initial value problem of PII, PIV, PV. Equation PIII is contained in PV for a special choice of one of the parameters of PV, equation PVI has been solved by C. Cosgrove and PI remains open. The basic approach is that introduced in[55] although we have made certain simplifications and extensions. Here I briefly summarize the main results using PIV as an illustrative example.

5.1. The Lax Pair.

PIV equation

$$\frac{d^2y}{dt^2} = \frac{1}{2y}\left(\frac{dy}{dt}\right)^2 + \frac{3}{2}y^3 + 4ty^2 + 2(t^2+\alpha)y + \frac{\beta}{y}, \tag{5.1}$$

is the compatibility of the following linear problems

$$Y_x(x;t) = \left[\begin{pmatrix} 1 & 0 \\ 0 & -1 \end{pmatrix} x + \begin{pmatrix} t & u \\ \frac{2}{u}(z-\theta_0-\theta_\infty) & -t \end{pmatrix} + \right.$$

$$\left. + \begin{pmatrix} \theta_0 - z & -\frac{uy}{2} \\ \frac{2z}{uy}(z-2\theta_0) & -(\theta_0 - z) \end{pmatrix} \frac{1}{x} \right] Y(x;t), \tag{5.2a}$$

$$Y_t(x;t) = \left[\begin{pmatrix} 1 & 0 \\ 0 & -1 \end{pmatrix} x + \begin{pmatrix} 0 & u \\ \frac{2}{u}(z-\theta_0-\theta_\infty) & 0 \end{pmatrix}\right] Y(x;t). \tag{5.2b}$$

Indeed $Y_{xt} = Y_{tx}$ implies

$$\frac{dy}{dt} = -4z + y^2 + 2ty + 4\theta_0, \tag{5.3a}$$

$$\frac{dz}{dt} = -\frac{2}{y}z^2 + \left(\frac{4\theta_0}{y} - y\right)z + (\theta_0 + \theta_\infty)y, \tag{5.3b}$$

$$\frac{du}{dt} = -u(y + 2t), \tag{5.3c}$$

where,

$$\alpha = 2\theta_\infty - 1, \quad \beta = -8\theta_0^2 . \tag{5.4}$$

Eliminating from (5.3) we obtain PIV.

5.2. Analytic Eigenfunctions and their Relationship.

We recall that in studying the initial value problem of an equation in 1+1 or in 2+1, one uses the time-independent part of the Lax pair to define an inverse problem, in terms of certain scattering or (more generally) inverse data. Then one uses the time dependent part to find the time evolution of these data. Similarly, here one uses (5.2a) to define an inverse probelm in terms of certain monodromy data; then one uses (5.2b) to find the time evolution of these data. To define an inverse problem one needs to consider the analyticity properties of $Y(x;t)$ in the whole complex x-plane. Since Y satisfies a linear ODE, its analyticity properties are completely determined from the singular points of (5.2a). Indeed, performing an analysis around $x = 0$, $x = \infty$, and introducing different solutions Y_j in different sectors S_j (so that Y_j's are normalized at ∞) it follows that (we assume $0 < \theta_0 < 1$, $0 \leq \theta_\infty < 1$, $\theta_0 \neq \frac{1}{2}$):

i) The Y_j's, $j = 1,\dots,5$ defined in the sectors S_j's, where the S_j's are given below and each S_j contains the intial boundary line, are related via

$$Y_{j+1}(x) = Y_j(x)G_j, \quad x \text{ on } C_{j+1}, \quad j = 1,2,3, \tag{5.5}$$

$$Y_1(x) = Y_4(xe^{2i\pi})\hat{G}_4, \quad x \text{ on } C_1.$$

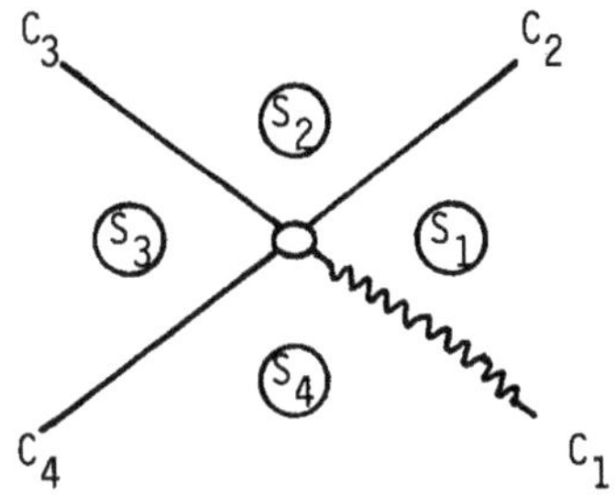

Figure 1.

where

$$G_1 = \begin{pmatrix} 1 & 0 \\ a & 1 \end{pmatrix}, \quad G_2 = \begin{pmatrix} 1 & b \\ 0 & 1 \end{pmatrix}, \quad G_3 = \begin{pmatrix} 1 & 0 \\ c & 1 \end{pmatrix}, \quad G_4 = \begin{pmatrix} 1 & d \\ 0 & 1 \end{pmatrix}, \tag{5.6}$$

$\hat{G}_4 = G_4 M_\infty$, $M_\infty = \text{diag}(e^{2i\pi\theta_\infty}, e^{-2i\pi\theta_\infty})$.

ii) $$Y_j(x) \sim Y_\infty(x) = \hat{Y}_\infty(x) e^{Q(x)} (1/x)^{D_\infty}, \quad \text{as } |x| \to \infty, \ x \text{ in } S_j, \tag{5.7}$$

where $\hat{Y}_\infty(x)$ is holomorphic at $x = \infty$,

$$Q(x) = \text{diag}[\tfrac{x^2}{2} + xt, -(\tfrac{x^2}{2} + xt)], \ D_\infty = \text{diag}(\theta_\infty, -\theta_\infty). \tag{5.8}$$

iii) $$Y_0(x) \sim \hat{Y}_0(x) x^{D_0}, \text{ as } x \to 0, \ x \text{ in } S_1, \tag{5.9}$$

where $\hat{Y}_0(x)$ is holomorphic at $x = 0$,

$$M_0 = \text{diag}(e^{2i\pi\theta_0}, e^{-2i\pi\theta_0}), \quad Y_0(xe^{2i\pi}) = Y_0(x) M_0. \tag{5.10}$$

iv) The connection matrix E_0 is defined by

$$E_0 \doteq \begin{pmatrix} \alpha_0 & \beta_0 \\ \gamma_0 & \delta_0 \end{pmatrix} \quad Y_1(x) = Y_0(x) E_0, \quad x \text{ in } S_1, \ \det E_0 = 1. \tag{5.11}$$

From the above it follows that Y is a sectionally holomorphic function. Its behavior at $x = 0$ and $x = \infty$ is determined from the monodromy matrices M_0 and M ; its jumps across the Stokes lines defined in Figure 1, are given by the Stoke's matrices $G_1, \ldots, G_4$. Hence its entire behavior is determined from the following data:

$$\text{Monodromy data} = \{a, b, c, d, \alpha_0, \beta_0, \gamma_0, \delta_0\}. \tag{5.12}$$

5.3. Properties of the Monodromy Data.

It turns out that all of the monodromy data can be expressed in terms of two. In particular

$$(\prod_{j=1}^{4} G_j)M_\infty = E_0^{-1}M_0^{-1}E_0, \tag{5.13}$$

$$(1+bc)e^{2i\pi\theta_\infty} + [a + (c+1)(ab+1)]e^{-2i\pi\theta_\infty} = 2\cos 2\pi\theta_0. \tag{5.14}$$

Furthermore, using (5.2b) it follows that the mondromy data are time invariant.

5.4. The Inverse Problem.

The inverse problem consists of reconstructing Y, or more precisely Ψ where $\Psi = Ye^{-Q(x)}$, in terms of the above monodromy data. The inverse problem can be formulated as a Riemann-Hilbert (RH) matrix problem along the contour $C_1 + C_2 + C_3 + C_4$. This RH problem is discontinuous both at the origin and at infinity. Using the method of[55] we separate this RH problem into a sum of two simpler RH problems, one defined on $C_2 + C_4$ and the other on $C_1 + C_3$. Furthermore, it is remarkable that the RH problem defined on $C_2 + C_4$ can be solved in closed form. Hence solving the initial vlaue problem of PIV is equivalent to solving a RH problem discontinuous at $x = 0$ and $x = \infty$ and defined on $C_1 + C_3$. This RH problem can be mapped to a continuous one using appropriate auxiliary functions. Hence its solution can be obtained in terms of a Fredholm integral equation. Having obtained Y it is straightforward to compute y, i.e. the solution of PIV.

5.5. Some Special Solutions.

For certain choices of the parameters α, β, PIV admits one parameter family of solutions expressible rationally in terms of the Weber-Hermite functions. Such solutions can also be obtained from the inverse problem. Let $\theta_\infty = \pm\theta_0 + \frac{n}{2}$, $n \in \mathbb{Z}$, then (5.14), (5.13) imply $a = c = 0$ and $b = -d$ respectively. Let us consider for concreteness $\theta_\infty = -\theta_0$, $0 < \theta_\infty < 1/2$. Then one is lead to the following RH problem:

$$\Psi^+(x)=\Psi^-(x)g_\Psi(x), \text{ on } C_1+C_3, \ g_\Psi=g_2^{-1} \text{ on } C_3; \ g_4 \text{ on } C_1, \tag{5.15}$$

$$\Psi \to I \text{ as } |x| \to \infty,$$

where

$$g_2 = e^{Q(x)} G_2 e^{-Q(x)}, \quad g_4 = e^{Q(x)} \hat{G}_4 e^{-Q(x)}.$$

This RH problem can be solved in closed form:

$$\Psi = \begin{pmatrix} x^{-\theta_\infty} & x^{\theta_\infty} F(x) \\ 0 & x^{\theta_\infty} \end{pmatrix}; \; F(x) \doteq \frac{d}{2\pi i} \int_L \hat{x}^{-2\theta_\infty} \frac{e^{2q(\hat{x})}}{\hat{x}-x} d\hat{x}, \; q(x) = \frac{x^2}{2} + xt, \tag{5.16}$$

where the contour L is given

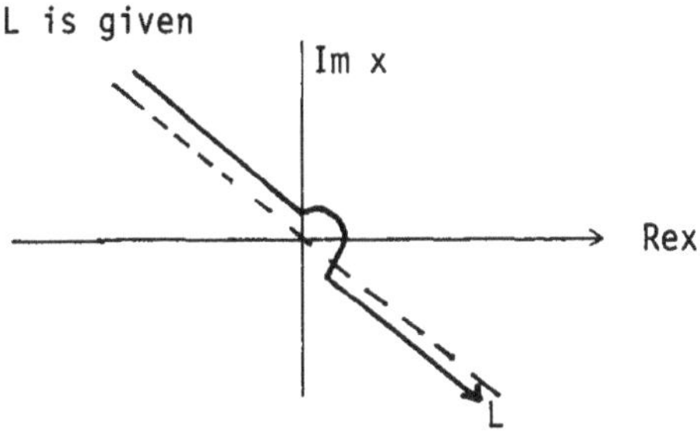

Note that

$$F(x) = -\frac{1}{x} f(t) + O\left(\frac{1}{x^2}\right), \; f(t) = He_{2\theta_\infty - 1}(it), \; He \doteq \text{Weber-Hermite}$$

function.

5.6. Schlesinger Transformations.

These transformations change θ_0, θ_∞ to θ_0', θ_∞', where

$$\theta_0' = \theta_0 \pm \frac{m}{2}, \quad \theta_\infty' = \theta_\infty \pm \frac{n}{2}, \quad m, n \in \mathbb{Z}. \tag{5.17}$$

To obtain these transformations let $Y \doteq R(x,t)Y$ correspond to θ_0', θ_∞' but to the same monodromy data. Then it can be shown that R satisfies a very simple RH problem, which can actually be sovled in closed form: For example

$$\begin{matrix} \theta_0' = \theta_0 - \frac{1}{2}, \\ \theta' = \theta_\infty + \frac{1}{2}, \end{matrix} \qquad R = \begin{pmatrix} 0 & 0 \\ 0 & 1 \end{pmatrix} x^{\frac{1}{2}} + \begin{pmatrix} 1 & \frac{uy}{2(z-2\theta_0)} \\ -\frac{z-\theta_0-\theta_\infty}{u} & -\frac{y(z-\theta_0-\theta_\infty)}{2(z-2\theta_0)} \end{pmatrix} x^{-\frac{1}{2}}. \tag{5.18}$$

6. SOLUTIONS TO A CLASS OF NONLINEAR SINGULAR INTEGRO-DIFFERENTIAL EQUATIONS

This is joint work with M.J. Ablowitz and M.D. Kruskal[56] and was motivated by some recent results of Constantin, Lax, and Majda[57]; these authors used the transformation $w = u + iHu$, where H denotes the Hilbert transform, to map the equation $u_t = uHu$ to the ODE $w_t = -\frac{i}{2}w^2$.

Using the fact that $w = u + iHu$ is the boundary value of a function analytic in the lower half complex x-plane it follows that $Hw = -iw$, and more generally $Hw^n = -iw^n$, $He^w = -ie^w$ etc. This enables us to map a large class of nonlinear singular integro-differential equations, via explicit transformations, to either ordinary differential equations or to linearizable partial differential equations. Conversely, given a linearizable PDE there is an algorithmic way of finding its singular integro-differential analogue. Examples include:

(a) ODE's	(b) Singular Integro-Differential Equations	
$w_t = -\frac{i}{2}w^2$	$u_t = uHu$	(6.1)
$w_t = w^3$	$u_t = u^3 - 3u(Hu)^2$	(6.2)
$w_t = ie^{-iw}$	$u_t = e^{Hu}\sin u.$	(6.3)

(a) PDE's	(b) Singular INtegro-Differential Equations	
$w_t = w_{xx} - i(w^2)x$	$u_t = u_{xx} + 2(uHu)_x$	(6.4)
$w_t + w_{xxx} - i\alpha(w^2)_x + \beta(w^3)_x = 0$	$u_t + u_{xxx} + 2\alpha(uHu)_x + \beta(u^3 - 3u(Hu)^2)_x = 0$	(6.5)
$w_{xt} = ie^{-iw}$	$u_{xt} = e^{Hu}\sin u$	(6.6)
$w_t + i(w_{xx} + (w^2)_x) = 0$	$u_t = (Hu)_{xx} + 2(uHu)_x.$	(6.7)
$\frac{\partial}{\partial x}(w_t + w_{xxx} - i(w^2)_x) = -3\sigma^2 w_{yy}$	$\frac{\partial}{\partial x}(u_t + u_{xxx} + (2uHu)_x) = -3\sigma^2 u_{yy}.$	(6.8)

Equation (6.4a) is essentially Burgers equation and can be linearized via the Cole-Hopf transformation $w = -i(\ln f)_x$. Equation (6.4b) arises in various population ecological models and to our knowledge, was first considered and solved via a dependent variable transformation and splitting into upper and lower functions by J. Satsuma[58]. In equations (6.5) α, β are real constants, and (6.5b) is an analog of the Gardner equation (a combination of KdV and modified KdV). Equation (6.6b) is related to the Liouville equation (6.6a) and is known to be linearizable.

A (3+1)-dimensional equation can also be linearized via (6.8a). Namely let $H_z u$ denote the Hilbert transform of $u(x,y,z,t)$ with respect to the variable z, i.e.,

$$H_z u = \frac{1}{\pi} \int_{-\infty}^{\infty} \frac{u(x,y,\xi,t)}{\xi - z} d\xi.$$

Then instead of K-P: we may consider a multi-dimensional analog of (6.8b)

$$\frac{\partial}{\partial x}(u_t + u_{xxx} + 2(uH_z u)_x) = -3\sigma^2 u_{yy}, \tag{6.9}$$

and it is also mapped to the KP equation (6.8a), via $w = u + iH_z u$.

7. HODOGRAPH TRANSFORMATIONS OF LINEARIZABLE PDE'S

This is joint work with P. Clarkson and M.J. Ablowitz[59]. Since P. Clarkson will present these results in a separate contribution of these proceedings, I will only make a few comments.

We call two PDE's <u>equivalent</u> if one can be obtained from the other by a transformation involving the dependent variables $u = \phi(v)$ and/or the introduction of a potential variable ($u = v_x$ or $u_x = v$).

It is well known[60] that the most general <u>semilinear</u> equation of the form

$$u_t = u_{xx} + f(u,u_x) \tag{7.1}$$

which is linearizable, is either linear or the Burger's equation (which with the above definition, is equivalent to a linear equation). Fokas and Yortsos[61] have shown that the most general <u>quasilinear</u> equation of the form

$$u_t = g(u)u_{xx} + f(u,u_x) \tag{7.2}$$

which is linearizable, is equivalent to the equation

$$u_t = (u^{-2}u_x)_x + \alpha u^{-2}u_x, \quad \alpha \text{ is an arbitrary constant.} \tag{7.3}$$

The above equation can be mapped to the Burger's equation via an <u>extended Hodograph</u> transformation, i.e. a transformation of the form

$$\tau = t, \quad \xi = \int^x u(x',t)dx'. \tag{7.4}$$

It is also well known that the Harry-Dym[62] equation, can be mapped to a modified Korteweg-deVries (MKdV) equation via an extended Hogograph[63]. In that sense, the Harry-Dym equation is a quasilinear analogue of the MKdV equation.

One is naturally motivated to ask the following questions:

i) Is there an algorithmic method of finding a quasilinear analogue of any semilinear equation?

ii) Is the associated quasilinear equation unique?

iii) Conversely, given a quasilinear equation, is there an algorithmic method of finding whether it can be mapped to a semilinear equation as well as finding this semilinear equation?

In the above paper we consider the above questions for semilinear and quasilinear equations of the type

$$u_t = u_{nx} + f(u,u_x,\ldots,u_{(n-1)x}),\ n \geq 2,\ u_{nx} \doteq \frac{\partial^n u}{\partial x^n}\,. \tag{7.5}$$

and

$$u_t = g(u)u_{nx}+f(u,u_x,\ldots,u_{(n-1)x}),\ n \geq 2,\ \frac{dg}{du} \neq 0 \tag{7.6}$$

respectively. The answer to question i) above is affirmative. Also, the associated quasilinear equation is unique, in the sense that extended and pure hodograph transformations yield equivalent quasilinear equations. (By pure hodograph we mean transformations of the form $\tau = t,\ \xi = u(x,t)$). Furthermore, we find the most general equation of the form (7.6) which can be mapped via an extended hodograph transformation to a semilinear form.

The above results are of some interest in establishing whether an equation is a candidate for linearization. Suppose that one is interested in investigating whether a given quasilinear equation is linearizable. We propose the following algorithmic procedure:

1. Put the equation into its potential canonical form

$$v_t = v_x^{-n} v_{nx} + H(v_x,v_{xx},\ldots,v_{(n-1)x}), \tag{7.7}$$

by using the transformation $v_x = g^{-1/n}(u)$.

2. Apply a pure hodograph transformation to equation (7.7). If equation (7.7) is transformable to a semilinear equation, it will

become

$$u_t = u_{nx} + F(u_x, u_{xx}, \ldots, u_{(n-1)x}). \tag{7.8}$$

3. Investigate whether equation(7.8) is linearizable. This is easier than investigating whether (7.6) is linearizable directly. The reason for this is twofold. First, for at least third order equations there is a complete classification of all linearizable equations. Within equivalence, there exist only six such equations[64]. Hence one simply needs to study if there exists an equivalence transformation to map equation (7.8) with $n = 3$, to one of these six canonical equations. Second, for equations with $n \geq 4$ one may investigate the question of linearization via the Painlevé test. We point out that quasilinear partial differential equations do not appear suitable for applying the Painlevé test. Ramani, Dorizzi and Grammaticos[65] (see also[67] and the references therein), introduced the notion of "<u>weak-Painlevé</u>" in order to deal with equations such as the Harry-Dym equation which are linearizable after a change of variables. However, the higher KdV equation

$$u_t = u_{xxx} + u^3 u_x ,$$

although it is thought to be nonlinearizable (it has only three independent polynomial conservation laws of a certain type[66]) and is also "weak-Painlevé"[67]. Therefore the "weak-Painlevé" concept does not seem to distinguish between a linearizable and a nonlinearizable equation.

We point out that one often finds in the literature claims of "new" third order linearizable equations. These equations, using the notion of equivalence can be mapped via a pure hodograph transformation to one of the six canonical equations mentioned above.

The above algorithmic approach is useful provided that a given linearizable quasilinear equation <u>can</u> be mapped to a semilinear form. The above approach will fail if there exist linearizable quasilinear equations which can not be mapped to a semilinear form. It is shown in[61] that such equations do not exist for at least $n = 2$. The question of whether such equations exist for $n \geq 3$ remains open.

8. THE SCALING REDUCTION OF THE THREE-WAVE RESONANT SYSTEM AND THE PAINLEVE VI EQUATION

This is joint work with R.A. Leo, L. Martina, and G. Soliani[68].

It is well known that for a large class of equations, the large time asymptotic limit is governed by certain similarity solutions of the underlying PDE. If this PDE is an exactly solvable equation in 1+1 (i.e. in one spatial and in one temporal dimension) one expects[45] that the similarity solutions satisfy an ODE of the Painlevé type.

In the above paper we considered the three-wave resonant interactions in the case of explosive instability[69-70].

$$u_{j_t} + c_j u_{j_x} - i u_\ell^* u_k^* = 0, \quad j,k,\ell = 1,2,3, \quad j \neq k \neq \ell, \tag{8.1}$$

where $u_j(x,t)$ are the complex amplitudes of the wave parameters, c_j are their group relations and * denotes complex conjugate. Assuming $c_1 < c_2 < c_3$ and using invariance under x-translation, t-translation and appropriate scaling we are lead to consider a system of three first order nonlinear ODE's.

This system via a series of transformations, can be mapped to a single second order ODE, which is quadratic in the second derivative. Such equations are outside the class investigated by Painlevé and his school[44], however the equation obtained above is a particular case of an equation recently studied by Fokas and Yortsos[75], in their investigation of exact trasnformations of Painleve VI equation:

A fundamental role in the exact treatment of the Painlevé equations is played by certain transformations which map solutions of a given Painlevé to solutions of the same Painlevé but with different choice of the parameters. Such transformations for PII-PVI were given in[71-74] respectively. Finding such a transformation for PVI necessitated the introduction of an auxiliary equation which is quadratic in the second derivative. However, it was shown in[75] that this equation can be mapped to PVI.

The second order ODE obtained from the similarity reduction of the three-wave resonant interactions is a special case of the above auxiliary equation (see equation (2) of[75]).

REFERENCES

[1] Fokas, A.S., J. Math. Phys. 27 (7), (1986).

[2] Fokas, A.S. and Ablowitz, M.J., Stud. Appl. Math. 69, 211 (1983).

[3] Ablowitz, M.J., Bar Yaacov, D., and Fokas, A.S., Stud. Appl. Math. 69, 135 (1983).

[4] Fokas, A.S., Phys. Rev. Lett. 51 (1983).

[5] Fokas, A.S., and Ablowitz, M.J., J. Math. Phys. 25, 2505 (1984).

[6] Fokas, A.S. and Ablowitz, M.J., Lectures on the Inverse Scattering Transform for Multidimensional (2+1) Problems, edited by K.B. Wolf (Springer, Berlin, 1983), pp. 137-183.

[7] Ablowitz, M.J. and Haberman, R., Phys. Rev. Lett. 35, 1125 (1975).

[8] Faddeev, L.D., J. Sov. Math. 5, 334 (1976).

[9] Newton, R.G., J. Math. Phys. 21, 1698 (1980); 2191 (1981); 23, 594, 693 (1982).

[10] Nachman, A. and Ablowitz, M.J., Stud. Appl. Math. 71, 243 (1984).

[11] Beals, R. and Coifman, R., "Multidimensional inverse scattering", preprint, Yale University, 1984.

[12] Newton, R.G., Scattering Theory of Waves and Particles (Springer, Berlin, 1982), 2nd ed.

[13] Nachman, A.I., Fokas, A.S., and Ablowitz, M.J., "Note on the inverse problem for first order multidimensional systems", Lectures in Applied Mathematics, 1984 AMS-SIAM Summer Seminar (Am. Math. Soc., Providence, RI, 1986).

[14] Nachman, A.I. and Ablowitz, M.J., Stud. Appl. Math. 71, 251 (1984).

[15] Fokas, A.S., Phys. Rev. Lett., 57, (2), 159-162, (1986).

[16] Konopelchenko, B.G., Phys. Lett., 93A, 442 (1983).

[17] Manakov, S.V., has also noted this fact (private communication).

[18] Fokas, A.S. and Santini, P.M., Stud. Appl. Math. 75, 179 (1986).

[19] Santini, P.M. and Fokas, A.S., Recursion Operators and Bi-Hamiltonian Structures in Multidimensions I, INS #65, preprint, 1986.

[20] Fokas, A.S. and Santini, P.M., Recursion Operators and Bi-Hamiltonian Structures in Multidimensions II, INS #67, preprint, 1986.

[21] Gardner, C.S., Greene, J.M., Kruskal, M.D. and Miura, R.M., Phys.

Rev. Lett. 19, 1095 (1967).

[22] Ablowitz, M.J. and Segur, H., Solitons and the Inverse Scattering Transform, SIAM, Philadelphia, 1981; Calogero, F., and Degasperis, A., Spectral Transform and Solitons I, North Holland, 1982.

[23] Ablowitz, M.J., Kaup, D.J., Newell, A.C., and Segur, H., Stud. Appl. Math. 53, 249 (1974).

[24] Olver, P.J., J. Math. Phys. 18, 1212 (1977).

[25] Fuchssteiner, B., Nonlinear Anal. 3, 849 (1979).

[26] Magri, F., J. Math. Phys. 19, 1156 (1978); in Nonlinear Evolution Equations and Dynamical Systems, M. Boiti, F. Pempinelli, and G. Soliani (Eds.), Lect. Notes in Phys., Vol. 120, Springer, New York, 1980, p. 233.

[27] Gel'fand, I.M. and Dorfman, I.Ya., Funct. Anal. Appl. 13, 13 (1979); 14, 71 (1980).

[28] Fokas, A.S. and Fuchssteiner, B., Lett. Nuovo Cimento 28, 299 (1980); Fuchssteiner, B. and Fokas, A.S., Symplectic Structures, their Backlund Transformations and Hereditary Symmetries, Phys. D 4, 47 (1981).

[29] Gerdjikov, V.S., Ivanov, H.I., and Kulish, P.P., Teor. Mat. Fyz. 44, 342 (1980).

[30]Fokas, A.S. and Anderson, R.L., J. Math. Phys. 23, 1066 (1982).

[31] Manakov, S.V., Phys. D, 3, 420 (1981).

[32] Fokas, A.S. and Ablowitz, M.J., Stud. Appl. Math. 68 (1983).

[33] Fokas, A.S. and Fuchssteiner, B., Phys. Lett. A, 86, 341 (1981).

[34] Oevel, W. and Fuchssteiner, B., Phys. Lett. A, 88A, 323 (1982).

[35] Gel'fand, I.M., private communication.

[36] Konopelchenko, B.G. and Dubrovsky, V.G., Phys. D, 16D, 79 (1985) and references therein.

[37] Zakharov, V.E. and Konopelchenko, B.G., Comm. Math. Phys. 94, 483 (1984).

[38] Papageorgiou, V. and Fokas, A.S., The Kadomtsev-Petviashvili Equation Periodic in One Spatial Dimension and Decaying in the Other", (preprint), Clarkson University, 1986.

[39] Papageorgiou, V., Ph.D. Thesis, Clarkson University, January 1987.

[40] Zakharov, V.E. and Manakov, S.V., Funct. Anal. and its Appl., 19 11 (1985).

[41] Caudrey, P.J., Discrete and Periodic Spectral Transforms Related to the Kadomtsev-Petviashvili Equation (preprint) 1985.

[42] Fokas, A.S., Mugan, U. and Ablowitz, M.J., The Inital Value Problem of the Second, Fourth and Fifth Painlevé Equations", (preprint), Clarkson University, 1986.

[43] Mugan, U., Ph.D. Thesis, Clarkson University, 1986.

[44] Ince, E.L., Ordinary Differential Equations (1927), New York, Dover, 1956.

[45] a) Ablowitz, M.J., Segur, H., Phys. Rev. Lett. 38, 1103 (1977);
b) Ablowitz, M.J., Ramani, M.J., and Segur, H., Lett. Nuovo Cimento 23, 333 (1978).
c) Ablowitz, M.J., Ramani, A., and Segur, H., J. Math. Phys., 21, 715 (1980).
d) Ablowitz, M.J., Ramani, A., and Segur, H., J. Math. Phys. 21, 1006 (1980).

[46] Barouch, E., McCoy, B.M., and Wu, T.T., Phys. Rev. Lett. 31, 1409 (1973).
Wu, T.T., McCoy, B.M., Tracy, C.A., and Barouch, E., Phys. Rev. B 13, 316 (1976).

[47] Jimbo, M., Miwa, T., Mori, Y., and Sato, M., Physica 1D, 80 (1980).
Jimbo, M., Miwa, T., Proc. Jpn. Acad. 56, A, 405 (1980).

[48] Miles, J.W., Proc. R. Soc. London, A 361, 277 (1978).

[49] McCoy, B.M., Perk, J.H.H., and Schrock, R., Nucl. Phys. B (1983) and Correlation functions of the transverse Ising chain at the critical field for large temporal and spatial separations.

[50] Fokas, A.S., Leo, R.A., Martina, L., and Soliani, G., Phys. Lett. A, Vol. 115, No. 7, (1986).

[51] Sato, M., Miwa, T., and Jimbo, M., Publ. RIMS, Kyoto Univ. 15 201 (1979); Proc. Jpn. Acad. 55 A, 267 (1979); In: Lect. Notes in Phys., Vol. 126, Berlin, Heidelberg, New York: Springer 1980, p. 429.

[52] Ueno, K., Master's thesis, Kyoto Univ. (1979). Proc. Jpn. Acad. 56 A, 97 (1980); 56 A, 103 (1980).

[53] Flaschka, H., and Newell, A.C., Commun., Math. Phys. 76, 67 (1980).

[54] Jimbo, M., Miwa, T., and Ueno, K., Physica 2D, 306 (1981).

[55] Fokas, A.S. and Ablowitz, M.J., Commun. Math. Phys. 91, 381-403 (1983).

[56] Ablowitz, M.J., Fokas, A.S. and Kruskal, M.D., Note on Solutions to a Class of Nonlinear Singular Integro-Differential Equations" to appear in Physics Lett. A., (1986).

[57] Constantin, P., Lax, P.D., and Majda, A., Comm. Pure Appl. Math. 38, p. 715 (1985).

[58] Satsuma, J., J. Phys. Soc. Japan, 50, 1423, (1981).

[59] Clarkson, P., Fokas, A.S. and Ablowitz, M.J., "Hodograph Transformations of Linearizable PDE's", preprint, Clarkson University (1986).

[60] Fokas, A.S., J. Math. Phys. 21, 1318 (1980).

[61] Fokas, A.S. and Yortsos, Y.C., SIAM J. Appl. Math., 42, 318-332, (1982).

[62] Kruskal, M.D., Nonlinear Wave Equations, in Dynamical Systems, Theory and Applications, ed. j. Moser, Lect. Notes Phys., 38, 310-254, Springer-Verlag, New York, (1975).

[63] Wadati, M., Konno, K., and Ichikawa, Y.H., J. Phys. Soc. Japan, 47, 1698 (1979).
Shimuzu, T. and Wadati, M., Prog. Theor. Phys., 63, 808 (1980).

[64] Svinolupov, S.I. and Sokolov, V.V., Func. Anal. Appl., 16, 317 (1982).
Svinolupov, S.I., Sokolov, V.V. and Yamilov, R.I., Sov. Math. Dokl., 28, 165-168, (1983).

[65] Ramani, A., Dorizzi, B., and Grammaticos, B., Phys. Rev. Lett., 49, 1539 (1982).
Ranada, A.F., Ramani, A., Dorizzi, B. and Grammaticos, B., J. Math. Phys. 26, 708 (1985).

[66] Miura, R.M., SIAM Rev., 18, 412 (1976).

[67] Weiss, J., J. Math. Phys., 1293 (1986).

[68] Fokas. A.S., Leo, R.A., Martina, L. and Soliani, G., Phys. Lett. A., 115, (7), 329 (1986).

[69] Zakharov, V.E. and Manakov, S.V., Sov. Phys. JETP 42, 842 (1976).

[70] Kaup, D., Reiman, A.H., and Bers, A., Rev.Mod. Phys. 51, 275 (1979).

[71] Erugin, N.P., Dokl. Akad. Nauk. BSSR 2 (1958).
Lukashevich, N.A., Diff. Urav. 7, 1124 (1971).

[72] Lukashevich, N.A., Diff. Urav. 3, 994 (1967); Gromak, V.I., DIff. Urav. 11, 373 (1975).

[73] Lukashevich, N.A., Diff. Urav. 3, 395 (1967); Abstract of Doctoral Thesis, Kiev (1971).

[74] Lukashevich, N.A., Diff. Urav. 4, 732 (1968); Gromak, V.I., Diff. Urav. 12, 740 (1976).

[75] Fokas, A.S. and Yortsos, Y.C., Lett. Nuovo Cimento 30, 539 (1981).

OLD AND NEW RESULTS ON RECURSION OPERATORS: AN ALGEBRAIC APPROACH TO KP EQUATION

F. Magri*, C. Morosi**

* Dipartimento di Matematica, Università di Milano
** Dipartimento di Matematica, Università di Perugia

ABSTRACT. We present a purely algebraic scheme for constructing hierarchies of commuting non linear evolution equations for field functions u(x,y,t) defined on R^3 and taking values in any associative algebra $\mathscr{A}$ with unity. By suitable choices of the algebra $\mathscr{A}$ one recovers several classes of equations well-known in the literature. So, if $\mathscr{A}$ is the algebra of real numbers one obtains the usual KdV hierarchy; if $\mathscr{A}$ is the algebra of n×n matrices one obtains the non-abelian KdV hierarchy; finally, if $\mathscr{A}$ is the algebra of linear differential operators with constant coefficients in one space variable one recovers the KP hierarchy. The relation between this algebraic scheme and the theory of bi-Hamiltonian manifolds is also emphasized.

1. INTRODUCTION

This paper aims to contrast the common opinion according to which KP theory is an *extension* of KdV theory. Indeed, the opposite statement can also be maintained: namely, that KP equation is a particular *reduction* of KdV equation. This point of view automatically solves the problem of finding the "recursion operator" for KP equation [1] [2]; indeed, one can state that KP and KdV equations have the *same* recursion operator.

The precise meaning of these statements will be clarified in Sec.4 after a critical review of KdV theory done in Secs. 2-3.
The discovery that the standard KdV hierarchy is not complete leads to introduce, in Sec. 2, the new concept of *"KdV bichain"* and to develop, in Sec. 3, the "KdV theory in two-space dimensions". The standard KP theory is then obtained, in Sec. 4, as a particular reduction of the last theory. All the results stated in Secs. 2-4 are proved in the final section of the paper, in the framework of the theory of bi-Hamiltonian manifolds with symmetries [3-4] .

2. KdV THEORY IN ONE SPACE DIMENSION

A critical review of KdV theory in one space dimension is essential for a right analysis of KP equation. The main remark is that the standard KdV hierarchy is merely a part of a more general chain, henceforth referred to as *"KdV bichain"*. The new chain is presently introduced from an axiomatic point of view, based on three results. Each result concerns one of the three main structures entering the theory, namely: the "recursion operator", the Hamiltonian structures and the symmetry algebra.

(i) <u>The Nijenhuis tensor</u>. Let A be any algebra with unit, $X: A \to A$ a derivation on A

$$X(a_1 \cdot a_2) = (Xa_1) \cdot a_2 + a_1 \cdot (Xa_2) \tag{2.1}$$

and $U \subset A$ a subalgebra which is stable with respect to X:

$$X(U) \subset U \tag{2.2}$$

We assume that the restriction of X to U is invertible. By X, X^{-1}, the commutator $[\ ,\ ]$ and the anticommutator $\{,\}$, we define the following map $N:(u,\varphi) \rightarrow N_u\varphi$ from $U \times U$ to U :

$$4\, N_u\varphi := -X^2\varphi + \{Xu,\, X^{-1}\varphi\} + 2\,\{u,\varphi\} + [u,\, X^{-1}\,[X^{-1}\varphi,u]] \tag{2.3}$$

Regarding U as the phase space of an abstract dynamical system, u as a point of this space, φ as a tangent vector to U, the mapping (2.3) defines a tensor field of type (1,1) on U. As such, N is a Nijenhuis tensor ([3], and Sec. 5), fulfilling the condition

$$N'_u(\varphi,N_u\psi)-N'_u(\psi,N_u\varphi) = N_u(N'_u(\varphi,\psi)-N'_u(\psi,\varphi)) \tag{2.4}$$

It provides the "recursion operator" of the KdV theory.

ii) <u>The bi-Hamiltonian structure</u>. Assume that A is endowed with a Trace--form tr: $A \rightarrow R$ fulfilling the conditions

$$tr(\varphi\cdot\psi) = tr(\psi\cdot\varphi) \tag{2.5}$$

$$tr(\varphi\cdot\psi) = 0 \quad \forall\, \psi \implies \varphi = 0 \tag{2.6}$$

If the derivation X is skew-symmetric with respect to this Trace-form

$$tr(\varphi\cdot X\psi) + tr(\psi . X\varphi) = 0 \tag{2.7}$$

then the mappings $P: (u,\alpha) \rightarrow P_u\alpha$ and $Q: (u,\alpha) \rightarrow Q_u\,\alpha$ given by

$$2\, P_u\,\alpha := -X^3\alpha + \{Xu,\alpha\} + 2\,\{u,X\alpha\} + [u,X^{-1}\,[\alpha,u]] \tag{2.8}$$

$$Q_u\alpha := 2X\alpha$$

define two Poisson tensors on U, which are related to the Nijenhuis tensor N by:

$$N = P\cdot Q^{-1} \tag{2.9}$$

They provide the "bi-Hamiltonian structure" of KdV theory ([3], and Sec.5).

iii) The symmetry algebra. Let $\hat{G}$ to be the Lie bi-algebra of the pairs of elements $c=(c_1,c_2)$ of Ker X in A, endowed with the double commutator

$$[(c_1,c_2),\ (d_1,d_2)]' := ([c_1,d_1],[c_1,d_2]+[c_2,d_1]) \qquad (2.10)$$
$$[(c_1,c_2),\ (d_1,d_2)]'' := (\ 0\ ,\ [c_2,d_2])$$

and let G be any abelian subalgebra of $\hat{G}$ (with respect to both commutators). Under the assumption that

$$U \cdot \mathrm{Ker}\, X \subset U \qquad \mathrm{Ker}\, X \cdot U \subset U \qquad (2.11)$$

the mapping $\varphi : G \times U \to U$ given by

$$2\varphi_c(u) := [u,c_1] + \{Xu,\ c_2\} + [\ u, X^{-1}[c_2,u]\ ] \qquad (2.12)$$

defines a family of vector fields on U. These vector fields give a representation of G on U

$$[\varphi_c,\ \varphi_d\ ] = 0 \qquad (2.13)$$

and they leave N, P and Q invariant

$$L_{\varphi_c}(N) = 0, \quad L_{\varphi_c}(P) = 0, \quad L_{\varphi_c}(Q) = 0 \qquad (2.14)$$

Consequently, the algebra G can be referred to as the *symmetry algebra* of KdV theory. These results allow the following construction. Let G_c be the abelian subalgebra of $\hat{G}$ engendered by any element $c=(c,0) \in G$ and by its powers $c^j=(c^j,0)$ (or by $c^j=(0,c^j)$). By means of the corresponding vector fields (2.12) and of the Nijenhuis tensor (2.3), we construct the bichain φ_j^k defined by:

$$\varphi_j^k(u;c) := N_u^k \cdot \varphi_{cj}(u) \qquad (2.15)$$

All these fields commute in pairs, by known results on Nijenhuis tensors [3] . The corresponding family of evolution equations

$$\frac{du}{dt_{kj}} = \varphi_j^k(u;c) \tag{2.16}$$

will be referred to as the KdV bichain on the algebra A, associated with the generator c $\in$ G. Indeed, the standard KdV hierarchy is recovered from (2.16) by setting j = 0 and by choosing $A = C^\infty(R,\mathscr{A})$ (the algebra of C^∞ functions on the real axis taking values in any finite-dimensional associative algebra $\mathscr{A}$); in this case X is the x-derivative and U the subalgebra of rapidly decreasing functions. Furthermore, under the assumption of the existence of a Trace-form making X skew-symmetric, the evolution equations (2.16) are bi-Hamiltonian, on account of the invariance condition (2.14) and of the Hamiltonian property of the vector fields (2.12). So, we can state that:

Proposition 2.1 (KdV bichain). In any associative algebra A with unit, endowed with a derivation fulfilling (2.2), one can define a bichain (2.15) of commuting vector fields, for every generator c of the symmetry algebra. Moreover, if A has the structure of bi-Hamiltonian manifold, these vector fields are bi-Hamiltonian.

3. KdV THEORY IN TWO SPACE DIMENSIONS

Now we assume that the algebra A is endowed with a second derivation Y, commuting with X and leaving U invariant:

$$[X,Y] = 0 \qquad\qquad Y(U) \subset U \tag{3.1}$$

These further assumptions are fulfilled, for example, by replacing the previous algebra $A = C^\infty(R,\mathscr{A})$ by $A = C^\infty(R^2,\mathscr{A})$. In this case Y is the partial derivative with respect to the second space variable y. One easily check that the algebra $A' = A[Y]$ of polynomials in Y with coefficients in A

$$a' \in A' : \quad a' = \sum_{i \geq 0} a_i Y^i \tag{3.2}$$

endowed with the derivation $X' : A' \to A'$ given by

$$X'(a') := \sum_{i \geq 0} X(a_i) Y^i \tag{3.3}$$

fulfils all the assumptions listed in the previous section. Consequently, a KdV bichain can be defined on U', the subalgebra of the polynomials with coefficients in U, by formally using the *same* Nijenhuis tensor N and the *same* symmetry algebra as before. Differences arise due to different multiplication rules to be used in A and A'. So, for example, the first few equations of the bichain engendered by the element $c = (0,Y)$ of the algebra G' are given by the vector fields

$$\begin{aligned}
\varphi_o^o &= u_x \\
\varphi_o^1 &= -\frac{1}{4} u_{xxx} + \frac{3}{4}\{u,u_x\} \\
\varphi_1^o &= u_x \cdot Y + \frac{1}{2} u_{xy} + \frac{1}{2}\left[u, X^{-1}u_y\right]
\end{aligned} \tag{3.4}$$

where u has been used instead of u' to denote the elements of U' and $u_x, u_y, \ldots$ denote Xu, Yu,... .

Two simple remarks are in order. The first one concerns the quite unsymmetric appearance of X and Y in these fields, reflecting the different role played by the two derivations in this theory; indeed, Y is used to define the phase space $A' = A[Y]$, while X is used to define the geometric structure on this space. The second remark concerns the number of the equations; indeed, since A' is in general an infinite-dimensional algebra, each equation (3.4) actually represents an infinite number of equations in an infinite number of field functions. Explicitly, these equations are obtained by expanding $\dot{u}$ and $\varphi^j_k(u;c)$ in powers of Y

$$\dot{u} = \sum_{i\geq 0} \dot{u}_i Y^i \qquad \varphi^j_k(u;c) = \sum_{i\geq 0} \varphi^j_{ki}(u;c)Y^i \tag{3.5}$$

and by equating the coefficients of the same powers:

$$\dot{u}_i = \varphi^j_{ki}(u_o,u_1,\ldots)$$

So, for example, the first equation (3.4) obviously gives

$$\dot{u}_i = u_{ix} \tag{3.7}$$

while the second one corresponds to the system

$$\dot{u}_i = -\frac{1}{4} u_{ixxx} + \frac{3}{4} \sum_{\substack{j,k\geq 0\\ 0\leq j-k\leq i}} \binom{j}{k} \left[u_j \frac{\partial^{k+1} u_{i+k-j}}{\partial y^k \partial x} + u_{jx} \frac{\partial^k u_{i+k-j}}{\partial u^k} \right] \tag{3.8}$$

As such, these equations are clearly of difficult use. However, simple hierarchies of equations in a *finite* number of field functions can easily be extracted from them by means of a reduction technique, which will be explained in the next section.

4. THE KP HIERARCHY

The idea of the reduction simply amounts to look for those linear combinations (with constant coefficients), if any, of the vector fields of the bichain

$$\varphi^{\rho} := \sum_{\substack{k\geq 0 \\ j\geq 0}} c^{\rho j}_{k} \, \varphi^{k}_{j} \tag{4.1}$$

which are tangent to a suitably chosen finite-dimensional submanifold S in U'. The restriction of these fields to S give then rise to evolution equations in a finite number of field functions.

To clarify this idea, let us first consider the vector fields $\varphi^k_o := N^k_u \cdot \varphi^o_o$, corresponding to the linear combinations

$$\varphi^{\rho} = \sum_{\substack{k\geq 0 \\ j\geq 0}} \delta^{\rho}_{k} \, \delta^{j}_{o} \, \varphi^{k}_{j} \tag{4.2}$$

Observe that these vector fields, restricted to the submanifold $S_o = \{u' \in U' : u' = u_o\}$ of the differential operators of order zero, are differential operators of order zero.

Namely, they are tangent to S_o. Indeed: (i) φ^o_o is a differential operator of order zero on S_o and ii) $N_u : \varphi \to N_u \varphi$ does not increase the order of φ on S_o. Obviously it is:

$$\varphi^{i+1} = N_u \cdot \varphi^{i} \tag{4.3}$$

This allows to recover the one-dimensional KdV theory as a restriction of the two-dimensional KdV theory to the submanifold S_o.

In the same way, let us consider the submanifold

$$S_1 = \{u' \in U' : u' = u+Y, \quad u \in U\} \tag{4.4}$$

of monic first-order differential operators. In this case, no vector fields of the chain (3.4) is tangent to S_1. However, if we form the linear combinations:

$$\varphi^0 = \varphi^0_0$$

$$\varphi^1 = \varphi^0_1 - \varphi^1_0 \tag{4.5}$$

$$\varphi^2 = \varphi^0_2 - 2\varphi^1_1 + \varphi^2_0$$

we obtain (by assuming $A = C^\infty(\mathbb{R}^2, \mathscr{A})$ with $\mathscr{A}$ any commutative algebra) the following vector fields

$$\varphi^0(u) = u_x \tag{4.6}$$

$$\varphi^1(u) = \frac{1}{2}(u_{xy} + X^{-1} u_{yy} + 2u_x Y) - \frac{1}{4}(-u_{xxx} + 6uu_x + \tag{4.7}$$

$$+ 2u_{xy} - X^{-1} u_{yy} + 4u_x Y)$$

$$= \frac{1}{4}(u_{xxx} - 6uu_x - 3X^{-1} \cdot u_{yy})$$

$$\varphi^2(u) = \left[(\text{Terms of order zero}) + (u_{xy} + X^{-1} \cdot u_{yy})Y + u_x Y^2\right] + \tag{4.8}$$

$$- 2\left[(\text{Terms of order zero}) + \frac{1}{4}(-u_{xxx} + 6uu_x + 4u_{xy} + \right.$$

$$\left. + X^{-1} u_{yy})\ Y + u_x Y^2\right] +$$

$$+ \left[(\text{Terms of order zero}) + \frac{1}{2}(-u_{xxx} + 6uu_x + 2u_{xy} + \right.$$

$$\left. - X^{-1} u_{yy})\ Y + u_x Y^2\right]$$

$$= \text{Terms of order zero}$$

Namely, these linear combinations are tangent to S_1. More in general, this is true for the linear combination of the form

$$\varphi^l = \sum_{\substack{i\geq 0\\ j\geq 0}} (-1)^i \binom{l}{i} \delta_i^{l-j} \varphi_j^i \tag{4.9}$$

as it will be proved in Sec. 5.

The restrictions of these vector fields to S_1 actually give the KP chain. In this sense the KP chain is a different reduction of the two-dimensional KdV bichain. In this case, the relation (4.3) is no longer true; from this restrictive point of view, one can correctly say that KP equation does not admit a recursion operator [1]. However, this is not true from the enlarged point of view of the "bichain theory", according to which KP has the *same* recursion operator as KdV, being engendered by the same bichain.

In order to give the previous results a more concise and clear form, it is suitable to introduce the following symbol

$$\varphi_a(\lambda,\mu) := \exp(\lambda N_u)\ \varphi_{\exp\mu a}(u) = \sum_{\substack{k\geq 0\\ j\geq 0}} \frac{\lambda^k}{k!} \frac{\mu^j}{j!} \varphi_j^k(u;a) \tag{4.10}$$

which will be called the "symbol of the bichain" engendered by the generator a of the symmetry algebra. Then we can state that:

Proposition 4.1 (Reduction theorem). The one-parameter chains $\varphi_a(\lambda,o)$ and $\varphi_a(\lambda,-\lambda)$ extracted from the two-parameters KdV bichain (4.10) are tangent respectively to the submanifolds S_o and S_1 of the zero and first-order monic differential operators in U'. For $a = (0,Y)$ and $a = (Y,0)$, their restrictions to S_o and S_1 give rise to the standard one-dimensional KdV hierarchy and to the KP hierarchies (KP_1 and KP_2 [5]) respectively.

This result will be proved in the next section. The understanding of the connection between the choice of the reduction submanifold S and the kind of constraints to be imposed to the parameters (λ,μ) of the bichain ($\mu = 0$ and $\lambda+\mu = 0$ in the cases previously dealt with) is at present an open problem.

5. THE PROOFS OF THE RESULTS

In this final section we sketch the proofs of the results previously given in Secs. 2-4 for the KdV and KP chains. They can be divided in two parts, concerning respectivley:

i) The bi-Hamiltonian structure and the symmetry algebra.

ii) The KP chain.

i) Results from the theory of bi-Hamiltonian manifolds with symmetries. Here we limit ourselves to recall some basic definitions and to give merely an outline of the reduction process giving rise to the Nijenhuis and bi-Hamiltonian structures introduced in Sec. 2; for more details, we refer to the literature, and in particular to [3]. A bi-Hamiltonian manifold M is a differentiable manifold endowed with two Poisson brackets $\{,\}_P$, $\{,\}_Q$ such that their linear combination still fulfils Jacobi's identity. They can be associated with two Poisson tensors P and Q given by

$$P: \mathscr{X}^*(M) \to \mathscr{X}(M) \quad : \qquad \{f,g\}_P = \langle df, P\, dg \rangle \tag{5.1}$$

$$Q: \mathscr{X}^*(M) \to \mathscr{X}(M) \quad : \qquad \{f,g\}_Q = \langle df, Q\, dg \rangle \tag{5.2}$$

for any pair of functions $f,g \in C^\infty(M)$. If P is regular, then the tensor N defined by

$$N : \mathscr{X}(M) \to \mathscr{X}(M) \; : \qquad N = Q.P^{-1} \tag{5.3}$$

is torsionless

$$T(N)(\varphi,\psi) := L_{N\varphi}(N)\psi - N\, L_{\varphi}(N)\psi = 0 \tag{5.4}$$

We will say that N is a Nijenhuis tensor (usually, it is said the recursion operator associated with P and Q).

The action of a Lie algebra G on the manifold M will be called bi-Hamiltonian if it keeps both Poisson tensors invariant, that is if its infinitesimal generators φ_b(b $\in$ G) are invariance fields for both P and Q

$$L_{\varphi_b}(P) = 0 \qquad\qquad L_{\varphi_b}(Q) = 0 \tag{5.5}$$

and consequently for N

$$L_{\varphi_b}(N) = 0 \tag{5.6}$$

Then the vector fields φ^j recursively associated with φ_b by

$$\varphi^{j+1} := N \cdot \varphi^j \qquad (\varphi^o = \varphi_b) \tag{5.7}$$

are Hamiltonian, commute in pairs and keep N invariant: this is the geometric interpretation of the so-called Lenard recursion scheme.

To construct the tensors N, Q, P and the symmetry algebra associated with the KdV theory, we follow a reduction method [3] .We start from the associative algebra gl(2,A) of 2×2 matrices with entries in A. On this manifold, we consider the following tensors:

$$P_\mu\, \xi := X\xi + [\xi,\mu] \qquad\qquad Q_\mu \xi := [\xi, a] \tag{5.8}$$

and the tensor

$$N_\mu := Q_\mu \cdot P_\mu^{-1} \tag{5.9}$$

where $\xi \in$ gl(2,U) and a $\in$ gl(2, KerX). It is easily checked that P and Q are Poisson tensors with respect to the Trace-form, if any, making X skew-symmetric, and that N is a Nijenhuis tensor, independent of the existence of a Trace-form [3] . Moreover, we remak that the Lie algebra

G_a defined by

$$G_a = \{ b \in G : [b,a] = 0 \} \tag{5.10}$$

is a symmetry algebra of the bihamiltonian manifold with respect to the action defined by the generators

$$\varphi_b(\mu) := [b,\mu] \tag{5.11}$$

In fact:

$$L_{\varphi_b}(P_\mu)\,\xi = P'_\mu\,(\xi,\varphi_b(\mu)) - \varphi'_b\,P_\mu\xi - P_\mu\,\varphi'^{*}_b\xi \overset{(5.8)}{=} [\xi,[b,\mu]] + [b,[\mu,\xi]] + [\mu,[\xi,b]] = 0 \tag{5.12}$$

on account of the Jacobi's identity, and similarly

$$L_{\varphi_b}(Q_\mu)\xi \overset{(5.8)}{=} [\xi,[a,b]] \overset{(5.10)}{=} 0. \tag{5.13}$$

Next, we apply the reduction method described in [3], [6]. Essentially, the process consists in a sequence of restrictions and projections. The restrictions are on the characteristic leaves of the Poisson tensor Q; the projections concern the null foliation of the presymplectic tensor obtained by restricting P^{-1} to the previous leaves. By choosing

$$a = \begin{pmatrix} 0 & 1 \\ 0 & 0 \end{pmatrix} \tag{5.14}$$

the process ends to the first step, giving rise to the tensor (2.8) we are looking for. The required properties of the reduced tensors follow automatically from the corresponding properties of the tensors (5.8) and of the tensorial character of the reduction process. This avoids a cumbersome direct inspection.

The symmetry algebra (2.12) is obtained, by the same reduction, from the algebra (5.11) . To prove this result, we recall from [3] that in correspondence with the choice (5.14) of the matrix a the integral leaf S of Q is given by

$$S = \{ \mu : \mu = \begin{pmatrix} p & q \\ 1 & -p \end{pmatrix} \} \qquad (5.15)$$

and that the projection onto the reduced phase space $M = S/\mathrm{Ker}\, P^{-1}\big|_S$ is

$$\pi : S \to M' = S/\mathrm{Ker}\, P^{-1}\big|_S \qquad u = Xp - p^2 - q \qquad (5.16)$$

The subalgebra G_a is

$$b \in G_a : \quad b = b_1 e + b_2 a \quad (b_1, b_2 \in A) \qquad (5.17)$$

so that the generators φ_b at any point of S are

$$\varphi_{b_1} = \left(\begin{array}{c|c} [b_1,p] & [b_1,q] \\ \hline & -[b_1,p] \end{array} \right) \qquad \varphi_{b_2} = \left(\begin{array}{c|c} b_2 & -\{p,b_2\} \\ \hline & -b_2 \end{array} \right) \qquad (5.18)$$

The first generator φ_{b_1} is tangent to S, its restriction φ'_{b_1} being

$$\varphi'_{b_1} = ([b_1,p] \ ; \ [b_1,q]) \qquad (5.19)$$

On account of (5.16), it follows that φ'_{b_1} is projectable on M' and that the projected vector field φ''_{b_1} is

$$\varphi''_{b_1} = [b_1,u] \qquad (5.20)$$

This is the first generator (2.12) of Sec. 2.

The second generator φ_{b_2} is not tangent to S, due to the presence of b_2 on the diagonal. However, the vector fields

$$X_{b_2} = N^{-1} \cdot \varphi_{b_2} \tag{5.21}$$

are tangent to S; their restrictions to S are given by

$$X'_{b_2} = (X\xi_1+\xi_2-q\xi_3+[\xi_1,p]\ ;\ X\xi_2+\xi_1 q-q\xi_4-\{p,\xi_2\}) \tag{5.22}$$

where ξ_2 is arbitrary and ξ_1,ξ_3,ξ_4 are determined by the relations

$$\xi_3=-b_2;\ \xi_1-\xi_4 = -\{p,b_2\}\quad ;\quad X(\xi_1+\xi_4) = [b_2,\ p^2+q] \tag{5.23}$$

Independent of the particular choice of ξ_2, they project onto a *unique* vector field $\hat{\varphi}_{b_2}$ on the reduced phase space

$$\hat{\varphi}_{b_2} = -\frac{1}{2}\{b_2,\ Xu\} - \frac{1}{2}[[b_2,\ X^{-1}u],\ u] \tag{5.24}$$

It is the second generator (2.12) of Sec. 2. The results of Sec. 2 are thus completely proved in the context of the reduction theory of bi-Hamiltonian manifolds with symmetries.

ii) The KP chain. The reduction theorem of Sec. 4 can be easily proved taking into account some remarkable properties of the Nijenhuis tensor N and of the action of the symmetry algebra. To put into evidence these algebraic properties, we consider the set $B \subset \mathrm{End}(A')$ defined as follows. On the natural basis of End(A'), the elements $T \in B$ have the matrix triangular form

$$T \in B: \qquad T_{ik} = \begin{cases} 0 & k < i \\ \binom{k}{i} T_{k-i} & k \geq i \end{cases} \qquad (i,k \geq 0) \tag{5.25}$$

with $T_k \in \mathrm{End}(A)$. By using this matrix form, one directly verifies that B is an associative algebra with unit (an intrinsic definition of B will be given in the following).

The reason to introduce such an algebra is that the left and right-translations L_a and R_a ($a \in A$) defined on A' belong to B, as well as X, X^{-1}. This follows from the matrix representation of X, X^{-1}, L_a, R_a:

$$(X)_{ik} = X \quad \delta_{ik} \qquad (X^{-1})_{ik} = X^{-1} \delta_{ik} \tag{5.26}$$

$$(L_a)_{ik} = a\, \delta_{ik} \qquad (R_a)_{ik} = \begin{cases} 0 & k<i \\ \binom{k}{i} a_{k-i} & k\geq i \end{cases} \tag{5.27}$$

where $a_j := Y^j(a)$. Thus since the Nijenhuis tensor N (2.3) is defined in terms of X, X^{-1}, and left and right-translations by u, Xu,... it follows that N itself, restricted to U, belongs to B.

The next step is to define, in a quite natural way, an extension $B' = B[\Delta]$ constructed by means of the right-translation $\Delta := R_y : A' \to A'$

$$\Delta_{ik} = \delta_{i,k+1} \qquad (i,k\geq 0) \tag{5.28}$$

and defined as the algebra of the polynomials in Δ with coefficients in B:

$$T' \in B': \qquad T' = \sum_{k\geq 0} T_k \Delta^k \qquad (T_k \in B) \tag{5.29}$$

The main point is now the remark that although $\Delta \notin B$, it fulfils the condition

$$[B,\Delta] \subset B \tag{5.30}$$

(Thus, the algebra B can be intrinsically characterized as the sub-algebra of the endomorphisms of A' which maintains the order and is stable with respect to the derivation $[\cdot,\Delta]$).

On account of these properties once can then show that $L_{a'}, R_{a'}$ ($a' \in A'$) as well as X, X^{-1} belong to B', so that the Nijenhuis tensor N itself, restricted to U', belongs to B'. Moreover, the mapping N enjoies the following two properties:

I) N is *homogeneous of degree zero*. This means that if $u \in A_k[Y]$ is of the form

$$u = \sum_{0}^{k} {}_j \, u_j \, Y^j \qquad (u_j \in A) \tag{5.31}$$

then N_u belongs to $B_k[\Delta]$, namely it can be written as

$$N_u = \sum_{0}^{k} {}_j \, N_{ju} \, \Delta^j \qquad (N_{ju} \in B) \tag{5.32}$$

II) N *maps monic polynomials into monic polynomials*. This means that if u is of the form

$$u = \sum_{0}^{k-1} {}_j \, u_j \, Y^j + Y^k \tag{5.33}$$

then N_u takes the form

$$N_u = \sum_{0}^{k-1} {}_j \, N_{uj} \, \Delta^j + \Delta^k \tag{5.34}$$

If we introduce the infinitesimal generator $\phi_u : G \to T_u U$ of the action of the symmetry algebra

$$\phi_u(a) := \varphi_a(u) \tag{5.35}$$

one can directly verify from (2.12) that

III) ϕ_u *is homogeneous of degree zero.*

On account of these properties, we can now prove that the submanifold S_0 of the differential operators of degree zero in U' is a reduction submanifold for the one-parameter chain $\varphi_a(\lambda,0)$, with $a=a_0$. Indeed from (4.10) it follows that

$$\varphi_a(\lambda,0) = \exp(\lambda N_u)\cdot\Phi_u(a) \tag{5.36}$$

so that the chain (5.36), restricted to S_o, is tangent to S_o, on account of the previous properties I and III for N and ϕ respectively. If we consider the submanifold S_1 of first-order monic differential operators, S_1 is a reduction submanifold for the one-parameter chain $\varphi_a(\lambda,-\lambda)$. Indeed, let us write the evolution equations in the form

$$\frac{du}{dt} = \Phi_u^\lambda(a) \tag{5.37}$$

where the "modified" generator Φ_u^λ is defined by

$$\Phi_u^\lambda(a) := \exp(\lambda N_u)\cdot\Phi_u(\exp(-\lambda a) \tag{5.38}$$

Then the problem is reduced to show that Φ^λ, restricted to S_1, maps S_1 into operators of degree zero

$$\Phi^\lambda \Big|_{S_1} = \Phi_o^\lambda \tag{5.39}$$

To this end, it suffices to observe that

$$\frac{d\phi_u^\lambda(a)}{d\lambda} = N_u\cdot\phi_u^\lambda(a) - \phi_u^\lambda(a)\cdot\Delta \tag{5.40}$$

and that on account of the previous properties I, II it is

$$N\Big|_{S_1} = N_o + \Delta \qquad (N_o \in B) \tag{5.41}$$

Restricting (5.40) to the submanifold S_1 one has

$$\frac{d\phi_u^\lambda(a)}{d\lambda}\Bigg|_{S_1} = N_o\cdot\Phi_u^\lambda(a) + \left[\Delta,\ \phi_u^\lambda(a)\right] \tag{5.42}$$

so that the required result follows from the previous properties of N and ϕ and from the fact that $\phi_u\Big|_{S_1} = \phi_o$.

In conclusion, the standard KdV and KP hierarchies are obtained by reducing to S_o and S_1 the two one-parameter chains $\varphi_a(\lambda,0)$ and

$\varphi_a(\lambda,-\lambda)$, which are themselves different restrictions of the same KdV bichain $\varphi_a(\lambda,\mu)$. In this precise sense, we conclude that KP can actually be seen as a reduction of KdV.

ACKNOWLEDGMENTS.

This work has been partially supported by the Italian M.P.I. (Project "Geometria e Fisica") and by the G.N.F.M. of the Italian C.N.R.

REFERENCES

[1] Zakharov V.E. and Konopelchenko B.G., "On the theory of recursion operator", Comm. Math. Phys., 94 (1984), 483-509.

[2] Santini P. and Fokas A.S., "The recursion operator of the Kadomtsev-Petviashvili equation", preprint, Postdam (1986).

[3] Magri F. and Morosi C., "A geometrical characterization of integrable Hamiltonian systems through the theory of Poisson-Nijenhuis manifolds", Quaderno S/19 (1984), Dpt. Matematica Università di Milano.

[4] Olver P.J., "Application of Lie groups to differential equations", Graduate Texts in Mathematics, Vol. 107, Springer-Verlag, 1986.

[5] Oevel W. and Fuchssteiner B., "Explicit formulas for symmetries and conservation laws for the Kadomtsev-Petviashvili equation", Phys. Lett. 88A (1982), 323-327.

[6] Magri F., Morosi C. and Ragnisco O., "Reduction techniques for infinite-dimensional Hamiltonian systems: some ideas and applications", Comm. Math. Phys., 99 (1985), 115-140.

Factorization of the Recursion Operators and the Integrability of Nonlinear Evolution Equations

H. H. Chen
Department of Physics and Astronomy
University of Maryland
College Park, MD. 20742 U. S. A.
and
J. E. Lin
Department of Mathematical Sciences
George Mason University
Fairfax, VA. 22030 U. S. A.

ABSTRACT

We show that a slight modification of the integrability-test scheme proposed previously by Chen, Lee, and Liu leads to the factorization of recursion operators as a consequence for the integrability of KdV and MKdV equations.

1. INTRODUCTION

One of the most interesting problems concerning the nonlinear evolution equations[1-7] is the testing of their integrability and the subsequent construction of the Lax pairs[2] for their solutions. One of such schemes was proposed sometime ago by Chen, Lee, and Liu[8]. They argued that a natural candidate of the linear Lax scattering equations is the linearized equation (or its adjoint) of the original nonlinear evolution equation. Considerable amount of information are contained in the linearized equation (or its adjoint). With some ingenuity, these information could be extracted to help decide on the issue of the integrability of the nonlinear evolution equation. For example, a WKB-type expansion of the solution of the adjoint of the linearized equation yields a recursion formula that could generate an infinite set of conservation laws for the integrable equation. On the other hand, if the nonlinear evolution equation is not integrable, the recursion formula would be disrupted at a certain stage and thus provides a crucial integrability test of the nonlinear evolution equation. To insure that the recursive generation of the constants of motion not be disrupted, the existence of a linear recursion operator[8,9] for solutions of the linearized equation (or its adjoint) would be sufficient. In fact, the linear recursion operator for solutions of the linearized equation and the linearized equation together constitute a Lax pair[8] for the integrable nonlinear evolution equation (see (14) and (15) below) and thus provides the basis for its complete solution by the inverse scattering method[1-7]. We remind the readers that a Lax pair associated with an integrable equation $u_t = K(u)$ is a pair of operators $A = A(u)$ (temporal part) and $L = L(u)$ (spectral part) which satisfy the Lax condition[2]:

$$L_t = [A,L] = AL - LA. \tag{1}$$

In this work, we will show that a slight modification of the Chen-Lee-Liu scheme will yield the recursion operators of the solutions of the linearized equation and its adjoint if the original equation is integrable. The Korteweg- de Vries (KdV) equation and the modified Korteweg- de Vries (MKdV) equation will be used as prototypes of our presentation.

In Section 2, we shall give the Chen-Lee-Liu scheme in detail for the KdV equation. The modification that leads to the factorization and construction for the recursion operator will be presented in Section 3. All the solutions of the KdV or MKdV equation are assumed to decay sufficiently fast at $|x| = \infty$. In what follows, we shall use ∂_x to denote $\partial/\partial x$ and ∂_x^{-1} to denote $\int_{-\infty}^{x}\cdot dx$.

<u>Definition 1</u>. (See ref. [10]) Given a functional $F = \int_{-\infty}^{\infty} f(u)dx$, where $u = u(x,\alpha)$ is its argument function, depending on a parameter α as well as on the variable x, and decays sufficiently fast at $|x| = \infty$, then the functional derivative $\delta F/\delta u$ is defined by

$(d/d\alpha)F\{u\} = \int_{-\infty}^{\infty}(\delta F/\delta u)(\partial u/\partial\alpha)dx$.

<u>Definition 2</u>. Given a function $G(u) = G(x,t,u,u_x,\cdots)$ which depends on u and its partial derivatives and possibly on the independent variables x and t,

$$G'[h] = \partial G(u+\varepsilon h)/\partial\varepsilon|_{\varepsilon=0} \tag{2}$$

is the Gateaux derivative of G in the direction h with respect to u. Therefore, the operator G' (depending on u) acts on the function h.

<u>Definition 3</u>. Given an evolution equation $u_t = K(u)$, where K(u) depends on u and its partial derivatives and possibly on the independent variables x and t. A function $\sigma(x,t,u,u_x,\cdots)$ is called a symmetry (generator) of the equation if σ satisfies the linearized equation of $u_t = K(u)$:

$$\partial\sigma/\partial t = K'[\sigma], \tag{3}$$

where $K'[\sigma]$ is defined as in (2).

<u>Remark 1</u>. These symmetries are the infinitesimal symmetry generators of one-parameter groups of invariant transformations of the equation. A classical theorem of Noether[11] states that constants of motion can usually be constructed from the symmetries of the Lagrangian.

We will use M^* to denote the adjoint of the operator M in the following presentation.

2. THE CHEN-LEE-LIU TEST SCHEME

We shall take the Korteweg- de Vries (KdV) equation as an example to demonstrate the integrability-test scheme proposed previously by Chen, Lee, and Liu[8]. The KdV equation is given by

$$u_t = K(u) = 6uu_x + u_{xxx}. \tag{4}$$

We linearize Eq. (4) and obtain

$$\phi_t = K'[\phi] = 6u_x\phi + 6u\phi_x + \phi_{xxx}. \tag{5}$$

This equation or its adjoint

$$\psi_t = -K'^*[\psi] = 6u\psi_x + \psi_{xxx} \tag{6}$$

can be taken as a member of the Lax pair. Note that ψ and ϕ are related by

$$\psi_x = \phi. \tag{7}$$

Suppose we take Eq.(6) as the Lax scattering equation. The wave function ψ at large wave numbers can be represented as

$$\psi(k,x) = \exp(kx + \omega t + \partial_x^{-1}T), \tag{8}$$

where T is the phase function. Substitute this into Eq.(6), we obtain

$$\partial_x^{-1}T_t = 6u(k+T) + T_{xx} + 3(k+T)T_x + T^3 + 3kT^2 + 3k^2T, \tag{9}$$

where we have set $\omega = k^3$ to eliminate secularity. Eq.(9) can be interpreted as a conservation law: If T could be represented as a polynomial of u and its partial derivatives, then, as $x \to \infty$, the right-hand side approaches zero, and therefore, the left-hand side says that $\int_{-\infty}^{\infty} T_t dx = 0$, that is, $\int_{-\infty}^{\infty} T dx$ is a constant of motion. Furthermore, an asymptotic expansion of T,

$$T = \sum_{n=0}^{\infty} k^{-n} T_n , \tag{10}$$

yields an infinite number of T_n's that are conserved in time. These T_n's can be constructed recursively from a formula obtained from (9) and (10), (Note that we use $\delta_{i,j}$ to denote the Kronecker delta in (11))

$$\partial_x^{-1}(T_n)_t = 6u\delta_{n+1,0} + 6uT_n + \partial_x^2 T_n + 3\partial_x T_{n+1} + 3\sum_{j=0}^{n} T_j \partial_x T_{n-j}$$
$$+ \sum_{j+p=0}^{n} T_j T_p T_{n-j-p} + 3\sum_{j=0}^{n+1} T_j T_{n+1-j} + 3T_{n+2} . \tag{11}$$

Eq.(11) can be used to generate T_{n+2} (the last term on the right-hand side of Eq.(11)) from the knowledge of T_j's, $j = 0,1,2,\cdots,n+1$. Furthermore, if the right-hand side approaches zero as $x \to \infty$, then $\int_{-\infty}^{\infty} (\partial/\partial t) T_n dx = 0$, that is, $\int_{-\infty}^{\infty} T_n dx$ is a constant of motion. From Eq.(11) we find that, assuming $T_{-2} = 0 = T_{-1}$,

$$T_0 = 0,$$
$$T_1 = -2u,$$
$$T_2 = 2u_x,$$
$$T_3 = -2u^2 - 2u_{xx},$$
$$T_4 = (4u^2 + 2u_{xx})_x,$$
$$T_5 = -2u_{xxxx} - 10u_x^2 - 12uu_{xx} - 4u^3,$$

etc.

It is interesting to note that the explicit construction of T_{n+2} depends on the fact that $\partial_x^{-1}(T_n)_t$ can be expressed in terms of differential polynomials of u. In fact, for general nonlinear evolution equations, this fact serves as a test to distinguish an integrable system from a nonintegrable one. For example, the quartic KdV equation

$$w_t = 6w^3 w_x + w_{xxx}$$

would not pass this test. We can show immediately that

$$T_1 \text{ (quartic KdV equation)} = -2w^3$$

but w^3 is not a conserved density at all. The recursive formula therefore is disrupted at n = 1. On the other hand, for the integrable system like the KdV equation (4), we can go on to find the higher members of the conserved densities without disruption. However, from Eq.(11) alone, we can not establish the fact that the recursive operation could go on forever. More informations are needed to do that. It is at this stage that the role of the linear recursion operator comes into play. Actually, the linear recursion operator does not act directly on the conserved densities but on the solutions of Eq.(6). It is well known[2] that the functional derivative of constants of motion C_n satisfy Eq.(6), that is,

$$\delta C_n/\delta u = \psi_n \tag{12}$$

and ψ_n satisfies Eq.(6). A linear recursion operator R^a would act on ψ_n and generate ψ_{n+1}, that is,

$$\psi_{n+1} = R^a \psi_n, \tag{13a}$$

$$(\partial/\partial t)\psi_n = -K'^*[\psi_n], \tag{13b}$$

$$\text{and } (\partial/\partial t)\psi_{n+1} = -K'^*[\psi_{n+1}]. \tag{13c}$$

The consistency of Eq.(13a), (13b), and (13c) requires that

$$(\partial/\partial t)R^a = [-K'^*, R^a] = -K'^*R^a + R^a K'^* \tag{14}$$

which is exactly the Lax condition (1) for the integrability of nonlinear evolution equations. Thus the linear recursion operator R^a plays the role of the spectral operator while -K'* plays the role of the temporal evolution operator! Note that this is one of the most important properties of the recursion operator.

Remark 2. There is also a recursion operator R^s acting on the symmetries of the equation $u_t = K(u)$ and this operator together with K' also form a Lax pair[8,12],

$$(\partial/\partial t)R^s = [K', R^s] = K'R^s - R^s K'. \tag{15}$$

Note that R^s plays the role of the spectral operator and K' plays the role of the temporal evolution operator.

Remark 3. The existences and applications of recursion operators for the nonlinear integrable evolution equations with one-spatial dimension have been discussed in many literatures, see, for example, ref. [8,9,12-18].

3. CONSTRUCTION OF RECURSION OPERATORS

Take a look at (8) again, this says that

$$(\partial_x - v)\psi = 0, \tag{16}$$

where $v = k + T$. Now we modify the right-hand side of the above equation by replacing 0 with an arbitrary constant ξ,

$$(\partial_x - v)\psi = \xi. \tag{17}$$

Requesting that (17) be compatible with Eq.(6), that is, differentiating (17) and (6) with respect to t and x, respectively, and equating ψ_{xt} with ψ_{tx}, we get

$$(v_t - 6u_x v - 6uv_x - v_{xxx} - 3v_{xx}v - 3v_x^2 - 3v^2v_x)\psi$$
$$- 3(2u_x + v_{xx} + vv_x)\xi = 0.$$

Separately equating the coefficients of ψ and ξ to be zero, we obtain

$$2u_x + v_{xx} + vv_x = 0$$

and

$$v_t - 6u_x v - 6uv_x - v_{xxx} - 3v_{xx}v - 3v_x^2 - 3v^2v_x = 0, \tag{18}$$

therefore,

$$u = -(1/2)v_x - (1/4)v^2, \tag{19}$$

and replacing (19) into (18), we get

$$v_t + (3/2)v^2v_x - v_{xxx} = 0. \tag{20}$$

Note that (19) is the Miura transformation[19] and (20) is the modified Korteweg- de Vries (MKdV) equation[19]. In fact, we have

$$u_t - 6uu_x - u_{xxx} = -(1/2)(\partial_x + v)(v_t + (3/2)v^2v_x - v_{xxx}). \tag{21}$$

Associated with Eq.(20) there are two equations

$$(\phi^M)_t = (K^M)'[\phi^M] = (-3vv_x - (3/2)v^2\partial_x + \partial_x^3)[\phi^M] \tag{22}$$

and

$$(\psi^M)_t = -(K^M)'^*[\psi^M] = (-(3/2)v^2\partial_x + \partial_x^3)[\psi^M] \tag{23}$$

just as Eqs.(5) and (6) are associated with Eq.(4). Note that Eq.(22) is the linearized equation of the MKdV equation (20) and (23) is the adjoint of Eq.(22). Hence, ϕ^M and ψ^M are the symmetries and the functional derivative of the constants of motion of the MKdV equation,

Eq.(20), respectively. Note also that K^M in (22) and (23) is

$$K^M(v) = -(3/2)v^2 v_x + v_{xxx} \tag{24}$$

and ψ^M and ϕ^M are related by

$$(\psi^M)_x = \phi^M. \tag{25}$$

It is rather surprising that the addition of an arbitrary constant ξ to Eq. (16) could yield the Miura transformation. This actually is due to the fact that ξ is a solution of Eq.(23). We now ready to construct the recursion operators for KdV equation (4). Let

$$J = \partial_x - v \tag{26}$$

and

$$G = 6u\partial_x + \partial_x^3. \tag{27}$$

Using Miura transformation (19), we find that,

$$J_t = HJ - JG, \tag{28}$$

where $H = -(3/2)v^2\partial_x + \partial_x^3$. Note that $\underline{H = -(K^M)'*!}$

Lemma. If $L_t = BL - LA$ and $\eta_t = A\eta$, then $L\eta$ satisfies $(L\eta)_t = B(L\eta)$.

Proof: $(L\eta)_t = L_t\eta + L\eta_t = (BL - LA)\eta + LA\eta = BL\eta$.

Corollary 1. J maps the solution ψ of Eq.(6) to the solution ψ^M of Eq.(23), that is, J maps the functional derivative of the constants of motion of the KdV equation (4) to the functional derivative of the constants of motion of the MKdV equation (20).

Proof: This follows directly from (6), (23), (28), and the above lemma.

Corollary 2. J^* maps the symmetries of the MKdV equation (20) to the symmetries of the KdV equation (4).

Proof: Taking the adjoint of Eq.(28), we get $(J^*)_t = J^*H^* - G^*J^* = (-G^*)J^* - J^*(-H^*)$. Hence, using the lemma again, we see that $J^*\rho$ satisfies $(J^*\rho)_t = -G^*(J^*\rho)$ if ρ satisfies $\rho_t = -H^*\rho$. Note that $-G^* = 6u_x + 6u\partial_x + \partial_x^3$ and $-H^* = -(3/2)v^2\partial_x - 3vv_x + \partial_x^3$, therefore, ϕ of Eq.(5) and ϕ^M of Eq.(22) satisfy $(\phi)_t = -G^*\phi$ and $(\psi^M)_t = -H^*\psi^M$, respectively. Corollary 2 now follows from the definition of symmetries (Defnition 3) and the lemma.

We now use (7) and (25) to obtain

Corollary 3. $\partial_x^{-1}J^*\partial_x$ maps the functional derivative of the constants of motion of the MKdV equation (20) to the functional derivative of the constants of motion of the KdV equation (4).

Therefore, we have

Corollary 4.

$$R^a = -\partial_x^{-1}J^*\partial_x J = \partial_x^2 + 4u - 2\partial_x^{-1}\cdot u_x \tag{29}$$

is a recursion operator that maps the functional derivatives of the constants of motion of the KdV equation (4) to themselves.

Corollary 5.

$$R^s = -J^*\partial_x J\partial_x^{-1} = \partial_x^2 + 4u + 2u_x\partial_x^{-1} \tag{30}$$

is a recursion operator that maps the symmetries of the KdV equation (4) to themselves.

Similarly, we also have

Corollary 6.

$$R_M^a = -J\partial_x^{-1}J^*\partial_x = \partial_x^2 - v^2 + v\partial_x^{-1}\cdot v_x \tag{31}$$

is a recursion operator that maps the functional derivatives of the constants of motion of the MKdV equation (20) to themselves.

Corollary 7.

$$R_M^s = -\partial_x J\partial_x^{-1}J^* = \partial_x^2 - v^2 - v_x\partial_x^{-1}\cdot v$$

is a recursion operator that maps the symmetries of the MKdV equation (20) to themselves.

Remark 4. Similarly to (14) and (15) for the KdV equation, R_M^a together with $-(K^M)'^*$ and R_M^s together with $(K^M)'$ also form Lax pairs for the MKdV equation (20):

$$(R_M^a)_t = [-(K^M)'^*, R_M^a] = -(K^M)'^*R_M^a + R_M^a(K^M)'^*, \tag{32}$$

and

$$(R_M^s)_t = [(K^M)', R_M^s] = (K^M)'R_M^s - R_M^s(K^M)'. \tag{33}$$

The above result seems to be very general. For the MKdV equation (20) and the subsequent MKdV equation[20]

$$r_t = r_{xxx} + (1/2)r_x^3 + (3k^2/2)r_x\sin^2 r,$$

where k is a real constant, let

$v = -ir_x + ik\sin r,$

we have

$v_t + (3/2)v^2v_x - v_{xxx}$

$= -i(\partial_x - k\cos r)(r_t - r_{xxx} - (1/2)r_x^3 - (3/2)k^2r_x\sin^2 r).$

From Corollary 7, $R_M^S - k^2$ is a recursion operator of symmetries for the MKdV equation (20). Moreover, it can be factorized as

$R_M^S - k^2 = \partial_x^2 - v^2 - v_x\partial_x^{-1}\cdot v - k^2 = MHM^*\partial_x^{-1},$

where $M = \partial_x - k\cos r$ and $H = -\partial_x - r_x\partial_x^{-1}\cdot r_x$.

In summary, the recursion operators of the symmetries and the functional derivatives of the constants of motion for the KdV equation (4) and the MKdV equation (20) are constructed and their factorizations are given through the application of the lemma which uses the triad equation (28) with a more general form than that of the Lax pair (1). This type of triad-operator equation has been proposed by the authors [21,22] as an alternative condition for the integrability of nonlinear evolution equations with higher spatial dimensions.

AKNOWLEDGMENTS

The work of H. H. Chen is supported in part by the NSF Grant PHYS-80-14189 AOL. The second author is supported in part by the George Mason University 1986 Summer Stipend Grant.

REFERENCES

[1] Gardner, C., Greene, J., Kruskal, M., and Miura. R., Commun. Pure Appl. Math. 27, 97-133 (1974).

[2] Lax, P., Commun. Pure Appl. Math. 21, 467-490 (1968).

[3] Scott, A. C., Chu, F. Y. F., and McLaughlin, D. W., Proc. IEEE, 61, 1443-1483 (1973).

[4] Miura, R., SIAM Rev. 18, 412-459 (1976).

[5] Ablowitz, M. J. and Segur, H., Solitons and the Inverse Scattering Transform (SIAM, Philadelphia 1981).

[6] Newell, A. C., Solitons in Mathematics and Physics (SIAM, Philadelphia 1985).

[7] Novikov, S., Manakov, S., Pitaevskii, L., and Zakharov, V. E., Theory of Solitons (Consultants Bureau, New York 1984).

[8] Chen, H. H., Lee, Y. C., and Liu, C. S., Physica Scripta 20, 490-492 (1979).

[9] Lax, P., SIAM Rev. 18, 351-375 (1976).

[10] Gardner, C. S., J. Math. Phys. 12, 1548-1551 (1971).

[11] Arnold, V. I., Mathematical Methods of Classical Mechanics (Springer-Verlag, New York 1978).

[12] Chen, H. H., Lin, J. E., and Lee, Y. C., On the Direct Construction of the Inverse Scattering Operators of Integrable Nonlinear Hamiltonian Systems. Preprint (1985). To appear in Physica D.

[13] Fokas, A. S., J. Math. Phys. 21, 1318-1325 (1980).

[14] Fuchssteiner, B., Nonlinear Anal. 3, 849-862 (1979).

[15] Fuchssteiner, B. and Fokas, A. S., Physica D 4, 47-66 (1981).

[16] Ibragimov, N. Kh. and Shabat, A. B., Func. Anal. Appl. 14, 19-28 (1980).

[17] McKean, H. P., in Topics in Functional Analysis (Academic Press, New York 1978) 217-226.

[18] Zhu, G. C. and Chen, H. H., J. Math. Phys. 27, 100-103 (1986).

[19] Miura, R., J. Math. Phys. 9, 1202-1204 (1968).

[20] Chen, H. H., in Backlund Transformations (Lecture Notes in Mathematics, Vol.515. Springer-Verlag, New York 1976) 241-252.

[21] Chen, H. H. and Lin, J. E., On the Integrability of Multidimensional Nonlinear Evolution Equations. Preprint (1986). To appear in J. Math. Phys..

[22] Chen, H. H. and Lin, J. E., On the Infinite Hierarchies of Symmetries and Constants of Motion for The Kadomtsev-Petviashvili Equation. Preprint (1986). To appear in Physica D.

(January 13, 1987)

A GEOMETRICAL APPROACH TO INTEGRABLE SYSTEMS ADMITTING TIME DEPENDENT INVARIANTS

Walter Oevel

Universitaet Paderborn, D 4790 Paderborn,
Federal Republic of Germany

ABSTRACT

The structure of time dependent invariants for integrable systems is investigated. Using a purely geometrical framework a unified description of these quantities can be given. The crucial objects for the theory are master symmetries, i.e. vector fields giving rise to time dependent symmetries for the integrable system. It is shown how they fit into the algebraic structure, their close relationship to recursion operators and (bi-)hamiltonian formulations is revealed. The results are very general, they can be applied to most of the known soliton systems as well as to finite dimensional examples.

1. Introduction

In the last two decades much interest has been placed on the integrable nonlinear evolution equations for their rich and beautiful mathematical structure. The solvability of these systems refers to the striking feature that all these equations can be linearized up to a certain extent by various methods. In the classical (i.e. the finite dimensional) case of an integrable system this linearization is essentially done by means of action and angle variables. It is known that a hamiltonian machinery can be set up for most of the soliton systems as well (using IST one even gets explicit representations for action/angle coordinates[30)]), but obviously the integrability of an equation does not really depend on its hamiltonian structure. There are many other properties indicating solvability such as Direct Linearization, Baecklund transformations, Hirota representation,..., which are not directly connected to a symplectic structure. Even the existence of a large number of conservation laws is not necessary (although probably sufficient) for the integrability of an equation, dissipative systems such as Burgers equation are also linearizable and share many nice properties with the conservative soliton systems.

In the classical case the notion of "integrability" is well defined, it is designed for hamiltonian systems admitting a sufficiently large number of conserved functions in involution. For infinite dimensional systems the definition of "integrability" is still open, there is an overwhelming number of properties (many symmetries, many conservation laws, a recursion operator, a bi-hamiltonian formulation, a Lax representation, IST, Painleve property,...), which indicate nice features of the considered equation and might serve as an integrability definition. Most of these properties are related to some extent, a lot of techniques have been designed how to convert one of these structures into another one. Nevertheless, lacking rigorous and general theorems about the relations it is not clear which property is the basic one for the notion of integrability.

The existence of a large abelian symmetry group seems to be an essential feature, all the interesting examples (the soliton systems as well as the non conservative diffusion equations) do have this property. This obviously reflects the fact that the considered equation can be linearized after some transformation: linear equations always admit many commuting symmetries.

But this really is the weakest of the above structures, just the existence of a large symmetry group does not help much to find the linearizing transformation or to solve the equation.

Nevertheless, for the purpose of this paper we will understand an integrable system as a set of commuting vector fields on some manifold. This clearly is a substitute for the set of commuting functions (the action variables) on a symplectic manifold, which is the classical notion of integrability. The idea now is to copy the classical structure of an integrable system on the functions on a symplectic manyfold to the tangent bundle of an arbitrary manifold. Clearly we thus will loose a lot of structure, the result will be weaker, but it will be more general and apply to more examples. Having substituted the action variables by a set of commuting vector fields, we now have to find the substitute for the angle variables. This is the stage, where time dependent invariants and master symmetries[7),20)] arise.

In § 2 it will be motivated how time dependent invariants enter the theory in a very natural way. Looking at the classical case it is very simple to see what structure one has to look for.

In § 3 the geometrical framework will be set up. We will use the standard notation of differential geometry, the basic facts are reviewed briefly.

In § 4 the general structure of time dependent invariants will be discussed.

In § 5 Burgers equation will be investigated. It turns out that the "action/angle" structure of this particular example is typical for a huge class of integrable systems, in fact almost all systems admitting a scaling transformation and a recursion operator are included.

In § 6 the general theory is developped. For those systems admitting a bihamiltonian formulation and (or) a recursion operator a general theorem can be formulated granting the structure of the Burgers example. For those systems for which no recursion operator is known general statements cannot be made, nevertheless the concept of master symmetries provides a powerful working principle.

In § 7 these results are applied to some examples.

A more elaborate version of the theory presented here is found in[18)] and [19)]. Similar results were found by ten Eikelder[24)]. For technical details we refer to [25)]. A major part of this theory is based on the symmetry approach[10),12),13),17)] and Fuchssteiner's ideas of master symmetries[7)].

2. Motivation: angle coordinates and master symmetries

We first want to demonstrate that quantities with explicit time dependence arise in the theory of integrable systems in a very natural way. Let us have a look at the classical case: Consider a set of n independent functions $f_1,..., f_n$ in involution on a $2n$-dimensional symplectic manifold. A set of angle coordinates[2] $\phi_1,\dots,\phi_n$ for this integrable system consists of functions, whose Poisson brackets $\{f_i,\phi_j\}$ are conserved w.r.t. all the elements of the integrable system. Due to the completeness structure of a "completely integrable system" these Poisson brackets can be expressed in terms of the action variables f_i , i.e. there are functions $\omega_{i,j}$ such that

$$\{f_i,\phi_j\} = \omega_{i,j}(f_1,\dots,f_n). \tag{1}$$

Let us denote the hamiltonian vector fields associated to the action variables f_i by K_i, those associated to the angle variables ϕ_j by τ_j. Clearly the flow of K_i is linearized using the action/angle variables. The above Poisson bracket gives the time evolution of ϕ_j for the dynamic given by the vector field K_i, the functions $\omega_{i,j}$ are conserved w.r.t. the integrable system, hence (1) can be integrated immediatealy to $\phi_j(t) = t\,\omega_{i,j} + \phi_j(0)$. But this obviously implies that the functions

$$\phi_j - t\,\omega_{i,j} \tag{2}$$

are conservation laws of K_i with explicit time dependence. Mapping this structure to the vector fields we find, that

$$\tau_j - t\,[K_i,\tau_j,] \tag{3}$$

generates a symmetry transformation for K_i depending explicitly on the evolution parameter t. Note that the time dependent (3) is a symmetry only for the specific K_i, whereas the time independent part τ_j is the same for all members of the integrable system. Referring to this fact the notion **master symmetry**[7] was chosen for these objects. We remark that this notion is associated to an integrable system, i.e. a set of Hamiltonians, rather than to a single equation. The master symmetries may be regarded as vector field versions of the angle coordinates.

Our point now is that these vector fields might be easier to find and to investigate than the angle variables themselves.

To illustrate this we consider the trivial example of the harmonic oscillator with Hamiltonian $H = (p^2 + q^2)/2$, q and p being the space variable and the momentum. In phase space the evolution is restricted to a circle, on which the system moves with constant angular velocity. The angle coordinate is just the polar angle, i.e $\phi = arctan(q/p)$ in the first quadrant of phase space. Assuming the standard symplectic form we find

$$\tau = \frac{1}{2H}\begin{pmatrix} q \\ p \end{pmatrix} \tag{4}$$

as the associated vector field. The nice phenomenon now is the following: note that the above angle ϕ can be represented by the original coordinates p and q only locally, one has to patch several different charts. The associated τ is much nicer in this respect ! Of course, mathematically this does not make much of a difference, the angle coordinate is globally defined, it is just the representation with the original coordinates which can be done only locally. But one should bear in mind that the distinguished features of integrable systems arise from the fact that all the structures are defined globally (locally every hamiltonian system is integrable[2)]). And a straight forward way of proving global existence certainly is to write down the necessary structures explicitly, i.e. to find a nice representation for them. So, global representations do play an important role. The angle variables usually are cyclic (when the underlying Liouville manifold is compact and hence a torus), so they should not be expected to admit explicit global representations. Due to the above phenomenon the associated master symmetries might be easier to approach.

But let us do even one more step away from hamiltonian structure. Multiplication of (4) with the conservation law $2H$ will again yield the generator of a symmetry for the harmonic oscillator, we find the generator of a scaling transformation. This vector field now is not hamiltonian any more, nevertheless is still bears the information about the angle variable in it (via the integrating factor $1/(2H)$).

These elementary considerations lead to the following concept: Instead of considering an action/angle structure on the functions on a manifold we rather look for the corresponding structure on the vector fields. This structure obviously is a set of commuting vector fields (called the "Lax hierarchy", the substitute for the integrable system) and a set of master symmetries (the substitute for the angle variables). These two sets should be connected according to (1) in such a way that the commutator between a Lax member and a master symmetry is again in the integrable system, i.e. is another Lax member. We should not necessarily expect these master symmetries to be hamiltonian.

Indeed, it will turn out that for nearly all known soliton systems these objects can be found easily. In many cases these vector fields will turn out to be non hamiltonian. They also exist for non conservative systems without symplectic structures.

3. Notation

Let M be a manifold (of finite or infinite dimension), TM and T^*M be its tangent and cotangent bundle, respectively. For $u \in M$ the tangent and cotangent spaces are denoted by T_uM and T_u^*M. We will consider tensor fields in the following spaces:

$$\begin{aligned}
F(M) &:= \{ f : M \to \mathbf{R} \}, \\
X(M) &:= \{ K : M \to TM \}, \\
X^*(M) &:= \{ \gamma : M \to T^*M \}, \\
T^{(2,0)}(M) &:= \{ J : M \to L(TM,T^*M) \}, \\
T^{(0,2)}(M) &:= \{ \Theta : M \to L(T^*M,TM) \}, \\
T^{(1,1)}(M) &:= \{ \Phi : M \to L(TM,TM) \}.
\end{aligned}$$

By $L(\,.,.)$ we understand the space of fibre preserving linear maps between the corresponding bundles, e.g. $J(u)$ is a linear map $T_uM \to T_u^*M$. All fields considered here are supposed to be C^∞ w.r.t. a suitable topology. The elements of $F(M)$ and $X^*(M)$ are the 0- and 1-forms on M, $X(M)$ is the space of vector fields. The elements of $T^{(p,q)}(M)$ are regarded as sections in the tensor bundels by identifying $J \in T^{(2,0)}(M)$,say, with the tensor field T_J given by

$$T_J(u)(X_1, X_2) = < J(u)\, X_1 \,, X_2> \qquad \text{for all } X_1, X_2 \in T_u M. \tag{5}$$

The bracket <.,.> is to denote the application of a covector to a vector. All these tensor fields are called nondegenerate, if the maps between the fibres of the corresponding bundles are invertible. By the composition of two tensor fields we mean the composition of these maps, e.g. for $J \in T^{(2,0)}(M)$, $\Theta \in T^{(0,2)}(M)$ we have $\Theta J \in T^{(1,1)}(M)$.

The basic tool to handle these various sorts of tensor fields is the notion of Lie derivatives[4] into the direction of a vector field. They may be defined invariantly, here we just want to recall the local definition in a chart:

Definition 1: Let E be one of the above spaces. The Lie derivatives $L_\tau : E \to E$ of the following quantities into the direction of $\tau \in X(M)$ are:

a) for $f \in F(M)$: $L_\tau f := < df, \tau > := f'[\tau]$,
b) for $K \in X(M)$: $L_\tau K := [\tau, K] := K'[\tau] - \tau'[K]$,
c) for $\gamma \in X^*(M)$: $L_\tau \gamma := \gamma'[\tau] + \tau'^* \gamma \equiv (\gamma' - \gamma'^*)\, \tau + d < \gamma\,, \tau >$,
d) for $J \in T^{(2,0)}(M)$: $L_\tau J := J'[\tau] + \tau'^* J + J\, \tau'$,
e) for $\Theta \in T^{(0,2)}(M)$: $L_\tau \Theta := \Theta'[\tau] - \tau' \Theta - \Theta\, \tau'^*$,
f) for $\Phi \in T^{(1,1)}(M)$: $L_\tau \Phi := \Phi'[\tau] - \tau' \Phi + \Phi\, \tau'$,

where * stands for transposition w.r.t. <.,.>. The prime indicates the usual directional derivative of a tensor field $T(u)$, i.e.

$$T(u)'[X] := \frac{\partial}{\partial \varepsilon} T(u + \varepsilon X)\Big|_{\varepsilon=0} \quad , \quad X \in T_u M\,. \tag{6}$$

One checks that all these definitions are invariant w.r.t. the usual transformation laws[10].

It is well known that the commutator $[\tau, K] = L_\tau K$ of two vector fields satisfies the Jacobi identity such giving $X(M)$ a Lie algebra structure. The Lie derivative corresponds to the adjoint representation of this algebra and, more general, every Lie derivative is a representation of the vector field algebra, i.e. one has

$$L_{[\tau, K]} = L_\tau\, L_K - L_K\, L_\tau \tag{7}$$

for each of the cases in definition 1.

The Lie derivatives satisfy a general product rule

$$L_\tau (A \circledast B) = (L_\tau A) \circledast B + A \circledast (L_\tau B) \tag{8}$$

for any fields A and B in one of the spaces above and meaningful operations $\circledast$ between them. This operation $\circledast$ could be an inner product, the commutator of 2 vector fields or just the composition of 2 tensor fields, e.g.

$$L_\tau(\Theta J) = (L_\tau\Theta)\, J + \Theta\, (L_\tau J) \ , \tag{9}$$

where one has to choose the proper definition of L_τ according to the differentiated object.

4. Invariants

Let us consider a (time independent) vector field $K \in X(M)$, which gives rise to a dynamical system

$$u_t = K(u) \ . \tag{10}$$

A (time independent) field is called **invariant** w.r.t. this system if the Lie derivative of this field into the direction of K vanishes. The following terminology for the various sorts of invariants has been introduced:

An invariant scalar function $f \in F(M)$ is called a **conservation law**, an invariant vector field $\tilde{K}$ is called the generator of a symmetry group or **symmetry** for short. The importance of these notions is obvious[2)]. An invariant $\gamma \in X^*(M)$ shall be called a **conserved covariant**, typical expamples are the differentials of conservation laws. **Recursion operators** have been introduced by Olver[22)], they are invariant (1,1)-tensor fields. Using (8) it is easily seen that they map symmetries to symmetries, their adjoints map conserved covariants to conserved covariants. Typical examples for invariant (0,2) and (2,0)-tensors (sometimes called (inverse) **Noether** structures [10)]) are given by the hamiltonian operators of symplectic geometry, they map symmetries to conservation laws and vice versa. Actually, the fact that Lie derivatives are derivations, i.e. that one has a product rule (8), is the reason why the concept of recursion operators, Noether operators and so on works.

We now want to admit the tensor fields to depend explicitly on the evolution parameter of the considered dynamical system (10). This equation is still assumed to be autonomous ! The invariance notion now has to be modified. As the Lie derivatives is in a certain sense the time derivative along the integral curves of (10), we have to replace it by the total time derivative:

Definition 2: A tensor field $T(u,t)$ is called **invariant** w.r.t. $K \in X(M)$ iff

$$\left(\frac{\partial}{\partial t} + L_K\right) T = 0 \ . \tag{11}$$

As $\left(\frac{\partial}{\partial t} + L_K\right)$ is again a derivation, we still have that e.g. a time dependent recursion operator maps time dependent symmetries to new ones and so on. At first glance the above concept seems to be trivial, one immediately finds

Observation 1: For every time independent tensor field T_0 the formal Lie series

$$exp(-t\,L_K)T_0 \;:=\; \sum_{n=0}^{\infty}\frac{(-t)^n}{n!}\,(L_K)^n\,T_0 \tag{12}$$

is invariant w.r.t. K.

On the other hand for every invariant with analytic time dependence, i.e. $T(u,t) = \sum_{n=0}^{\infty}\frac{(-t)^n}{n!}\,T_n$, one finds $T_n = (L_K)^n\,T_0$, hence the invariants are in 1-1 correspondence with the time independent zero-order coefficients (which can be chosen arbitrarily). So a huge number of invariants is available for any dynamical system. But of course the formal series (12) will not make sense in general as it might not converge. To overcome these difficulties we will restrict ourselves to those T_0's, for which (12) truncates.

Definition 3: A time independent tensor field T_0 is called **invariant of degree** d w.r.t. $K\in X(M)$ iff $(L_K)^{d+1}\,T_0 = 0$.

Obviously an invariant of degree d is in 1-1 correspondence with an invariant time polynomial of degree d via the Lie series (12).

The degree 0 invariants are the usual time independent conservation laws, symmetry generators, recursion operators and so on. But we also know some examples of higher degree invariants: according to § 2 the angle variables of an integrable system are degree 1 conservation laws, the corresponding hamiltonian vector fields are symmetries of degree 1. The products of 2 angle variables are of degree 2: note that the product of two expressions (2) is a conserved time polynomial of degree 2, its zero order being the product of 2 angle variables. Obviously invariant functions of arbitrary degree can be constructed by multiplying angle variables, the corresponding vector fields being higher degree symmetries.

The important feature of angle variables is the fact that they yield higher invariants not only for one specific dynamical system, but for all the members of the integrable system. According to § 2 we substitute the integrable system by a set of commuting vector fields and introduce:

Definition 4: A sequence $\{K_0,K_1,..\}$ of independent commuting vector fields shall be called a **Lax hierarchy**[15]. A time independent tensor field T_0 is called a **master invariant** (of degree d) for this hierarchy, if it is an invariant (of degree d) for every member K_i.

This essentially coincides with the notion of master symmetries and master integrals introduced in [7] and [20]. Master invariants give rise to invariant time polynomials via (12). Note that for any K in the Lax hierarchy the Lie derivative L_K reduces the degree of an invariant by 1. This corresponds to the fact that for every time dependent invariant its explicit time derivative is again invariant.

5. A first example: Burgers equation.

As underlying manifold we choose $S(\mathbf{R})$, the Schwartz space of smooth functions on the real line vanishing rapidly at infinity. We will consider Burgers equation in the form

$$u_t = K_1 = u_{xx} + 2uu_x \tag{13}$$

with $u(.,t)\in S(\mathbf{R})$, $t\in \mathbf{R}$. Let $D = \frac{\partial}{\partial x}$ and $D^{-1} = \int_{-\infty}^{x} \ldots$ Then the associated Lax hierarchy $\{K_n := \Phi^n K_0\}$ may be generated by the recursion operator

$$\Phi(u) = D + u + u_x D^{-1} \tag{14}$$

(interpreted as (1,1)-tensor field) acting on the generator of space translation $K_0(u) = u_x$. It is easily shown, that all these K_n's commute. A set of master symmetries (of degree 1) can also be derived easily: Note, that (13) is left invariant w.r.t. the scaling transformation $u(x,t) \rightarrow e^{\varepsilon}u(e^{\varepsilon}x, e^{2\varepsilon}t)$, which is generated by the vector field $xu_x + u + 2tu_t$. Hence the zero order coefficient $\tau_0 := xu_x + u$ is an invariant of degree 1 for Burgers equation. Actually, it can easily be checked that it is a master invariant for all members of Burgers hierarchy. Applying the recursion operator we obtain a series $\tau_n := \Phi^n\tau_0$ of master symmetries for the hierarchy, one finds

$$[K_n, K_m] = 0 \ , \ [\tau_n, K_m] = (1+m)\, K_{n+m} \ , \ [\tau_n, \tau_m] = (m-n)\, \tau_{n+m} \ . \tag{15}$$

Using the Lie series (12) we obtain the time dependent symmetries $\tau_n + (1+m)\, t\, K_{n+m}$ for the equation $u_t = K_m$. The master symmetries do not leave Φ invariant: $L_{\tau_n}\Phi = \Phi^{n+1}$. An additional element $\tau_{-1} := 1$ (associated to Galilean invariance) might be added to the above set.

Note, that (15) looks like an algebra of action and angle variables, although no hamiltonian framework for (13) is given. We want to show that this "action/angle" structure is not special for this particular example: The distinguished role of the "angle"-type master symmetries of degree 1 will be investigated in the next sections.

Higher degree invariants for (13) can be deduced easily from the observation that the vector field $x + 2\,t\,(xu_x+u) + 2t^2u_t$ is a symmetry. Hence the constant vector field x is a symmetry of degree 2. Taking the Lie derivative of Φ into the direction of this vector field x we obtain the interesting operator $\Psi := x + D^{-1}$. Checking $L_{K_1}\Psi = -2\Phi$, we have found a degree 1 invariant which via the Lie series corresponds to the time dependent recursion operator $\Psi + 2\,t\,\Phi$ for Burgers equation. It can be shown that every polynomial in Φ and Ψ is a master recursion operator for Burgers hierarchy, the degree of which is given by the Ψ degree of the polynomial. The easiest way to do this is to use the results of § 7, where a multi dimensional hierarchy is presented which includes the potential form of (13).

6.1. How master symmetries act on invariants: a powerful principle

Before giving some rigorous statements we want to motivate what structure has to be expected for the "angle" type master symmetries of degree 1. It is well known for the hierarchies of soliton equations that all the equations in the hierarchy have the same symmetries, the same conservation laws, they admit the same recursion operator and the same hamiltonian formulation(s) (if present). Let us assume that this necessarily has to be the case, i.e. let us believe in the

Hypothesis : All the members of an integrable system share the same (time independent) invariants.

Now suppose, we have an "angle" type master symmetry τ, say, with the property that (in analogy to (1)) the commutator between the master symmetry and a member K, say, of the integrable system yields another member of the integrable system. For an arbitrary tensor field T we have the identity (7)

$$L_{[\tau,K]}\, T = L_\tau(L_K\, T) - L_K(L_\tau\, T) \ . \tag{16}$$

If T is an invariant of K and if according to the above hypothesis T is also left invariant by $[\tau,K]$, then this identity tells us that $L_\tau T$ must be a new invariant. Thus we have the powerful principle:

$$L_{master\ symmetry}\ invariant \ = \ new\ \ invariant \tag{17}$$

This includes: symmetries and conservation laws of integrable systems should be mapped to higher ones via Lie derivatives, recursion operators should yield new recursion operators, hamiltonian formulations should be mapped to new hamiltonian formulations (!). So all the algebraic structure of an integrable system may be recovered from master symmetries. In the presence of a recursion operator this will be demonstrated rigorously in the next sections.

6.2. Master symmetries and hereditary recursion operators

We want to show, that the "action/angle" structure of Burgers equation is just an example of a big class of Lax hierarchies generated by recursion operators. Two properties are needed for these results: the hereditary property of the recursion operator and the presence of a scaling transformation.

Definiton 5: A tensor field $\Phi \in T^{(1,1)}(M)$ is called **hereditary**, if

$$L_{\Phi\tau}\Phi = \Phi\, L_\tau \Phi \qquad \text{for all } \ \tau \in X(M). \tag{18}$$

This definition coincides with[10)], the notion of normal operators[17)] and is closely related to the notion of compatible Hamiltonian pairs[13)].

Definition 6: A vector field $\tau_0 \in X(M)$ is called a **scaling** (or a conformal invariance) for a tensor field T, if $L_{\tau_0} T = \kappa T$, $\kappa \in \mathbf{R}$.

With these notions one obtains

Theorem 1: Let $\Phi\in T^{(1,1)}(M)$ be a hereditary recursion operator for $K_0\in X(M)$. Let $\tau_0\in X(M)$ be a scaling for both K_0 and Φ, i.e. $L_{\tau_0}K_0 = \kappa K_0$, $L_{\tau_0}\Phi = \lambda\Phi$, $\kappa,\lambda\in\mathbb{R}$.

Defining $K_n := \Phi^n K_0$, $\tau_n := \Phi^n\tau_0$ one finds: $L_{K_n}\Phi = 0$, $L_{\tau_n}\Phi = \lambda\Phi^{n+1}$ and

$$[K_n, K_m] = 0 \quad , \quad [\tau_n, K_m] = (\kappa+m\lambda)K_{n+m} \quad ,[\tau_n,\tau_m] = \lambda(m-n)\tau_{n+m} \quad . \tag{19}$$

The **proof** is a straight forward application of the hereditary property and the product rule (8).

Remark 1: This result might lead to the intuition that a conformal invariance seems to be an essential structure of an integrable system. We do not believe that this is true. For the construction of the "angle type" master symmetries one can start with any time dependent symmetry (e.g. Galilean invariance, if present). The reasons, why we have formulated the above theorem using a scaling field, are: 1) the existence of a scaling is a ubiquituous phenomenon among integrable systems, 2) a scaling is the most obvious of the time dependent symmetries, it is detected at first sight.

It turns out that the above theorem can be applied to nearly all integrable systems generated by a recursion operator, a scaling transformation is indeed a very common feature.

6.3. The bi-hamiltonian case

Let us briefly recall the notion of (bi-)hamiltonian formulations. Let $J\in T^{(2,0)}(M)$ be antisymmetic, i.e. it corresponds to a 2-form. This form is closed if the bracket

$$\{X_1, X_2, X_3\} \ := \ < J'[X_1]\, X_2, X_3 > \tag{20}$$

satisfies the Jacobi identity for arbitrary $X_i\in X(M)$. Such a J shall be called **symplectic** (or **pre-symplectic,** if it is degenerate). Let $\Theta\in T^{(0,2)}(M)$ be a Poisson operator, i.e. it is antisymmetric and the bracket

$$\{X_1^*, X_2^*, X_3^*\} \ := \ < X_1^*,\Theta'[\Theta X_2^*]\, X_3^* > \tag{21}$$

satisfies the Jacobi identity for arbitrary $X_i^*\in X^*(M)$. Poisson operators are called **implectic** in[10]. A vector field $K\in X(M)$ is called **hamiltonian,** if $JK = df$ for some $f\in F(M)$, it is called **inverse hamiltonian** if $K = \Theta\, d\tilde{f}$ for some $\tilde{f}\in F(M)$. K is called **bi-hamiltonian,** if it admits both sorts of formulations. The concept of hamiltonian vector fields is included in the considerations about invariants, as (for invertible structures) a vector field is (inverse) hamiltonian iff it leaves the symplectic (implectic) vector field invariant.

Theorem 2: Let $J_1\in T^{(2,0)}(M)$ and $\Theta_0\in T^{(0,2)}(M)$ be a pair of (pre-)symplectic and implectic operators. Let $K_0\in X(M)$ be bi-hamiltonian w.r.t. J_1 and Θ_0, i.e.

$K_0 = \Theta_0\, df_0\;, J_1K = df_1\;, f_0, f_1 \in F(M)$.

Define $K_n{:=}\Phi^n K_0$, $J_{n+1}{:=}J_1\Phi^n$, $\Theta_n{:=}\Phi^n\Theta_0$.

i) If J_1 and Θ_0 are compatible (i.e. $\Phi{:=}\Theta_0 J_1$ is hereditary) then all the J_n are symplectic, all the θ_n are implectic. Every K_n leaves Φ as well as the J_n's and Θ_n's invariant. Hence all these vector fields are hamiltonian w.r.t. to all the above symplectic/implectic structures, there is a series of functions f_i such that $K_{n+m} = \Theta_n\, df_m\;, J_nK_m = df_{n+m}$. All the functions f_i are in involution w.r.t. all the Poisson brackets given by the Poisson operators Θ_n. The vectorfields K_n commute. Hence every f_n and every K_n is a conservation law and a symmetry, respectively, for every K_m.

ii) Let there be a $\tau_0 \in X(M)$ scaling K_0, Θ_0 and J_1:

$$L_{\tau_0}K_0 = \kappa\, K_0 \;\;,\;\; L_{\tau_0}\Theta_0 = \alpha\, \Theta_0 \;\;,\;\; L_{\tau_0}J_1 = \beta\, J_1 \;\;;\;\; \alpha,\beta,\kappa \in \mathbf{R}\;. \tag{22}$$

Define $\tau_n{:=}\Phi^n\tau_0$. Then one finds for all n, m :

$$\begin{aligned} L_{\tau_n}K_m &= (\kappa+m(\alpha+\beta))\; K_{n+m}\;, \\ L_{\tau_n}\tau_m &= (m-n)(\alpha+\beta)\; \tau_{n+m}\;, \\ L_{\tau_n}\Phi &= (\alpha+\beta)\Phi^{n+1}\;, \\ L_{\tau_n}\Theta_m &= (\alpha+(m-n)(\alpha+\beta))\Theta_{n+m}\;, \\ L_{\tau_n}J_m &= (-\alpha+(m+n)(\alpha+\beta))J_{n+m}\;, \\ L_{\tau_n}f_m &= (\kappa-\alpha+(m+n)(\alpha+\beta))f_{n+m}\;. \end{aligned} \tag{23}$$

With suitable invertability assumptions all the above indices could be negative as well.

Proof : The first part of the theorem is a meanwhile classical result[6),10),12),17)], the part about the master symmetries is just a straight forward application of the hereditary property of Φ and the product rule (8). For technical details we refer to [19),25)].

Remark 2: This theorem can be applied to nearly all known bi-hamiltonian systems. One usually finds that both hamiltonian formulations admit the same conformal invariance, a fact which to our opinion deserves further investigation.

Remark 3: The above results verify the principle of master symmetries as formulated in § 6.1.: all the master symmetries map invariants to new invariants. The scaling field τ_0 is a universal scaling, every quantity is mapped to itself. All the structure seems to be hidden in a non trivial master symmetry: with just τ_1 given one can construct all the higher invariants recursively. Note that even in the presence of a hamiltonian formulation the degree 1 master symmetries need not be directly connected to angle coordinates: if the coefficients α and β do not vanish then the time dependent symmetries arising from the master symmetries will not be hamiltonian and hence will not correspond to time dependent conservation laws.

Remark 4: In the infinite dimensional case one usually finds the generator of space translation $K_0(u) = u_x$, say, in the Lax hierarchy. But the commutator with space translation picks out the part of τ_n which depends explicitly on the space variable, so from $[\tau_n, K_0] = K_n$ one finds as typical form of the master symmetries $\tau_n = x K_n + R_n$, where the rest R_n does not contain the space variable explicitly. Note that in most examples this rest is nonlocal, i.e. it contains integrations.

Remark 5: The master symmetries seem to be deeply connected to multi hamiltonian formulations. The above result shows that the master symmetries connect the Poisson operators Θ_n via Lie derivatives. But independent of scaling properties (or whatever one wants to regard as basic) compatible Poisson structures always admit vector fields connecting them via Lie derivatives. Indeed, up to certain ambiguities the master symmetries may be recovered just from the knowledge of the Poisson operators[19].

6.4. Lax hierarchies without recursion operator

In the preceding sections we have discussed the structure of degree 1 master symmetries in the presence of a recursion operator. The principle of § 6.1. seems to indicate that one always finds a recursion operator when the considered system is hamiltonian: using a master symmetry one can map this first hamiltonian structure to a new one. But this might be satisfied in a very trivial way: if the master symmetry itself is hamiltonian, then it leaves the corresponding symplectic or implectic operator invariant, hence the "new" hamiltonian operator is zero and no bi-hamiltonian structure is available ! This might happen and it does happen in various cases such as the Benjamin-Ono, the Kadomtsev-Petviashvili, or the Calogero-Moser system. But in these cases we are provided with the hamiltonian functions of the master symmetries, which are angle variables for the integrable system. Hence the master symmetries either yield a bi-hamiltonian formulation or explicit representations for the angle variables.

New results (cf. P. Santini's lecture in this book) indicate that a bi-hamiltonian approach to the above systems can be done using extensions of the original system.

Consider an evolution equation $u_t = K_1(u)$, whose higher invariants are wanted. What can master symmetries do for us if no recursion operator is known ?

Step 1: look for a non trivial master symmetry τ_1, say, which maps K_1 to a symmetry of K_1, i.e. look for a vector field satisfying $[[\tau_1, K_1], K_1] = 0$.

Step 2: If such a τ_1 is found, then $K_{n+1} := [\tau_1, K_n]$ are candidates for a Lax hierarchy. In general L_{τ_1} should map any invariant to higher ones.

Of course, step 1 is the crucial one. Note, that "trivial" master symmetries such as a scaling will not be useful to obtain higher invariants, at least one higher master symmetry is needed. As demonstrated in the last sections such a τ_1 can be achieved cheaply from simple transformations (Galilean or scaling invariance), if a recursion operator is known. Without a recursion operator one really has to look for τ_1. A useful hint as to the explicit form is given by remark 4, i.e. a typical ansatz for τ_1 would be something like $\tau_1 = x K_1 + R_1$. If the structure of the rest R_1 was understood, then this approach would algorithmically tell us the integrability of the equation.

As to step 2: Note that the structure of master symmetries is a very weak (and hence very general) one. In general there is no structural argument, why the sequence K_n should automatically yield symmetries of K_1. One likes to think of τ_1 as a substitute for the recursion operator, mapping symmetries to symmetries. But the linear map L_{τ_1} is not a tensor field, it is not clear how to substitute the essential hereditary property of recursion operators for this map. In [5),21)] a "highest order analysis" for the above candidates K_n was needed to show that they really are symmetries for the considered equation (using the Jacobi identity this immediately implies that all the K_n's commute). In[20)] a lot of additional structure was needed to obtain rigorous statements. In [18)] criteria were derived how to obtain commutativity results in a purely algebraic way, i.e. without using the explicit form of the K_n's. It turned out that 2 non trivial master symmetries are needed to obtain such statements, the criteria are sufficient but technically involved and by no means beautiful. For certain restricted cases[29)] commutativity is guaranteed by general theorems, but again these results do depend on the explicit form of the equations and do not seem to arise from a geometrical background.

With these problems still open the principle of master symmetries remains somewhat heuristic. Nevertheless, it is extremely powerful, as all the higher invariants of an integrable system seem to be hidden in one vector field. This approach has successfully applied to various systems[5),7),8),9),21)].

7.1. Examples of bi-hamiltonian systems with scaling

As underlying manifold we choose the Schwartz space $S(\mathbf{R})$ of rapidly decreasing functions (coordinates u) or $S(\mathbf{R}) \oplus S(\mathbf{R})$ for the coupled systems (coordinates (u,v)). The tangent and cotangent spaces shall be identified with the manifold, the duality shall be given by

$$< X^* , X > := \int_{-\infty}^{\infty} X^*(x)\, X(x)\, dx \tag{24}$$

or

$$< \begin{pmatrix} X_1^* \\ X_2^* \end{pmatrix} , \begin{pmatrix} X_1 \\ X_2 \end{pmatrix} > := \int_{-\infty}^{\infty} \left(X_1^*(x)\, X_1(x) + X_2^*(x)\, X_2(x) \right)\, dx \tag{25}$$

for the coupled systems. We directly refer to theorem 2 and just state the first vector field K_0, the scaling τ_0, the hamiltonian operators Θ_0, J_1, and the scaling factors α,β,κ. All other quantities are given by the theorem.

The hierarchy of the **Korteweg-de Vries** equation:

$K_0(u) = u_x, \quad \tau_0(u) = xu_x + 2u, \quad \Theta_0 = D^3 + 2Du + 2uD, \quad J_1 = D^{-1},$

$\alpha = 0, \quad \beta = 2, \quad \kappa = 1.$

The hierarchy of the **CDGSK** equation[11)]:

$K_0(u) = u_{xxx} + 5uu_{xxx} + 5u_xu_{xx} + 5u^2u_x, \quad \tau_0(u) = xu_x + 2u, \quad \Theta_0 = D^3 + 2Du + 2uD,$

$J_1 = D^3 + 2D^2uD^{-1} + 2D^{-1}uD^2 + u^2D^{-1} + D^{-1}u^2, \quad \alpha = 0, \quad \beta = 6, \quad \kappa = 5.$

The **AKNS**-hierarchy[1]:

$$K_0(u,\mathrm{v}) = \begin{bmatrix} -u \\ \mathrm{v} \end{bmatrix}, \quad \tau_0(u) = \begin{bmatrix} xu_x + u \\ x\mathrm{v}_x + \mathrm{v} \end{bmatrix}, \quad \Theta_0 = \begin{bmatrix} 2uD^{-1}u & D - 2uD^{-1}\mathrm{v} \\ D - 2\mathrm{v}D^{-1}u & 2\mathrm{v}D^{-1}\mathrm{v} \end{bmatrix},$$

$$J_1 = \begin{bmatrix} 0 & 1 \\ -1 & 0 \end{bmatrix}, \quad \alpha = 0, \quad \beta = 1, \quad \kappa = 0.$$

The **Hirota-Satsuma-hierarchy**[14),6)]:

$$K_0(u,\mathrm{v}) = \begin{bmatrix} \frac{1}{2}u_{xxx} + 3uu_x - 6\mathrm{v}\mathrm{v}_x \\ -\mathrm{v}_{xxx} - 3u\mathrm{v}_x \end{bmatrix}, \quad \tau_0(u) = \begin{bmatrix} xu_x + 2u \\ x\mathrm{v}_x + 2\mathrm{v} \end{bmatrix},$$

$$\Theta_0 = \begin{bmatrix} \frac{1}{2}D^3 + Du + uD & D\mathrm{v} + \mathrm{v}D \\ D\mathrm{v} + \mathrm{v}D & \frac{1}{2}D^3 + Du + uD \end{bmatrix}, \; J_1 = \begin{bmatrix} \frac{1}{2}D + D^{-1}u + uD^{-1} & -2D^{-1}\mathrm{v} \\ -2\mathrm{v}D^{-1} & -2D \end{bmatrix},$$

$$\alpha = 0, \quad \beta = 4, \quad \kappa = 3.$$

Another part of the Lax hierarchy of the CDGSK and the Hirota Satsuma equation originates from space translation, some of these hierarchies admit a further master symmetry given by Galilean invariance. Many more examples have been investigated [3),16),18),19),23),26),27),28)].

7.2. An example for higher degree invariants: a multi-dimensional Burgers hierarchy

The well known Cole-Hopf transformation relates the potential form of Burgers equation with the heat equation. This can trivially be generalized considering the heat equation in an arbitrary number of space dimensions. For simplicity we restrict ourselves to 2 variables x_1 and x_2, but obviously everything can be done with an arbitrary number of variables. Being linear the heat equation

$$\tilde{u}_t = \tilde{u}_{x_1x_1} + \tilde{u}_{x_2x_2} \tag{26}$$

admits many commuting symmetries, two hereditary recursion operators are given by $\tilde{\Phi}_1 := \partial_{x_1}$ and $\tilde{\Phi}_2 := \partial_{x_2}$. The multiplication operators $\tilde{\Psi}_1 := x_1$ and $\tilde{\Psi}_2 := x_2$ (interpreted as (1,1)-tensor fields) are easily checked to be invariants of degree 1. We construct the vector fields

$$\tilde{m}_p^q = \tilde{\Psi}_1^{q_1}\,\tilde{\Psi}_2^{q_2}\,\tilde{\Phi}_1^{p_1}\,\tilde{\Phi}_2^{p_2}\,\tilde{m}_0^0 \quad ; \quad \tilde{m}_0^0(\tilde{u}) := \tilde{u} \tag{27}$$

(using multi indices). Obviously the vector fields $\tilde{K}_p := \tilde{m}_p^0$ form a Lax hierarchy, they all commute. It is easily checked that each $\tilde{m}_p^q$ is a master symmetry of degree

$|q| = q_1 + q_2$ for this hierarchy, they form a closed Lie algebra the structure constants of which can be calculated easily. The operators $\Psi_1{}^{q_1}\,\Psi_2{}^{q_2}\,\Phi_1{}^{p_1}\,\Phi_2{}^{p_2}$ are all master invariants of degree $|q|$ for the Lax hierarchy. It is a simple exercise to write down explicitly the time dependent symmetries and recursion operators arising from the Lie series (12).

These results get a less trivial shape using the Cole-Hopf transformation $\tilde{u} = e^u$. Using the proper transformation laws for tensor fields[10] we obtain in the u-chart hereditary operators

$$\Phi_1 := \partial_{x_1} + u_{x_1}\ ,\ \Phi_2 := \partial_{x_2} + u_{x_2}\ ,\ \Psi_1 := x_1\ ,\ \text{and}\ \ \Psi_2 := x_2\ . \tag{28}$$

The vector field $\tilde{m}_0^0$ becomes $m_0^0 = 1$, the others may again be represented by

$$m_p^q = \Psi_1^{q_1}\,\Psi_2^{q_2}\,\Phi_1^{p_1}\,\Phi_2^{p_2}\ m_0^0\ . \tag{29}$$

The first elements of the Lax hierarchy $K_p := m_p^0$ are

$K_{(0,0)} = 1,\ K_{(1,0)} = u_{x_1},\ K_{(0,1)} = u_{x_2},\ K_{(2,0)} = u_{x_1x_1} + u_{x_1}u_{x_1},\ K_{(1,1)} = u_{x_1x_2} + u_{x_1}u_{x_2},$

$K_{(0,2)} = u_{x_2x_2} + u_{x_2}u_{x_2},\ K_{(3,0)} = u_{x_1x_1x_1} + 3u_{x_1}u_{x_1x_1} + u_{x_1}u_{x_1}u_{x_1}\ , \ldots$

Every linear combination of elements of the above Lax hierachy is integrable (linearizable), it admits master symmetries and hereditary recursion operators of arbitrary degree. The algebraic structure of this integrable system can be written down simply using the results in the heat equation chart.

Master symmetries of higher degree also exist for the Benjamin-Ono and the Kadomtsev-Petviashvili equations[7] as well as for the Calogero Moser system[20]. Santini's extended recursion operators seem to give a straight forward approach to these quantities.

References

1) Ablowitz, M.J., Kaup, D.J., Newell, A.C., Segur,H.: "*The Inverse Scattering Transform - Fourier Analysis for Nonlinear Problems*", Studies in Applied Mathematics 53, 249-316 (**1974**).
2) Abraham, R., Marsden, J.E. : "*Foundations of Mechanics*", Benjamin Publ., London (**1978**).
3) Chen, H.H., Lee, Y.C., Lin, J.E.: "*On a new hierarchy of symmetries for the integrable nonlinear evolution equations*", preprint (**1982**).
4) Choquet-Bruhat, Y., de Witt-Morette, C., M. Dillard-Bleick, M.: "*Analysis, Manifolds and Physics*", North-Holland, Amsterdam (**1977**).
5) Fokas, A.S., Fuchssteiner, B: "*The Hierarchy of the Benjamin-Ono Equation*", Phys.Letters 86A 341-345 (**1981**).

6) Fuchssteiner, B.: "*The Lie-Algebra structure of Degenerate Hamiltonian and bi-Hamiltonian Systems*", Progr. Theor. Phys. 68, 1082-1104 (**1982**).

7) Fuchssteiner, B.: "*Mastersymmetries and higher order time-dependent symmetries and conserved densities of nonlinear evolution equations*", Prog. Theor. Phys. 70, 1508-1522 (**1983**).

8) Fuchssteiner, B.: "*On the hierarchy of the Landau-Lifshitz equation*", Physica 13D (**1984**).

9) Fuchssteiner, B.: "*Master symmetries for completely integrable systems in statistical mechanics*", Proc. Sitges Conference 1984 (in Springer Lecture Notes in Physics 216,), 305-315 (**1984**).

10) Fuchssteiner, B.: Fokas, A.S.: "*Symplectic structures, their Baecklund transformations, and hereditary symmetries*", Physica 4D, 47-66 (**1981**).

11) Fuchssteiner, B., Oevel, W.: "*The bi-Hamiltonian structure of some nonlinear fifth and seventh order differential equations and recursion formulas for their symmetries and conserved covariants*", J. Math. Phys. 23, 358-363 (**1982**).

12) Gel'fand, I.M., Dorfman, I.Y.: "*Hamiltonian Operators and Algebraic Structures related to them* ", Funct. Analy. i.Eqs. Priloz. 13, 248-262 (**1979**).

13) Gel'fand, I.M., Dorfman, I.Y.: "*The Schouten Bracket and Hamiltonian Operators*"", Funct. Analy. i.Eqs. Priloz. 14, 223-226 (**1980**).

14) Hirota, R., Satsuma, J.: "*Soliton solutions of a coupled Korteweg-de Vries equation*", Phys. Letters 85A 407-408 (**1981**).

15) Lax, P. D.: "*Integrals of Nonlinear Equations of Evolution and Solitary Waves*", Comm. Pure Appl. Math. 21, 467-490 (**1968**).

16) Li Yi-shen, Zhu Guo-Cheng: "*New set of symmetries of the integrable equations,..* ", USTC preprint, Hefei, Anhui People's republic of China (**1985**).

17) Magri, F.: "*A simple model of the integrable Hamiltonian equation*", J. Math. Phys. 19, 1156-1162 (**1978**).

18) Oevel, W.: *Dissertation*, Paderborn (**1984**).

19) Oevel, W.: "*A Geometrical Approach to Integrable Systems admitting Scaling Symmetries*", preprint, Paderborn, (**1985**).

20) Oevel, W., Falck, M.: "*Master Symmetries for Finite Dimensional Integrable Systems: The Calogero-Moser System*", Prog. Theor. Phys. 75, 1328-1341 (**1986**).

21) Oevel, W., Fuchssteiner, B: "*Explicit Formulas for Symmetries and Conservation Laws of the Kadomtsev-Petviashvili equation*", Phys. Letters 88A 323-327 (**1982**).

22) Olver, P.J.: "*Evolution equations possessing infinitely many symmetries*", J. Math. Phys. 18, (**1977**).

23) Strampp, W., Oevel, W.: "*A Nonlinear Derivative Schroedinger-Equation: Its Bi-Hamilton Structures, Their Inverses, Nonlocal Symmetries and Mastersymmetries*", Prog. Theor. Phys. 74, (**1985**). 922-925

24) Ten Eikelder, H.M.M.: *Ph.D. thesis*, Eindhoven (**1984**).

25) Ten Eikelder, H.M.M.: "*Symmetries for dynamical and Hamiltonian systems*", CWI tract 17, CWI, Amsterdam (**1985**).

26) Ten Eikelder, H.M.M.: "*Symmetries of the massive Thirring model*", J.Math.Phys 27, (**1986**).

27) Ten Eikelder, H.M.M., Broer, L.J.F.: "*Non-hamiltonian symmetries of a Boussinesq equation* ", to appear in J.Math.Phys (**1986**).

28) Tian Chou: "*Symmetries and a hierarchy of general KdV equation*", ICTP, Trieste, preprint (**1985**).

29) Tu, G.Z.: "*A commutativity theorem of partial differential operators*", Comm. Math. Phys. 77, 289-297 (**1980**).

30) Zakharov, V.E., Manakov, S.V., Novikov, S.P., Pitayevsky, L.P.: "*Theory of Solitons. The Method of the Inverse Scattering Problem*", Nauka, Moskau (**1980**).

DIRECT LINEARIZATION METHOD AND INTEGRABLE LATTICE SYSTEMS

H.W. Capel
Instituut voor Theoretische Fysica - Universiteit van Amsterdam
Valckenierstraat 65, 1018 XE Amsterdam
THE NETHERLANDS

ABSTRACT

Recently, many integrable difference equations on two- and three-dimensional lattices have been investigated on the basis of the direct linearization method (DLM) introduced by Fokas and Ablowitz. The DLM is based on a linear integral equation containing an arbitrary measure and contour. For every measure and contour the integral equation yields the basis functions of a Lax representation and the potential is obtained integrating the basis functions over the same contour with the same measure. The integrable lattice equations are obtained in the form of Bianchi identities expressing the commutativity of Bäcklund transformations (BT's) of the integral equation. As an example the integrable lattice versions of the Korteweg-de Vries (KdV) and the Kadomtsev-Petviashvili (KP) equations are treated. Hierarchies of integrable systems with one or more continuous coordinates, as well as the associated hamiltonian structures can be obtained applying appropriate continuum limits to the lattice equation, as well as to the linearizing integral equation. This is illustrated using the KdV and its lattice analogues as an example.

1. INTRODUCTION

Integrable systems may be characterized by the property that rather general solutions of these systems can be obtained solving linear problems only. An important formalism in the theory of integrable systems is the inverse scattering transform[1)-3)]. An other formalism is the method of direct linearization (DLM). It is based on an integral equation containing an integration over an arbitrary contour with an arbitrary measure. For every choice of measure and contour one can solve a wave function depending on a complex spectral parameter as well as on the coordinates of the system. By integrating the wave function over the same contour with the same measure one obtains a potential depending only on the coordinates of the system. The important feature is now that the potential satisfies one and the same integrable system, independently of the choice of measure and contour.

As an example we consider the integral equation proposed by Fokas and Ablowitz[4)] for the solutions of the Korteweg-de Vries (KdV) equation:

$$u_k + \rho_k \int_C d\lambda(\ell) \frac{u_\ell}{k+\ell} = \rho_k \quad . \tag{1.1}$$

In eq. (1.1) C denotes a contour in the plane of the complex spectral parameter k, $d\lambda(k)$ is a measure depending on k, ρ_k is a free-wave function and u_k is the wave function to be solved from the integral equation. Integrating u_k over the same contour C with the same measure $d\lambda(k)$ one obtains a potential

$$u = \int_C d\lambda(k) u_k \quad , \tag{1.2}$$

depending only on the coordinates of the system. In this paper we shall restrict ourselves to contours and measures satisfying the <u>uniqueness condition</u>, i.e. at given ρ_k the solution u_k of the integral equation is unique.

Starting from a suitable choice of the free-wave function ρ_k one can derive the various linear relations of a Lax representation by showing that the left-hand side minus the right-hand side satisfies a homogeneous integral equation. The uniqueness condition then implies that l.h.s. - r.h.s. is zero. The integrable equation can be obtained by integrating over the contour C with the measure $d\lambda(k)$ or by working out the compatibility of the linear relations. Choosing ρ_k to be a plane wave with dispersion $\omega(k) = k^3$, i.e. $\rho_k = \exp(kx + k^3 t)$, it can be shown that[4)]

$$\partial_t u - \partial_x^3 u - 3(\partial_x u)^2 = 0 \quad , \tag{1.3}$$

i.e. $\partial_x u$ satisfies the KdV.

Since then various generalizations have been obtained:

i) Including a factor $k^{-\alpha}$, (α), in the r.h.s. of (1.1) and a factor $k^{-\beta}$, (β), in the integral at the r.h.s. of (1.2), one can define[5)] a matrix of potentials $u^{(\alpha,\beta)}$. In this way one can derive some modified equations[5)] involving matrix elements $(\alpha,\beta) \neq (0,0)$ as well as the Miura transformations (MT's) relating different matrix elements.

ii) By studying two coupled equations for wave functions ϕ_k, ψ_k, one obtains the direct linearization (DL)[6)] of the AKNS system[7)], see also ref. 5 for a treatment of special cases.

iii) By including polynomials in k and ℓ in the integral equation one can obtain the DL[6),8)] of finite-matrix extensions[9)] of existing PDE's.

iv) By replacing the contour C and the measure $d\lambda(\ell)$ by a hypercontour D and a measure $d\zeta(\ell,\ell')$ in terms of two complex variables ℓ,ℓ' one can establish the DL[10),11)] of PDE's in 2+1 dimensions.

v) The integral equation can be regarded as a transformation mapping a free-wave function ρ_k into a new solution u_k of a Lax representation and is therefore called a free-reference integral equation. One can also investigate a general-reference integral equation mapping an arbitrary solution u_k^0 of a Lax representation into a new solution of the Lax representation[8),10)-13)].

vi) On the basis of the integral equation one can investigate Bäcklund transformations (BT's)[14),15)] and integrable lattice systems[16)]. Here one should keep in mind that from all lattice equations changing into one integrable PDE in a continuum limit, only few will be integrable.

In this paper we shall focus on the last type of generalization. In section 2 the BT's associated with the integral equation (1.1) are reviewed and in section 3 we derive the lattice KdV together with its linearizing integral equation. The lattice KP and its DL can be obtained from an integral equation of the type (1.1), but with C and $d\lambda(\ell)$ replaced by D and $d\zeta(\ell,\ell')$. This will be shown in section 4. By applying suitable continuum limits to the lattice equations and the linearizing integral equations one can derive[8),11),16)] the continuum versions of the KdV and KP with the associated DL.

In the continuum limits treated so far only the dominating terms in a small lattice parameter are taken into account. One can do better and also treat higher-order terms in the lattice parameter using suitable vertex operators. Such vertex operators are reminiscent of the ones that have been used in the τ-function approach[17)-19)]. In section 5 we perform a first continuum limit on the lattice KdV and derive an infinite hierarchy of integrable higher-order discrete KdV equations for time-dependent fields $u_n(t)$ at sites n of a one-dimensional chain, together with their DL. The equations can also be derived from a hamiltonian structure via a Poisson bracket[20)] which is a discrete analogue of the Gardner bracket[21)]. Finally, in section 6 a second continuum limit is treated. Starting from the discrete KdV, we derive the usual hierarchy of higher-order KdV's with the associated hamiltonian structure[22),23)]. A more detailed treatment containing also analogous results related to the modified Korteweg-de Vries (mKdV) equation can be found in ref. 24. Furthermore, it will be of interest to establish similar relations between the lattice KP and recent results[25),26)] on the hierarchy and (bi-)hamiltonian structure of the KP.

2. BÄCKLUND TRANSFORMATION

Before introducing the BT we first indicate how the soliton solutions[4)] arise in the DLM. For this purpose we consider a measure $d\lambda(k)$ which is a linear combination of simple poles located at $k_1, k_2, \ldots, k_N$, i.e.

$$d\lambda(k) = \frac{1}{2\pi i} \sum_{\alpha=1}^{N} \frac{c_\alpha}{k-k_\alpha} dk \quad , \tag{2.1}$$

and we choose a contour C surrounding the N poles. From Cauchy's theorem for the contour integration, i.e.

$$\frac{1}{2\pi i} \int_C \frac{f(k)}{k-k_\alpha} dk = f(k_\alpha) \quad , \tag{2.2}$$

it is clear that (1.1) reduces to N algebraic linear equations

$$u_{k_\alpha} + \rho_{k_\alpha} \sum_{\beta=1}^{N} \frac{c_\beta u_{k_\beta}}{k_\alpha + k_\beta} = \rho_{k_\alpha} \quad , \tag{2.3}$$

and eq. (1.2) yields

$$u = \sum_{\alpha=1}^{N} c_\alpha u_{k_\alpha} \quad . \tag{2.4}$$

The potential u in (2.4) is the general N-soliton solution, independently of the specific features of the free-wave function ρ_k .

Let us consider a transformation of the measure $d\lambda(\ell)$ in (1.1) and a new measure $d\bar{\lambda}(\ell)$ which is given by

$$d\lambda(\ell) \to d\bar{\lambda}(\ell) = \frac{p-\ell}{p+\ell} d\lambda(\ell) \quad . \tag{2.5}$$

From (1.1) with the new measure $d\bar{\lambda}(\ell)$ one solves a new wave function $\bar{u}_k$ and the new potential $\bar{u}$ is obtained by (1.2) with $d\lambda(k)$ and u_k replaced by $d\bar{\lambda}(k)$ and $\bar{u}_k$. We anticipate $\bar{u}$ to be a solution of the same integrable system as u and therefore the transformation (2.5) is called a BT. This way of introducing BT's has some analogy with the transformation of the reflection coefficient in the inverse scattering transform considered e.g. in ref. 3. It is easy to see that the transformation (2.5) can increase the number of solitons by 1. In fact, consider the case (2.1) with C surrounding the poles $k_1,k_2,\dots,k_N$ corresponding to an N-soliton solution u. Then from partial fractions it follows that $d\bar{\lambda}(k)$ is a linear combination of N+1 poles and when C surrounds the extra pole at $k = -p$, the new solution is an (N+1)-soliton solution.

Instead of (2.5) we can apply a similar transformation to the free-wave function ρ_k, i.e.

$$\rho_k \to \tilde{\rho}_k = \rho_k \frac{p-k}{p+k} \quad . \tag{2.6}$$

Using the new $\tilde{\rho}_k$, eq. (1.1) can be solved to find a new wave function $\tilde{u}$ and using (1.2) with the old $d\lambda(k)$ and the new $\tilde{u}_k$ one obtains the new potential $\tilde{u}$. It is easy to show that $\tilde{u} = \bar{u}$. In the following the BT's will be investigated on the basis of (2.6).

To work out (2.6) let us consider the integral equation (1.1) with $\tilde{\rho}_k$ and the new solution $\tilde{u}_k$. Multiplying this equation by (p+k), we obtain the relation

$$(p+k)\tilde{u}_k + \rho_k \int_C d\lambda(\ell) \frac{(p+\ell)\tilde{u}_\ell}{k+\ell} = (p-k)\rho_k + \tilde{u}\rho_k \quad . \tag{2.7}$$

In fact, after multiplication by (p+k) the factor $(p-k)\rho_k$ in front of the integral in the l.h.s. is decomposed into a term $(p+\ell)\rho_k$ which is needed to have an integral equation for $(p+k)\tilde{u}_k$ and a term $-(k+\ell)\rho_k$ which cancels the denominator so that the integral can be worked out leading to the term $\tilde{u}\rho_k$ in the r.h.s. Introducing now a wave function y_k satisfying the integral equation

$$y_k + \rho_k \int_C d\lambda(\ell) \frac{y_\ell}{k+\ell} = k\rho_k \tag{2.8}$$

it is clear that the r.h.s. of (2.7) can be identified as the source term in an integral equation for $(p + \tilde{u})u_k - y_k$. The uniqueness condition implies the linear relation

$$(p+k)\tilde{u}_k = (p + \tilde{u})u_k - y_k \quad . \tag{2.9}$$

Eq. (2.9) will be used in section 3 to derive the Bianchi identity expressing the commutativity of two BT's. To conclude this section we shall work out the spatial part of the BT[15] of the KdV in the case that $\rho_k = \exp[kx + \omega(k)t]$ for an arbitrary dispersion $\omega(k)$. From (1.1) with $\partial_x \rho_k = k\rho_k$ one finds

$$(2\partial_x - k)u_k + \rho_k \int_C d\lambda(\ell) \frac{(2\partial_x - \ell)u_\ell}{k+\ell} = (k-u)\rho_k \quad , \tag{2.10}$$

and the uniqueness condition implies that

$$(2\partial_x - k)u_k + (p+k)\tilde{u}_k = (p + \tilde{u} - u)u_k \tag{2.12}$$

for the BT (2.6).

From (2.6) it is clear that the inverse BT, i.e. $\tilde{\rho}_k \to \rho_k$, can be obtained changing $p \to -p$ and interchanging the wiggles and no wiggles. From (2.12) we then obtain for the inverse BT

$$(2\partial_x - k)\tilde{u}_k - (p-k)u_k = -(p + \tilde{u} - u)u_k \quad . \tag{2.13}$$

Adding (2.12) and (2.13) we see that the terms with k cancel. The result can be integrated with (1.2) to yield

$$\partial_x\left(\frac{\tilde{u}-u}{2}\right) + p\left(\frac{\tilde{u}-u}{2}\right) + \left(\frac{\tilde{u}-u}{2}\right)^2 + \partial_x u = 0 \quad , \tag{2.14}$$

which is the spatial part of the BT of the KdV-hierarchy.

3. LATTICE KdV

In section 2 we have derived eq. (2.9) for the BT (2.6). Since the inverse BT $\tilde{\rho}_k \to \rho_k$ is generated by $p \to -p$ and interchanging the wiggles and no wiggles, we also have a second relation, i.e.

$$(p+k)\tilde{u}_k = (p+\tilde{u})u_k - y_k$$

$$(p-k)u_k = (p-u)\tilde{u}_k + \tilde{y}_k \quad . \tag{3.1}$$

We also consider a second BT which is generated by

$$\rho_k \rightarrow \hat{\rho}_k = \rho_k \frac{q-k}{q+k} \tag{3.2}$$

and for this BT and its inverse we obtain

$$(q+k)\hat{u}_k = (q+\hat{u})u_k - y_k$$

$$(q-k)u_k = (q-u)\hat{u}_k + \hat{y}_k \quad , \tag{3.3}$$

which is eq. (3.1) with p and the wiggle replaced by q and the hat.

Since (2.6) and (3.2) are scalar multiplications, it is clear that the two BT's must commute with each other. The commutativity can be expressed in terms of a <u>Bianchi identity</u> which is an algebraic relation involving only the potential u and its Bäcklund transforms. Using the commutativity, i.e. $\hat{\tilde{y}}_k = \tilde{\hat{y}}_k$, and eliminating the y_k from (3.1) and (3.3), we obtain the linear relations

$$(q+k)\hat{u}_k - (p+k)\tilde{u}_k = (q + \hat{u} - p - \tilde{u})u_k \quad ,$$

$$(p-k)\hat{u}_k - (q-k)\tilde{u}_k = (p - \hat{u} - q + \tilde{u})\hat{\tilde{u}}_k \quad , \tag{3.4}$$

$$(p+k)\tilde{\tilde{u}}_k + (p-k)u_k = (2p + \tilde{\tilde{u}} - u)\tilde{u}_k \quad . \tag{3.5}$$

Adding both relations (3.4) and integrating the result over the contour C with the measure $d\lambda(k)$ we obtain the Bianchi identity

$$(q + p + \hat{\tilde{u}} - u)(q - p + \tilde{u} - \hat{u}) = q^2 - p^2 \quad . \tag{3.6}$$

We are now going to interpret this Bianchi identity as an integrable lattice equation. In order to do so we identify the two BT's generated by (2.6) and (3.2) as the two primitive translations over $\vec{e}_1$ and $\vec{e}_2$ respectively of a two-dimensional lattice. Then, for a function f at the lattice site $\vec{r} = n\vec{e}_1 + m\vec{e}_2$, which can be either the free-wave function ρ_k , the wave function u_k or the potential u, the transformed function $\tilde{f}(\vec{r})$ is equal to $f(\vec{r} + \vec{e}_1)$ and similarly $\hat{f}(\vec{r}) = f(\vec{r} + \vec{e}_2)$. For the free-wave function ρ_k at the lattice site (n,m) we can take

$$\rho_k(n,m) = \left(\frac{p-k}{p+k}\right)^n \left(\frac{q-k}{q+k}\right)^m \quad , \tag{3.7}$$

as this expression is multiplied by (p-k)/(p+k) by a translation over $\vec{e}_1$ and by (q-k)/(q+k) by a translation over $\vec{e}_2$.

In the Bianchi identity (3.6) we can now take u as u(n,m), $\tilde{u}$ as u(n+1,m), $\hat{u}$ as u(n,m+1) and $\hat{\tilde{u}}$ as u(n+1,m+1) to obtain the equation

$$\{q+p+u(n+1,m+1)-u(n,m)\}\{q-p+u(n,m+1)-u(n+1,m)\} = q^2 - p^2 \tag{3.8}$$

at the sites (n,m) of a two-dimensional lattice. Eq. (3.8) is an integrable equation in the sense that solutions can be obtained from the linearizing integral equation (1.1) with $\rho_k = \rho_k(n,m)$ given by (3.7). This integral equation has the solution $u_k(n,m)$ and the potential u(n,m) is obtained integrating $u_k(n,m)$ over the contour C with the measure $d\lambda(k)$. Eq. (3.8) can be shown to be a lattice version of the KdV by two consecutive continuum limits[6),8),16)], in which only dominating terms in the lattice parameter are considered. Eq. (1.1) with $\rho_k(n,m)$ given by (3.6) is a free-reference integral equation, a general-reference equation for the lattice KdV has been derived in ref. 8.

4. LATTICE KP

In section 3 we derived a Bianchi identity from the relations (3.1) and (3.3) arising from the BT's (2.6) and (3.2) and their inverses. Let us now suppose that we only have the relations for $(p+k)\tilde{u}_k$ and $(q+k)\hat{u}_k$, but not the inverse relation for $(p-k)u_k$ and

$(q-k)u_k$. In order to have an algebraic identity for the potential u and its transforms, we need a third transformation characterized by a parameter r and e.g. an open dot instead of a wiggle or a hat. From three such relations we can derive a Bianchi identity which can be interpreted as an integrable lattice equation in 3 dimensions.

To be specific let us consider the integral equation

$$u_k + \rho_k \iint_D d\zeta(\ell,\ell') \frac{\sigma(\ell,\ell')}{k+\ell'} u_\ell = \rho_k \tag{4.1}$$

which is (1.1), apart from the following two differences:

i) The integral equation contains an integration over a hyper contour D in the space of two complex variables ℓ,ℓ' with a measure $d\zeta(\ell,\ell')$ depending on ℓ and ℓ' as well, rather than a contour C and a measure $d\lambda(\ell)$ in the space of one complex variable.

ii) The integral equation contains a factor $\sigma(\ell,\ell')$ depending on ℓ and ℓ' as well. Of course, we could have included this factor in the measure, but we prefer to keep the measure invariant under the BT's which will be introduced.

From the integral equation with given free-wave function ρ_k, $\sigma(\ell,\ell')$, D and $d\zeta(\ell,\ell')$ one can solve the wave function u_k and the potential u is obtained integrating u_k multiplied by σ over the same hyper contour D with the measure $d\zeta(\ell,\ell')$

$$u = \iint_D d\zeta(k,k')\sigma(k,k')u_k \quad . \tag{4.2}$$

Also here we restrict ourselves to hyper contours and measures satisfying the <u>uniqueness condition</u>, i.e. at given ρ_k and $\sigma(\ell,\ell')$ the solution u_k of the integral equation is unique.

In order to define a BT for the integral equation (4.1) we introduce a <u>simultaneous transformation</u> for ρ_k and $\sigma(\ell,\ell')$, i.e.

$$\rho_k \to \tilde{\rho}_k = \rho_k \frac{p-k}{p+k}$$

$$\sigma(\ell,\ell') \to \tilde{\sigma}(\ell,\ell') = \sigma(\ell,\ell') \frac{p+\ell}{p+\ell'} \quad . \tag{4.3}$$

The integral equation (4.1) with $\tilde{\rho}_k$ and $\tilde{\sigma}(\ell,\ell')$ has a new solution $\tilde{u}_k$, and the new potential $\tilde{u}$ is obtained from (4.2) with $\tilde{\sigma}(k,k')$ and $\tilde{u}_k$ instead of $\sigma(k,k')$ and u_k .

Multiplying (4.1) for $\tilde{u}_k$ by p+k and using the relation

$$\tilde{\sigma}(\ell,\ell')\,\frac{p-k}{p+k} = \sigma(\ell,\ell')\,\frac{p+\ell}{k+\ell'} - \tilde{\sigma}(\ell,\ell') \tag{4.4}$$

it is straightforward to derive the relation

$$(p+k)\tilde{u}_k + \rho_k \iint_D d\zeta(\ell,\ell')\,\frac{\sigma(\ell,\ell')}{k+\ell'}\,(p+\ell)\tilde{u}_\ell = (p-k)\rho_k + \tilde{u}\rho_k\ , \tag{4.5}$$

which is the analogue of (2.7). Introducing a wave function y_k defined by

$$y_k + \rho_k \iint_D d\zeta(\ell,\ell')\,\frac{\sigma(\ell,\ell')}{k+\ell'}\,y_\ell = k\rho_k \quad , \tag{4.6}$$

similarly to (2.8), we obtain the linear relation (2.9). However, due to the transformation of the σ the BT (4.3) cannot be inverted, so that we do not have an inverse relation for $(p-k)u_k$ as in (3.1).

In order to derive an algebraic identity in the potential u and its transforms, we need two other transformations

$$\rho_k \to \hat{\rho}_k = \rho_k\,\frac{q-k}{q+k}\ , \quad \sigma(\ell,\ell') \to \hat{\sigma}(\ell,\ell') = \sigma(\ell,\ell')\,\frac{q+\ell}{q+\ell'}\quad ,$$

$$\rho_k \to \mathring{\rho}_k = \rho_k\,\frac{r-k}{r+k}\ , \quad \mathring{\sigma}(\ell,\ell') \to \sigma(\ell,\ell') = \sigma(\ell,\ell')\,\frac{r+\ell}{r+\ell'}\quad , \tag{4.7}$$

leading to two linear relations which can be obtained from (2.9) replacing p and the wiggle by q and the hat, and by r and the open dot, respectively. We have

$$(p+k)\tilde{u}_k = (p+\tilde{u})u_k - y_k \quad ,$$

$$(q+k)\hat{u}_k = (q+\hat{u})u_k - y_k \quad ,$$

$$(r+k)\mathring{u}_k = (r+\mathring{u})u_k - y_k \quad . \tag{4.8}$$

The transformations introduced in (4.3) and (4.7) commute and therefore one can evaluate $(p+k)(q+k)\hat{\tilde{u}}_k$ in two different ways, yielding

$$\{(p+\hat{\tilde{u}})(q+\hat{u}) - (q+\hat{\tilde{u}})(p+\tilde{u})\}u_k = (q+k)\hat{y}_k - (p+k)\tilde{y}_k + (p-q)y_k \ , \quad (4.9)$$

in which $\hat{y}_k$ and $\tilde{y}_k$ are the solutions of (4.6) with $\hat{\rho}_k$, $\hat{\sigma}(\ell,\ell')$ and $\tilde{\rho}_k$, $\tilde{\sigma}(\ell,\ell')$ instead of ρ_k , $\sigma(\ell,\ell')$. Considering also the third equation (4.8) one can derive two other relations that follow from (4.9) by cyclic permutations of p, q and r and simultaneously of the wiggle, the hat and the open dot. Adding the three relations it is easy to see that all terms involving the y_k's cancel, and assuming $u_k \neq 0$, we immediately obtain the algebraic relation

$$\begin{aligned} &(p+\hat{\tilde{u}})(q+\hat{u}) - (q+\hat{\tilde{u}})(p+\tilde{u}) \\ &+ (q+\mathring{\hat{u}})(r+\mathring{u}) - (r+\mathring{\hat{u}})(q+\hat{u}) \\ &+ (r+\tilde{\mathring{u}})(p+\tilde{u}) - (p+\tilde{\mathring{u}})(r+\mathring{u}) = 0 \quad . \end{aligned} \quad (4.10)$$

Eq. (4.10) is the Bianchi identity expressing the commutativity of the 3 BT's (4.3) and (4.7). In order to derive an integrable lattice equation we interpret the 3 BT's generated by the wiggle, the hat and the open dot as the 3 primitive translations over $\vec{e}_1$, $\vec{e}_2$ and $\vec{e}_3$, respectively, of a three-dimensional lattice. For a function f, which can be ρ_k , $\sigma(\ell,\ell')$, u_k or u, we then have with $f \equiv f(\vec{r})$, $\tilde{f}(\vec{r}) = f(\vec{r}+\vec{e}_1)$, $\hat{f}(\vec{r}) = f(\vec{r}+\vec{e}_2)$, $\mathring{f}(\vec{r}) = f(\vec{r}+\vec{e}_3)$. The functions ρ_k and $\sigma(\ell,\ell')$ at the lattice site $\vec{r} = n_1\vec{e}_1 + n_2\vec{e}_2 + n_3\vec{e}_3$ are given by

$$\begin{aligned} \rho_k(n_1,n_2,n_3) &= \left(\frac{p-k}{p+k}\right)^{n_1} \left(\frac{q-k}{q+k}\right)^{n_2} \left(\frac{r-k}{r+k}\right)^{n_3} \\ \sigma(\ell,\ell')(n_1,n_2,n_3) &= \left(\frac{p+\ell}{p+\ell'}\right)^{n_1} \left(\frac{q+\ell}{q+\ell'}\right)^{n_2} \left(\frac{r+\ell}{r+\ell'}\right)^{n_3} \quad . \end{aligned} \quad (4.11)$$

From the Bianchi identity with the identifications for $f \equiv u$ we immediately obtain the partial difference equation

$$\{p+u(n_1+1,n_2+1,n_3)\}\{q+u(n_1,n_2+1,n_3)\}$$

$$-\ \{q+u(n_1+1,n_2+1,n_3)\}\{p+u(n_1+1,n_2,n_3)\}$$

$$+\ \{q+u(n_1,n_2+1,n_3+1)\}\{r+u(n_1,n_2,n_3+1)\}$$

$$-\ \{r+u(n_1,n_2+1,n_3+1)\}\{q+u(n_1,n_2+1,n_3)\}$$

$$+\ \{r+u(n_1+1,n_2,n_3+1)\}\{p+u(n_1+1,n_2,n_3)\}$$

$$-\ \{p+u(n_1+1,n_2,n_3+1)\}\{r+u(n_1,n_2,n_3+1)\} = 0 \quad . \tag{4.12}$$

Eq. (4.12) is an integrable equation for fields $u(n_1,n_2,n_3)$ at the sites $\vec{r} = n_1\vec{e}_1 + n_2\vec{e}_2 + n_3\vec{e}_3$ of a three-dimensional lattice in the sense that solutions can be obtained from the linearizing integral equation with $\rho_k \equiv \rho_k(n_1,n_2,n_3)$ and $\sigma(\ell,\ell') \equiv \sigma(\ell,\ell')(n_1,n_2,n_3)$, as given by (4.11). Eq. (4.12) is a lattice version of the KP, as can be shown by three consecutive continuum limits[6),8),16)] in which only dominating terms in the lattice parameter are considered. Eq. (4.1) with $\rho_k(n_1,n_2,n_3)$ and $\sigma(\ell,\ell')(n_1,n_2,n_3)$ given by (4.11) is a free-reference integral equation, a general-reference integral equation for the lattice KP has been derived in ref. 8.

5. THE DISCRETE KdV HIERARCHY

5.1 Continued-fraction expansion

In this section we shall work out a first continuum limit starting from the lattice KdV and taking into account all orders of a small lattice parameter. For reasons of convenience we shall use a slightly different interpretation of the BT's generated by (2.6) and (3.2). From now on these BT's will be identified with the translations over $\vec{e}_1$ and $\vec{e}_1 + \vec{e}_2$ respectively, in which $\vec{e}_1$ and $\vec{e}_2$ are the basis vectors of the two-dimensional lattice. Then for a function $f = \{\rho_k, u_k, u\}$, at $\vec{r} = n\vec{e}_1 + m\vec{e}_2$, we have $\tilde{f}(r) = f(\vec{r} + \vec{e}_1)$, $\hat{f}(\vec{r}) = f(\vec{r} + \vec{e}_1 + \vec{e}_2)$. This leads to the expression

$$\rho_k(\vec{r}) = \left(\frac{p-k}{p+k}\right)^n \exp\left\{m \ln\left(\frac{p+k}{p-k}\,\frac{q-k}{q+k}\right)\right\} \quad , \tag{5.1}$$

as $\rho_k(\vec{r})$ must be multiplied by $(q-k)/(q+k)$ by the transformation $(n,m) \to (n+1,m+1)$. The Bianchi identity (3.5) then leads to the lattice equation

$$q-p+u(n,m+1)-u(n,m) = \frac{q^2 - p^2}{v(n)+(q-p)+u(n+1,m+1)-u(n+1,m)} \tag{5.2}$$

with

$$v(n) \equiv 2p + u(n+1,m) - u(n-1,m) \quad . \tag{5.3}$$

Iterating (5.2) and assuming that u(n,m) tends to a value, independent of m, in the limit $n \to \infty$, we have

$$q-p+u(n,m+1)-u(n,m) = \cfrac{-(p^2-q^2)}{v(n) - \cfrac{(p^2-q^2)}{v(n+1) - \cfrac{(p^2-q^2)}{v(n+2) - \cdots}}} \tag{5.4}$$

which is an infinite continued-fraction representation for the solutions of the lattice KdV (5.2). One may regard (5.4) as a formal solution of (5.2). In fact, choosing the potentials u(n,m) at fixed m as an initial condition, we can obtain u(n,m+1) up to arbitrary order in $p^2 - q^2$ expanding the continued fraction. By repeating this one can obtain u(n,m') with m' > m up to arbitrary order in $p^2 - q^2$.

5.2 Continuum limit

To perform a continuum limit we introduce an infinite number of variables $\tau_1, \tau_2, \ldots$, which is reminiscent of the work of the Kyoto group[18),19)]. For an arbitrary function f(n,m) which can again be ρ_k, u_k or the potential u, we have $f(n,m) = f(n;\tau_1,\tau_2,\ldots)$, and the function f(n,m+1) is obtained from f(n,m) by a primitive translation in the m direction. Such a translation is expressed as a <u>vertex operator</u> $\exp\partial_m$, in which ∂_m is a linear combination of derivatives ∂_{τ_s} multiplied by the sth power of the lattice parameter. We take

$$f(n,m+1) = e^{\partial_m} f(n;\tau_1,\tau_2,\ldots) \quad ,$$

$$\partial_m = \sum_{s=1}^{\infty} \{(p^2 - q^2)^s / s\} \partial_{\tau_s} \quad . \tag{5.5}$$

In eq. (5.5) we have chosen $p^2 - q^2$ as the lattice parameter, as is suggested by the r.h.s. of (5.4), the factor s in the denominator is for convenience.

To evaluate the factor ρ_k as a function of the time variables $\tau_1,\tau_2,\ldots$, we use the expansion

$$\ln\left(\frac{p+k}{p-k}\,\frac{q-k}{q+k}\right) = \sum_{s=1}^{\infty} C_s(k)(p^2-q^2)^s/s \tag{5.6}$$

in which the coefficients $C_s(k)$ are determined by

$$(-1)^r\{(p+k)^{-r} -(p-k)^{-r}\}/r = \sum_{s=1}^{\infty} \{(-1)^s/s\}(2p)^{2s-r}\binom{s}{r-s}C_s(k) . \tag{5.7}$$

Using (5.1) is follows that the l.h.s. of (5.6) can be identified with $\partial_m \ln\rho_k$, implying with (5.5) that $\partial_{\tau_s} \ln \rho_k = C_s(k)$ and

$$\rho_k = \left(\frac{p-k}{p+k}\right)^n \exp\left[\sum_{s=1}^{\infty} C_s(k)\tau_s\right] \quad . \tag{5.8}$$

To obtain a hierarchy of integrable equations for time-dependent fields u_n at the sites n of a one-dimensional chain, we expand both members of (5.4) in powers of $p^2 - q^2$ using (5.5). Up to the order $(p^2 - q^2)^2$ we have

$$\partial_{\tau_1} u(n) = \frac{-1}{v(n)} + \frac{1}{2p} \quad , \tag{5.9}$$

$$\partial_{\tau_2} u(n) = - \partial_{\tau_1}^2 u(n) - \frac{2}{v^2(n)v(n+1)} + \frac{1}{4p^3}$$

$$= \frac{-1}{v^2(n)} \left\{\frac{1}{v(n+1)} + \frac{1}{v(n-1)}\right\} + \frac{1}{4p^3} \quad . \tag{5.10}$$

This procedure can be continued to derive the equations for all $\partial_{\tau_s} u(n)$. In fact, the term $\sim(p^2-q^2)^s$ in the l.h.s. of (5.4) contains $\partial_{\tau_s} u(n)$, but also products of lower order terms $\partial_{\tau_{s'}} u(n)$ with $s' < s$. Eliminating these terms with the help of lower-order equations that were already derived, we obtain an equation containing $\partial_{\tau_s} u$ and algebraic terms in the fields $v(n')$. Solutions of this equation can be obtained from the integral equation (1.1) with $\rho_k = (p-k)^n(p+k)^{-n}\exp[C_s(k)\tau_s]$. The equations for $\partial_{\tau_s} u(n)$ form the <u>discrete KdV</u>-hierarchy.

5.3 Hamiltonian structure

In order to establish the connection with a hamiltonian structure we start from the scattering problem for the one-dimensional chain. From (3.4), together with the identifications $f = f(\vec{r})$, $\tilde{f} = f(\vec{r}+\vec{e}_1)$, $\tilde{\tilde{f}} = f(\vec{r}+2\vec{e}_1)$, at the fixed value of m,

$$(p+k)u_k(n+1) - (p-k)u_k(n-1) = \{2p+u(n+1) - u(n-1)\}u_k(n) \quad . \quad (5.11)$$

From (5.8) and (5.11) it is straightforward to write down a linear relation connecting the u_k/ρ_k at the sites n-1,n,n+1. We consider the variable a(k), and an ansatz[27)] for the $(u_k/\rho_k)(n)$, i.e.

$$a(k) = \lim_{n\to\infty} \frac{u_k}{\rho_k}(n) \quad , \quad \frac{u_k}{\rho_k}(n) = \prod_{n'=-\infty}^{n} g_k(n') \quad . \quad (5.12)$$

From (5.11) and (5.12) we obtain

$$(p-k)g_k(n)g_k(n+1)+(p+k) = v(n)g_k(n) \quad . \quad (5.13)$$

We use the expansion

$$\ln g_k(n) = \ln(p+k)+H(k;n) \quad , \quad H(k;n) \equiv \sum_{r=0}^{\infty} (p^2-k^2)^r H_{r+1}(n) \quad , \quad (5.14)$$

leading to

$$\ln a(k) = H(k) + \text{constant} \quad , \quad H(k) \equiv \sum_{r=0}^{\infty} (p^2-k^2)^r H_{r+1} \quad ,$$

$$H_{r+1} = \sum_{n=-\infty}^{\infty} H_{r+1}(n) \quad , \quad H(k) = \sum_{n=-\infty}^{\infty} H(k;n) \quad . \tag{5.15}$$

In eq. (5.14) $\ln(p+k)$ arises from the second term in the l.h.s. of (5.13) and $H_1(n) = -\ln v(n)$, in agreement with the relation $g_p(n) = 2p/v(n)$. The $H(k;n)$ is the generating function of the $H_{r+1}(n)$. Similarly, in (5.15) we have the sum variables $H(k)$ and H_{r+1} which can be obtained from $H(k;n)$ and $H_{r+1}(n)$ summing over all sites n. The H_r in (5.15) are the constants of the motion of the discrete KdV-hierarchy. In fact, from eq.(3.3) together with the identifications $f \equiv f(\vec{r})$, $\tilde{f} = f(\vec{r}+\vec{e}_1)$, $\hat{f} = f(\vec{r}+\vec{e}_1+\vec{e}_2)$, and eq. (5.1) for $\rho_k(n,m)$ on the two-dimensional lattice, we have

$$(q-k)\,\frac{u_k}{\rho_k}(n+1,m+1) - (p-k)\,\frac{u_k}{\rho_k}(n+1,m)$$
$$= \{q - p + u(n+1,m+1) - u(n+1,m)\}\,\frac{u_k}{\rho_k}(n,m) \quad . \tag{5.16}$$

Assuming that $u(n,m)$ tends to a value independent of m in the limit $n \to \infty$ we find from (5.16) and (5.12) that $a(k,m) = a(k,m+1)$. Expanding in powers pf p^2-q^2 with the use of the vertex operator (5.5) yields the relation $\partial_{\tau_s} \ln a(k) = 0$. Expanding this relation in powers of p^2-k^2, we find $\partial_{\tau_s} H_r = 0$, implying that the H_r are constants of the motion under all time evolutions τ_s .

We shall now show that the constants of the motion H_r are hamiltonians generating the various equations of the discrete KdV-hierarchy. Inserting (5.14) in (5.13) yields the continued fraction expansion

$$\exp H(k;n) = \cfrac{1}{v(n) - \cfrac{(p^2-k^2)}{v(n+1) - \cfrac{(p^2-k^2)}{v(n+2) - \dots}}} \quad . \tag{5.17}$$

Comparing (5.3) and (5.17) we immediately obtain the relation

$$q-p+(e^{\partial_m}-1)u(n) = -(p^2-q^2)\exp H(q;n) \quad , \tag{5.18}$$

in which we have inserted the vertex operator $\exp\partial_m$. Eq. (5.18) connects the solution of the lattice equation (5.4) with the generating function H(q;n) of hamiltonian densities (5.14) on the chain.

Eq. (5.18) is not so convenient to investigate higher-order discrete KdV equations, but one can derive a relation which is linear in ∂_m. In fact, differentiating (5.18) with respect to q^2 gives

$$(\partial_{q^2}\partial_m)u(n) = \partial_{q^2}(p-q) - \partial_{q^2}\{(p^2-q^2)\exp H(q;n)\}$$
$$- (p^2-q^2)(\partial_{q^2}\partial_m)\exp H(q;n) \tag{5.19}$$

with

$$(\partial_{q^2}\partial_m) = -\sum_{s=1}^{\infty}(p^2-q^2)^{s-1}\,\partial_{\tau_s} \quad . \tag{5.20}$$

The r.h.s. of (5.19) can be simplified to give

$$(\partial_{q^2}\partial_m)u(n) = -\frac{\partial H(q)}{\partial v(n)} + \partial_{q^2}(p-q) \tag{5.21}$$

with H(q) defined by (5.15). Eqs. (5.21) and (5.3) imply that

$$(\partial_{q^2}\partial_m)v(n) = -\{v(n),H(q)\} \quad , \tag{5.22}$$

in which the Poisson bracket for two arbitrary functions A and B of the v(n') is the discrete analogue[20)] of the Gardner bracket[21)], i.e.

$$\{A,B\} = \sum_n \frac{\partial A}{\partial v(n)}\left\{\frac{\partial B}{\partial v(n+1)} - \frac{\partial B}{\partial v(n-1)}\right\} \quad . \tag{5.23}$$

Eq. (5.21) can be proved by induction taking into account the identity

$$(p^2-q^2)\{\exp H(q;n),H(q)\}+\partial_{q^2}\{(p^2-q^2)\exp H(q;n)\} = \frac{\partial H(q)}{\partial v(n)} \quad . \tag{5.24}$$

Eq. (5.24) can be shown inserting the derivatives which are evaluated from the continued-fraction expression (5.17). We have

$$\partial_{q^2}\{(p^2-q^2)\exp H(q;n)\} = -\sum_{s=0}^{\infty}(p^2-q^2)^s\, e^{2H(n)+\ldots+2H(n+s-1)+H(n+s)},$$

$$\frac{\partial H(q;n)}{\partial v(n+s)} = \begin{cases} -(p^2-q^2)^s\, e^{H(n)+2H(n+1)+\ldots+2H(n+s)} & , (s > 0)\ , \\ 0 & , (s < 0)\ , \end{cases} \tag{5.25}$$

in which the q dependence of H(q;n') has been omitted. All derivatives can be regarded as strings of exponential factors of H(q;n') at neighbouring sites. The Poisson bracket contains products of two derivatives and therefore two strings. Nearly all double string terms cancel with respect to each other, and we are only left with single string terms for which it is easy to see that (5.24) is satisfied.

To prove (5.21), let us assume that it is valid apart from a correction term of order $f p^2-q^2)^N$. With the explicit expressions $H_1(n) = -\ln v(n)$, $H_2(n) = \{v(n)v(n+1)\}^{-1}$ which follow from (5.17), it is easy to see that this assumption is correct for N = 1,2. From the assumption that (5.21) is valid apart from a term of order $(p^2-q^2)^N$ we have

$$(\partial_{q^2}\ \partial_m)\exp H(q;n) = -\ \{\exp H(q;n), H(q)\} + O\{(p^2-q^2)^N\}\ . \tag{5.26}$$

Inserting (5.26) into (5.19) and using also (5.24) one finds that (5.21) is valid apart from a term of the order $(p^2-q^2)^{N+1}$. This means that (5.21) has been justified up to infinite order in p^2-q^2.

Finally, expanding both members of (5.21) in powers of p^2-q^2, using also (5.20) and (5.15), we obtain

$$\partial_{\tau_r} u(n) = \frac{\partial H_r}{\partial v(n)} + \gamma_r \quad , \tag{5.27}$$

in which γ_r is the coefficient of $(p^2-q^2)^{r-1}$ in the expansion of $\partial_{q^2}(p-q)$. For the fields v(n) we have

$$\partial_{\tau_r} v(n) = \{v(n), H_r\} \tag{5.28}$$

with (5.23) for the Poisson bracket. With (5.27) and (5.28) we have established the hamiltonian structure for the discrete KdV hierarchy. From $\partial_{\tau_r} H_s = 0$ and (5.28) we have the involution property $\{H_s, H_r\} = 0$.

6. THE KdV HIERARCHY

6.1 Continuum limit

In this section we derive the KdV hierarchy by a second continuum limit starting from the discrete KdV

$$\frac{1}{p}\{1 + \partial_\tau u(n)\} = \frac{2}{2p+u(n+1)-u(n-1)} \quad , \tag{6.1}$$

cf. (5.9) with $\tau_1 = -2p\tau$. The solutions of the discrete KdV can be found from the integral equation (1.1) with

$$\rho_k = \exp\left[n \ \ell n \frac{p-k}{p+k} + \frac{2k\tau}{p^2-k^2}\right] \quad , \tag{6.2}$$

cf. (5.6) with $C_1(k) = -k/\{(p^2-k^2)p\}$ and $\tau_1 = -2p\tau$.

To perform a continuum limit we introduce an infinite number of variables $t_1, t_3, \ldots$. For an arbitrary function $f(n;\tau)$ which can be ρ_k , u_k , or u , we have $f(n;\tau) = f(t_1, t_3, \ldots)$ and $f(n+1;\tau)$ is obtained via a vertex operator $\exp\partial_n$, i.e.

$$f(n+1;\tau) = e^{\partial_n} f(t_1, t_3, \ldots) \ , \ \partial_n = \sum_{\ell=0}^{\infty} \frac{-2}{(2\ell+1)p^{2\ell+1}} \partial_{t_{2\ell+1}} \tag{6.3}$$

in which we have identified p^{-1} as the lattice parameter and the factor $-2/(2\ell+1)$ is for convenience. We have not included even powers of p^{-1}, since the odd powers will be sufficient for our purpose.

Using the expansion

$$n \ell n \frac{p-k}{p+k} + \frac{2k\tau}{p^2-k^2} = \sum_{\ell=0}^{\infty} \left\{\frac{-2n}{(2\ell+1)p^{2\ell+1}} + \frac{2\tau}{p^{2\ell+2}}\right\} k^{2\ell+1} \quad , \tag{6.4}$$

and eq. (6.3) we find $\partial_{t_{2\ell+1}} \ell n \ \rho_k = k^{2\ell+1}$, so that we can take

$$\rho_k = \exp[\sum_{\ell=0}^{\infty} k^{2\ell+1} \, t_{2\ell+1}] \quad . \tag{6.5}$$

From (6.4) we also have

$$\partial_\tau = \sum_{\ell=0}^{\infty} \frac{2}{p^{2\ell+2}} \partial_{t_{2\ell+1}} = (\partial_p \; \partial_n) \quad . \tag{6.6}$$

To obtain the KdV hierarchy for fields $u(x,t)$ with $x \equiv t_1$, $t \equiv t_{2\ell+1}$, $\ell = 1,2,\ldots$, we expand both members of (6.1) in powers of p^{-1}, using also (6.3) and (6.6). Up to the order p^{-7} we have

$$\partial_{t_3} u = \partial_x^{\,3} u + 3(\partial_x u)^2 \quad , \tag{6.7}$$

$$\begin{aligned} \partial_{t_5} u &= \frac{1}{6}\,\partial_x^{\,5} u + \frac{5}{6}\,\partial_{t_3}\partial_x^{\,2} u + \frac{5}{3}(\partial_x^{\,3} u)\,\partial_x u + \frac{10}{3}(\partial_{t_3} u)\,\partial_x u \\ &= \partial_x^{\,5} u + 10(\partial_x u)^3 + 10(\partial_x u)\,\partial_x^{\,3} u + 5(\partial_x^{\,2} u)^2 \quad . \end{aligned} \tag{6.8}$$

Eq. (6.7) with $t_3 = t$ is eq. (1.3) and with $v = \partial_x u$ one obtains from (6.7) and (6.8) the KdV and the next higher-order KdV. This procedure can be continued up to arbitrary order in p^{-1}. In fact, the coefficient of $p^{-(2\ell+3)}$ contains $\partial_{t_{2\ell+1}} u$, as well as products of terms involving $\partial_{t_{2\ell'+1}} u$ with $\ell' < \ell$, but these terms can be expressed in terms of $\partial_x u$. The solutions of the PDE for $\partial_{t_{2\ell+1}} u$ can be obtained from (1.1) with $\rho_k = \exp(k^{2\ell+1} \, t_{2\ell+1})$. The KdV hierarchy obtained from (6.1) satisfies the usual recursion relation[1),24)].

6.2 Hamiltonian structure

To connect with the hamiltonian structure we start from the scattering problem in the continuum case

$$\partial_x^{\,2} u_k - k\partial_x u_k = -(\partial_x u) u_k \quad , \tag{6.9}$$

which can be either derived directly from (1.1) with $\rho_k = \exp[kx+\omega(k)t]$, or from (5.11) with the vertex operator $\exp\partial_n = 1 - 2p^{-1}\partial_x - 2p^{-2}\partial_x^{\,2} + O(p^{-3})$. We introduce the scattering variable $a(k)$ and the function $q_k(x)$

$$a(k) = \lim_{x\to\infty} \frac{u_k}{\rho_k}(x) \quad , \quad \frac{u_k}{\rho_k}(x) = \exp \int_{-\infty}^{x} dx' \ \sigma_k(x') \quad . \qquad (6.10)$$

From (6.9) we then have the Riccati equation

$$\partial_x \sigma_k + \sigma_k^{\,2} + k\sigma_k + \phi = 0 \quad , \quad \phi \equiv \partial_x u \quad . \qquad (6.11)$$

For $\sigma_k(x)$ we use the expansion

$$\sigma_k(x) = \sum_{r=1}^{\infty} \left(\frac{-1}{k}\right)^r H_r(x) \qquad (6.12)$$

leading to

$$\ell n \ a(k) = \sum_{r=1}^{\infty} \left(\frac{-1}{k}\right)^r H_r \quad , \quad H_r = \int_{-\infty}^{\infty} dx \ H_r(x) \quad . \qquad (6.13)$$

From section 5 we have $\partial_\tau \ \ell n \ a(k) = 0$ and expanding in powers of p^{-1} with the vertex operator $\partial_\tau = (\partial_p \partial_n)$ of (6.6) we find $\partial_{t_{2\ell+1}} \ell n a(k)=0$, implying that $\partial_{t_{2\ell+1}} H_r = 0$, so that the H_r are constants of the motion under all time evolutions $t_{2\ell+1}$.

We finally show that the H_r generate the equations of motion of the KdV hierarchy, see also e.g. ref. 22 for a different treatment relating the hamiltonian structure and the recursion operator. We introduce first the odd and even parts of σ_k, i.e. $\sigma_k^{(0)} = \frac{1}{2}(\sigma_k - \sigma_{-k})$, $\sigma_k^{(e)} = \frac{1}{2}(\sigma_k + \sigma_{-k})$. From the Riccati equation (6.9) we find that $\sigma_k^{(e)}$ can be expressed as a total derivative and for $\sigma_k^{(0)}$ we find

$$\phi + \sigma_k^{(0)^2} + k\sigma_k^{(0)} = \frac{\partial_x^{\,2} \sigma_k^{(0)}}{k+2\sigma_k^{(0)}} - 3 \ \frac{(\partial_x \sigma_k^{(0)})^2}{(k+2\sigma_k^{(0)})^2} \quad . \qquad (6.14)$$

Subtracting (6.14) for ϕ and $\sigma_k^{(0)}$ from the same equation in terms of the functions $\phi + \delta\phi$, $\sigma_k^{(0)} + \delta\sigma_k^{(0)}$ leads in the case of infinitesimal variations δ to the relation

$$\frac{1}{k+2\sigma_k^{(0)}} \ \delta\phi + \delta\sigma_k^{(0)} = \partial_x \left\{ \frac{\partial_x \delta\sigma_k^{(0)}}{(k+2\sigma_k^{(0)})^2} - \frac{2(\partial_x \sigma_k^{(0)})\delta\sigma_k^{(0)}}{(k+2\sigma_k^{(0)})^3} \right\} \quad . \qquad (6.15)$$

Integrating from $-\infty$ to ∞ , we see that the r.h.s. does not contribute, and from the l.h.s. we obtain

$$\delta \int \sigma_k^{(0)} dx = - \int dx \frac{1}{k+2\sigma_k^{(0)}} \delta\phi \quad , \tag{6.16}$$

which is equivalent to the relation

$$\frac{-1}{k+2\sigma_k^{(0)}} = \frac{\delta\sigma_k^{(0)}}{\delta\phi} = \frac{\delta\sigma_k}{\delta\phi} \quad , \tag{6.17}$$

where $\delta/\delta\phi$ denotes the functional derivative with respect to ϕ and where we have used that $\sigma_k^{(e)}$ is a total derivative.

We finally consider eq. (6.1). Taking into account the interpretation of the BT (2.6) in terms of the primitive translation over $\vec{e}_1$, we can identify $u(n+1) = \tilde{u}$, $u(n-1) = \underset{\sim}{u}$, in which the subscript wiggle denotes the transform of u under the inverse BT. Comparing (2.14) for $\tilde{u}$ - u with the Riccati equation (6.9) we immediately have the relation $\tilde{u} - u = 2\sigma_p$. From the inverse BT we have a second relation changing $p \to -p$ and interchanging wiggles and no wiggles, i.e. $u - \tilde{u} = 2\tilde{\sigma}_{-p} \to \underset{\sim}{u} - u = 2\sigma_{-p}$.

From (6.1) and (6.17) we obtain the relation

$$\frac{1}{p}\left[1 + \partial_\tau u\right] = \frac{1}{p+2\sigma_p^{(0)}} = - \frac{\delta\sigma_p}{\delta\phi} \quad . \tag{6.18}$$

Expanding both members of (6.18) with (6.6) and (6.12) yields

$$2\partial_{t_{2\ell-1}} u = \frac{\delta H_{2\ell+1}}{\delta\phi} \quad , \quad 2\partial_{t_{2\ell-1}} \phi = \{\phi, H_{2\ell+1}\} \quad , \tag{6.19}$$

in which the Poisson bracket, for arbitrary functions A and B depending on ϕ , is the Gardner bracket[21)]

$$\{A,B\} = \int_{-\infty}^{\infty} dx \frac{\delta A}{\delta\phi} \partial_x \frac{\delta B}{\delta\phi} \quad . \tag{6.20}$$

Eqs.(6.19) and (6.20) give the well-known hamiltonian structure[22),23)] of the KdV hierarchy.

REFERENCES

1) Ablowitz, M.J. and Segur, H., "Solitons and the Inverse Scattering Transform", SIAM, Philadelphia (1981).
2) Novikov, S., Manakov, S.V.,Pitaevskii, L.P. and Zakharov, V.E., "Theory of Solitons, The Inverse Scattering Method", Plenum, New York (1984).
3) Calogero, F. and Degasperis, A., "Spectral Transform and Solitons", Vol. 1, North-Holland, Amsterdam (1982).
4) Fokas, A.S. and Ablowitz, M.J., Phys.Rev.Lett. 47,1096(1981).
5) Nijhoff, F.W., Quispel, G.R.W., Capel, H.W. and van der Linden, J. Physica 119A,101(1983).
6) van der Linden, J., Nijhoff, F.W., Capel, H.W. and Quispel, G.R.W. Physica 137A,44(1986).
7) Ablowitz, M.J., Kaup, D.J., Newell, A.C. and Segur, H., Stud.Appl. Math., 53,249(1974).
8) Capel, H.W., Wiersma, G.L. and Nijhoff, F.W., to be published in Physica A.
9) Wadati, M., in: "Solitons", edited by R.K. Bullough and P.J. Caudrey, Topics in Current Physics 17,287(1980), Springer (Berlin, Heidelberg, New York, Tokyo).
10) Fokas, A.S. and Ablowitz, M.J., in: "Nonlinear phenomena", edited by K.B. Wolf, Lecture Notes in Physica 189,167(1983), Springer (Berlin, Heidelberg, Tokyo).
11) Nijhoff, F.W., Capel, H.W. and Wiersma, G.L., in: "Geometric aspects of the Einstein equations and integrable systems", edited by R. Martini, Lecture Notes in Physics 239,263(1985), Springer (Berlin, Heidelberg, New York, Tokyo).
12) Santini, P.M., Fokas, A.S. and Ablowitz, M.J., J.Math.Phys. 25,261(1984).
13) Nijhoff, F.W., Capel, H.W., Wiersma, G.L. and Quispel, G.R.W., Phys.Lett. 103A,293(1984).
14) Nijhoff, F.W. and Capel, H.W., Phys.Lett. 91A,431(1982).
15) Quispel, G.R.W., Nijhoff, F.W., Capel, H.W. and van der Linden, J. Physica 123A,319(1984).

16) Nijhoff, F.W., Quispel, G.R.W., Capel, H.W. and van der Linden, J. Physica 125A,344(1984).

17) Lepowski, J. and Wilson, R.H., Commun.Math.Phys. 62,43(1978).

18) Date, E., Kashiwara, M., Jimbo, M. and Miwa, T., in: Proc. RIMS Symp. Kyoto (1981), M. Jimbo and T. Miwa, eds., World Scientific Publ. Co. (Singapore, 1983).

19) Jimbo, M. and Miwa, T., Publ. RIMS, Kyoto Univ. 19,943(1983).

20) Kako, F. and Mugibayashi, N., Progr.Theor.Phys. 60,975(1978).

21) Gardner, C.S., J.Math.Phys. 12,1548(1971).

22) Flaschka, H. and Newell, A.N., in: "Dynamical Systems. Theory and Applications", J. Moser ed., Lecture Notes in Physics 38, Springer (Berlin, 1975).

23) Kupershmidt, B.A. and Wilson, G., Invent.Math. 62,403(1981).

24) Wiersma, G.L. and Capel, H.W., to be published in Physica A.

25) Fokas, A.S., this volume.

26) Santini, P.M., this volume.

27) Ablowitz, M.J. and Ladik, J.F., J.Math.Phys. 17,1011(1976).

INTEGRABLE HIERARCHIES, LAGRANGIAN STRUCTURES AND NON-COMMUTING FLOWS¶

Frank W. Nijhoff†

Department of Physics, University of Lecce,

73100 Lecce, Italy

ABSTRACT

The direct linearization method is taken as a starting point to investigate hierarchies of partial differential equations (PDE's) in 1+1 and 2+1 dimensions. All relevant equations are deduced from an infinite-dimensional matrix system, in which the flows build a representation of a loop algebra. In general these flows correspond to non-commuting differentiations. The associated PDE's are derived, and for the the commuting case they are shown to posess a Lagrangian structure related to the Wess-Zumino model. Generalizations of the hierarchies to 2+1 dimensions are discussed.

¶) Work supported in part by MPI-Italy.

†) Present Address: Laboratoire de Physique Théorique et Hautes Energies, Université Pierre et Marie Curie, 4 Place Jussieu, F-75230 Paris Cedex 05.

1. INTRODUCTION

There exist nowadays a variety of methods to construct exactly integrable partial differential equations (PDE's), most of which can be brought together under the following headings:

- -Inverse Problems[1-4] (Inverse Scattering Transform, Riemann-Hilbert Problem, $\bar{\partial}$-problem)
- -Algebraic-geometric methods[5,6]
- -τ-function approach[7-10] (Grassmannian Manifolds)
- -Hamiltonian methods[11-14].

In most cases these approaches have led to an algebraïzation of the theory of integrable PDE's, and there is a clear message that emerges from this, namely that *there exists an intimate connection between infinite-dimensional nonlinear integrable systems on the one hand and the representation theory of infinite-dimensional Lie algebras (and groups!) on the other hand.* The infinite-dimensional Lie algebras which are meant here are the affine Kac-Moody Lie algebras[15], which over the last few years have gained much interest in many areas of mathematical physics. As for the application to the construction of integrable systems, there are some in principle straightforward procedures to build a hierarchy of nonlinear evolution equations starting from the representation theory of the algebra, cf. refs. 8,9,13,14,16-18. The other way around, i.e. the problem of recovering the algebra from a (large enough) system of nonlinear PDE's and their linearizations, is much less straightforward. Nevertheless, this is more in line with the physicist's way of thinking: given a principle to build a theory, find the most natural object (group, algebra) on which to apply it in order to get 'sensible' results. Adopting the ideas of earlier work by Drinfel'd and Sokolov[19], Reiman and Semenov-Tian-Shanskii[20] and Adler and van Moerbeke[21], Flaschka, Newell and Ratiu have shown in ref. 22, cf. also ref. 23, that sensible hierarchies of PDE's arise when one applies the Kostant-Adler-Symes construction to the loop algebras (i.e. affine Kac-Moody algebras without a central extension). It is not clear from this work, however, why it is necessary to consider the central extension to the loop algebra for a comprehensive description of the properties of the nonlinear evolution equations. Of course, a central term can be added by hand, and some authors[24,25] have indicated what modifications this gives rise to, e.g. in the Hamiltonian description of the flows, cf. refs. 13,14. Still, this seems rather ad hoc, and a satisfactory exlanation for the necessity of having a central term in the algebra has in my view yet to be given.

Another unresolved problem in the theory of integrable nonlinear evolution equations is the question how to construct *genuine* multi-dimensional generalizations. For 2+1 dimensions one has nowadays a good deal of insight, cf. e.g. refs. 26-32, and there are many new results on inverse scattering in multi-dimensional space (cf. the articles by M.J. Ablowitz, A.S. Fokas and A.I. Nachmann in this volume). Nevertheless, despite some few examples (among which the self-dual Yang-Mills equations, and the generalized sine-Gordon equation[33]), a systematic procedure for the construction of integrable PDE's in more than two space dimensions is still non-existent. It is my belief that a firm insight in the algebraic structure of the low-dimensional cases may eventually lead to the invention of some new structure to add to the theory, which is still lacking at present.

In this article I hope to give some novel insights in the algebraic structure of hierarchies of nonlinear PDE's in 1+1 and 2+1 dimensions. A convenient starting point for building these hierarchies is the direct linearizing transform[34-37] (DLT). Using the DLT it is shown that the hierarchies can be defined in a very concise form using infinitely sized (i.e. $\mathbb{Z}\times\mathbb{Z}$-dimensional) matrices, in which the algebraic structure is clearly recognized. In fact, in this picture the flows build directly (i.e. without having to take refuge to the use of vertex-operators[38-40] and τ-functions) a representation of a loop algebra, if one allows for the flows to be non-commuting corresponding to non-commutative differentiations. Nonetheless, hierarchies of nonlinear PDE's can be derived, which in the special case of commuting flows reduce to some well-known equations for integrable hierarchies of $N\times N$ matrix systems in 1+1 dimensions. The connection with a Kac-Moody algebra is obtained by looking at Lagrangian structures rather than Hamiltonian ones. The hierarchical structure in its primitive form (i.e. without explicitely eliminating auxiliary fields) lends itself very well to be cast into a Lagrangian form. Furthermore, by formally summing the whole hierarchy together one obtains a direct connection with a class of nonlinear σmodels for which an action principle can be written down immediately. These socalled Wess-Zumino-Witten chiral field models[41,42] belong to the class of models with multi-valued action[43,44] and their Poisson algebra of conserved currents is of Kac-Moody type.

Finally, I shall present the DLT for 2+1-dimensional matrix systems and show how the basic equations generalize to this case. In a formal way one can define hierarchies, which now form a lattice of equations rather than a linear sequence. The explicit form, however, of these equations is not easily determined, and I restrict myself here to a few examples.

2. DIRECT LINEARIZING TRANSFORMATION

As a starting point for building a hierarchy of integrable flows I will take the DLT, cf. refs. 34-37. This section outlines some of the results of a more detailed account[4,5] of the study of integrable hierarchies and their Hamiltonian structure on the basis of the DLT. The main idea is to start from a linear integral equation for the solution of a general type of linear spectral problems, viz. Riemann-Hilbert problems[3,4], without going into the details concerning specific choices of boundary conditions. The integral equation is chosen so general, containing integrations over arbitrary measures and contours in the complex plane of the spectral parameter k, as to cover a great many possibly interesting solutions. Some assumptions on these measures and contours have to be made in order that the integral equation is well-defined, and in order to be able to perform some manipulations on its solutions. Furthermore, we have to assume that for given measure and contour the solution of the integral equation is unique. The advantage of having a formulation independent of boundary conditions is that it enables one to investigate what freedom one has to impose additional dependences of the solution of the integral equation on certain auxiliary parameters. Another way of putting this is to say that we search for equations with as guiding principle the invariance under integral transformations of a certain type. For the sake of definiteness let us assume that we work in $\mathrm{Mat}_N(\mathbb{C})$, i.e. the space of N×N matrices with $\mathbb{C}$-valued coefficients. The linear spectral problem from which one may start is of the following type

$$\frac{\partial}{\partial x}\Phi_k(x) = (kJ + Q(x))\,\Phi_k(x) \tag{1}$$

where J, Q and Φ_k live in $\mathrm{Mat}_N(\mathbb{C})$, Q and Φ_k depending on x and Φ_k also depending on the spectral parameter k ($k\in\mathbb{C}$), the dependence on k being denoted by a lower-case index. The constant matrix J may for the time being be thought of as a diagonal matrix with distinct entries, $J=\mathrm{diag}(J_1,\cdots,J_N)$, $J_i\neq J_j$ if $i\neq j$, and the 'potential matrix' Q(x) being an off-diagonal matrix. The inverse scattering problem for eq. (1) has been studied by Beals and Coifman[4,6]. The DLT for eq. (1) can be formulated as follows. Suppose we have a known reference solution Φ_k^0 of (1) associated with a potential matrix Q^0, then we can find a new solution Φ_k of (1) associated with a new potential matrix Q by solving the following *linear* integral equation for Φ_k

$$\Phi_k(x) = \Phi_k^0(x) + \int_C \Phi_\ell(x)\, d\lambda(\ell)\, \frac{\Phi_\ell^0(x)^{-1}\Phi_k^0(x)}{k-\ell} \,. \tag{2}$$

In eq. (2) the integration is performed over a measure $d\lambda(\ell)$ and a contour C taking values in $Mat_N(\mathbb{C})$. This means that with every component $d\lambda_{ij}(\ell)$ there is associated a contour C_{ij} in the complex ℓ-plane. We assume furthermore that the above mentioned assumptions are valid. Integral equations like eq. (2) have been shown to arise as solutions of RH-problems and $\bar{\partial}$-problems associated with nonlinear evolution equations, cf. refs. 2-4,46. The solution Φ_k of (2) is again a solution of the spectral problem (1) provided that the new potential matrix Q is related to the reference-potential Q^0 by the formula

$$Q(x) = Q^0(x) + \left[\int_C \Phi_\ell \, d\lambda(\ell) \, {\Phi^0_\ell}^{-1} , J \right] . \qquad (3)$$

It is natural from eq. (3) to introduce a matrix H which is defined by its transformation rule under the DLT

$$H = H^0 + \int_C \Phi_\ell \, d\lambda(\ell) \, {\Phi^0_\ell}^{-1} \qquad (4)$$

in tems of which we may write $Q=[H,J]+$ inv. (inv. means an arbitrary matrix invariant under the DLT, which for simplicety we take to be zero). It is this matrix H which will play an important role in the construction of integrable hierarchies.

To proceed, we note that as Φ_k solves (2) with Φ^0_k as inhomogeneous term, then $(k-H)\Phi_k$ also solves (2) with the same kernel but with $(k-H^0)\Phi^0_k$ as inhomogeneous term. Secondly, there is a matrix H' say, such that $(k^2-kH-H')\Phi_k$ solves (2) but now with $(k^2-kH^0-H'^0)\Phi_k^0$ as inhomogeneous term. This procedure can be iterated, and it leads in a natural way to the introduction of quantities $\Phi_k^{(0)}\equiv\Phi_k$, $\Phi_k^{(1)}\equiv(k-H)\Phi_k$, $\Phi_k^{(2)}\equiv(k^2-kH-H')\Phi_k$,... which have a natural transformation property under the DLT. This can be extended also to the introduction of $\Phi_k^{(-1)}$, $\Phi_k^{(-2)}$, etc. corresponding to negative powers of k. Furthermore, one must take care to define properly adjoint solutions of the integral equation, which are of the form ${}^t\Phi_k^{(0)}\equiv\Phi_k^{-1}$, ${}^t\Phi_k^{(1)}\equiv\Phi_k^{-1}(k+H)$, etc. Without giving a detailed account of this construction, cf. ref. 45, I shall now introduce some notations in order to be able to present the whole structure in one stroke. Let us build from the objects $\Phi_k^{(i)}$, ${}^t\Phi_k^{(i)}$ infinite-component vectors $\underline{\Phi}_k=(\Phi_k^{(i)})_{i\in\mathbb{Z}}$ and ${}^t\underline{\Phi}_k=({}^t\Phi_k^{(i)})_{i\in\mathbb{Z}}$, i.e. a column- and a row-vector respectively. Let $\underline{\underline{\Lambda}}$ and ${}^t\underline{\underline{\Lambda}}$ denote index-raising operators acting from the left, from the right respectively, i.e. $(\underline{\underline{\Lambda}}\cdot\underline{\Phi}_k)^{(i)}=\Phi_k^{(i+1)}$, $({}^t\underline{\Phi}_k\cdot{}^t\underline{\underline{\Lambda}})^{(i)}={}^t\Phi_k^{(i+1)}$, (${}^t\underline{\underline{\Lambda}}$ being the matrix transposed of $\underline{\underline{\Lambda}}$), and let $\underline{\underline{O}}$ denote the projection matrix picking out the 0-elements, i.e. $(\underline{\underline{O}}\cdot\underline{\Phi}_k)^{(i)}=\delta_{i,0}\Phi_k^{(0)}$, $({}^t\underline{\Phi}_k\cdot\underline{\underline{O}})^{(i)}=\delta_{i,0}\,{}^t\underline{\Phi}_k^{(0)}$. Furthermore, I introduce a $\mathbb{Z}\times\mathbb{Z}$ matrix $\underline{\underline{H}}$, with components $H_{i,j}$, $\underline{\underline{\Lambda}}$ and ${}^t\underline{\underline{\Lambda}}$ acting on $\underline{\underline{H}}$ as follows: $(\underline{\underline{\Lambda}}^p\cdot\underline{\underline{H}}\cdot{}^t\underline{\underline{\Lambda}}^q)_{i,j}=H_{i+p,j+q}$. The integral transformation (2) can now be extended to a transformation for the vectors $\underline{\Phi}_k$ and ${}^t\underline{\Phi}_k$, namely

$$\underline{\Phi}_k = \underline{\Phi}_k^0 + \int_C \underline{\Phi}_\ell \, d\lambda(\ell) \, \frac{\Phi_\ell^{0\,-1}\Phi_k^0}{k-\ell} \tag{5a}$$

$${}^t\underline{\Phi}_k = {}^t\Phi_k^0 + \int_C \frac{\Phi_k^{-1}\,\Phi_\ell}{k-\ell}\, d\lambda(\ell)\, {}^t\underline{\Phi}_\ell^0 \quad . \tag{5b}$$

The $\mathbb{Z}\times\mathbb{Z}$ matrix $\underline{\underline{H}}$ transforms according to

$$\underline{\underline{H}} = \underline{\underline{H}}^0 + \int_C \underline{\Phi}_\ell \, d\lambda(\ell) \; {}^t\underline{\Phi}^0_\ell \quad , \tag{6}$$

the expression in the integral being a $\mathbb{Z}\times\mathbb{Z}$ dyadic matrix. The vectors $\underline{\Phi}_k$ and ${}^t\underline{\Phi}_k$ are defined by the following relations

$$k^p \underline{\Phi}_k = \underline{\underline{\Lambda}}^p \cdot \underline{\Phi}_k + \underline{\underline{H}} \cdot \underline{\underline{O}}_p \cdot \underline{\Phi}_k \tag{7a}$$

$$k^p \; {}^t\underline{\Phi}_k = {}^t\underline{\Phi}_k \cdot {}^t\underline{\underline{\Lambda}}^p - {}^t\underline{\Phi}_k \cdot \underline{\underline{O}}_p \cdot \underline{\underline{H}} \quad , (\forall p \in \mathbb{Z}), \tag{7b}$$

where the matrices $\underline{\underline{O}}_p$, $p\in\mathbb{Z}$, are given by

$$\underline{\underline{O}}_p = \begin{cases} \sum\limits_{j=0}^{p-1} {}^t\underline{\underline{\Lambda}}^j \cdot \underline{\underline{O}} \cdot \underline{\underline{\Lambda}}^{p-1-j} \quad , & p \geq 0 \quad , \\ -\sum\limits_{j=0}^{-p-1} {}^t\underline{\underline{\Lambda}}^{j+p} \cdot \underline{\underline{O}} \cdot \underline{\underline{\Lambda}}^{-1-j} \quad , & p < 0 \quad , \end{cases} \tag{8}$$

($\underline{\underline{\Lambda}}^{-1}$, ${}^t\underline{\underline{\Lambda}}^{-1}$ are index-lowering instead of -raising operators). Having now presented the material for defining a hierarchy of differential relations for the vectors $\underline{\Phi}_k$ and ${}^t\underline{\Phi}_k$, we introduce an infinite sequence of variables x_p, $p\in\mathbb{Z}$, in terms of which we can write down the following differential system

$$\frac{\partial}{\partial x_p} \underline{\Phi}_k = (\underline{\underline{\Lambda}}^p + \underline{\underline{H}} \cdot \underline{\underline{O}}_p) \cdot J \; \underline{\Phi}_k \tag{9a}$$

$$\frac{\partial}{\partial x_p} {}^t\underline{\Phi}_k = - {}^t\underline{\Phi}_k \; J \cdot ({}^t\underline{\underline{\Lambda}}^p - \underline{\underline{O}}_p \cdot \underline{\underline{H}}) \quad , (p\in\mathbb{Z}). \tag{9b}$$

The statement is now that eqs. (9) can be imposed *simultaneously* on $\underline{\Phi}_k$ and ${}^t\underline{\Phi}_k$, for all $p\in\mathbb{Z}$, and that they are preserved under the integral transformation (5). The matrix J in (9) is invariant under the DLT, $J=J^0$. For the proof of this statement we refer again to ref. 45. For the matrix H there are also differential relations, namely

$$\frac{\partial}{\partial x_p} \underline{\underline{H}} = \underline{\underline{\Lambda}}^p \cdot J \, \underline{\underline{H}} - \underline{\underline{H}} \, J \cdot {}^t\underline{\underline{\Lambda}}^p + \underline{\underline{H}} \cdot \underline{\underline{O}}_p \cdot J \, \underline{\underline{H}} \, , \tag{10}$$

which through (9) is compatible with the DLT, eq. (6), for $\underline{\underline{H}}$. The components of the $\mathbb{Z}\,\mathbb{Z}$ matrix H are not all independent, but they are related through a system of differential relations, namely

$$\underline{\underline{\Lambda}}^p \cdot \underline{\underline{H}} - \underline{\underline{H}} \cdot {}^t\underline{\underline{\Lambda}}^p + \underline{\underline{H}} \cdot \underline{\underline{O}}_p \cdot \underline{\underline{H}} = 0 \quad , \; \forall p \in \mathbb{Z} \quad , \tag{11}$$

which, in fact constitutes an infinite system of recursion relations. Eqs. (7)-(11) is an infinite coupled differential system for the matrix $\underline{\underline{H}}$ and the vectors $\underline{\Phi}_k, {}^t\underline{\Phi}_k$. In order to make some sense out of these equations one should like to reduce this infinite system to a finite one in terms of a finite number of quantities. This can indeed be achieved as I shall show in the next section. Furthermore, one can reduce the system by imposing additional symmetries on $\underline{\underline{H}}$, e.g. by the action of some automorphism on the matrix algebra in which we work. Such a reduction should be compatible with the DLT, and this leads to constraints on the measure in (5). I shall not go into such questions here, cf. ref. 45. Finally, it should bo noted that the use of index-raising operators to construct hierarchies of PDE's is also present in th work of the Kyoto school[17,47] .

3. INTEGRABLE HIERARCHIES AND LAGRANGIANS

In order to reduce the infinite system (7)-(11) we have to eliminate the matrices $\underline{\underline{\Lambda}}$ and ${}^t\underline{\underline{\Lambda}}$ from the equations, as they are responsible for the couplings between the different components. This can be done by combining differentiations of different order. In fact, from (9a) for p+q we have

$$\begin{aligned}\frac{\partial}{\partial x_{p+q}} \underline{\Phi}_k &= (\underline{\underline{\Lambda}}^{p+q} + \underline{\underline{H}}\cdot\underline{\underline{O}}_{p+q})\cdot J\ \underline{\Phi}_k \\ &= (\underline{\underline{\Lambda}}^{p} + \underline{\underline{H}}\cdot\underline{\underline{O}}_{p})\cdot(\underline{\underline{\Lambda}}^{q} + \underline{\underline{H}}\cdot\underline{\underline{O}}_{q})\cdot J\ \underline{\Phi}_k \\ &\quad - (\underline{\underline{\Lambda}}^{p}\cdot\underline{\underline{H}} - \underline{\underline{H}}\cdot{}^t\underline{\underline{\Lambda}}^{p} + \underline{\underline{H}}\cdot\underline{\underline{O}}_{p}\cdot\underline{\underline{H}})\cdot\underline{\underline{O}}_{q}\cdot J\ \underline{\Phi}_k \\ &= (\underline{\underline{\Lambda}}^{p} + \underline{\underline{H}}\cdot\underline{\underline{O}}_{p})\cdot\frac{\partial}{\partial x_q}\underline{\Phi}_k\ ,\end{aligned} \tag{12}$$

where in the second step use has been made of a relation for the matrices $\underline{\underline{O}}_p$, namely

$$\underline{\underline{O}}_{p+q} = \underline{\underline{O}}_p\cdot\underline{\underline{\Lambda}}^q + {}^t\underline{\underline{\Lambda}}^p\cdot\underline{\underline{O}}_q \quad , \tag{13}$$

which follows from the definition (8), and in the third step I have used (9a) and (11). From eq. (12), using (7a), we find

$$\frac{\partial}{\partial x_{p+q}}\underline{\Phi}_k = k^p\frac{\partial}{\partial x_q}\underline{\Phi}_k - \left(\frac{\partial}{\partial x_q}\underline{\underline{H}}\right)\cdot\ \underline{\underline{O}}_p\cdot\underline{\Phi}_k\ , \tag{14}$$

which for p=1 gives a one-step recursion relation. Furthermore, in that case we have in the second term on the right-hand side a matrix $\underline{\underline{O}}$, which enables us to project on the 0-component $\Phi_k^{(0)}=\Phi_k$. The equation only involves a matrix $H_{0,0}=H$, and upon iteration we obtain the following linear system

$$\frac{\partial \Phi_k}{\partial x_p} = Q_k^{(p)} \Phi_k \ , \ Q_k^{(p)} = \begin{cases} k^p J + \sum_{j=1}^{p} k^{p-j} Q_j & ,(p \geq 0), \\ k^p J - \sum_{j=0}^{-p-1} k^{p+j} Q_{-j} & ,(p<0), \end{cases} \tag{15}$$

in which the coefficients Q_j are given by $Q_j = -\partial H/\partial x_{j-1}$.

By a similar calculation as performed in (12) one can obtain recursion-like relations for $\underline{\underline{H}}$, starting from (10), and we thus obtain

$$\frac{\partial \underline{\underline{H}}}{\partial x_{p+q}} = (\underline{\underline{\Lambda}}^p + \underline{\underline{H}} \cdot \underline{\underline{O}}_p) \cdot \frac{\partial \underline{\underline{H}}}{\partial x_q} \quad . \tag{16}$$

To eliminate the $\underline{\underline{\Lambda}}$-term we differentiate again with respect to another x_r, leading to

$$\begin{aligned} \frac{\partial}{\partial x_r}\left(\frac{\partial \underline{\underline{H}}}{\partial x_{p+q}} \right) &= (\underline{\underline{\Lambda}}^p + \underline{\underline{H}} \cdot \underline{\underline{O}}_p) \cdot \frac{\partial^2 \underline{\underline{H}}}{\partial x_r \partial x_q} + \frac{\partial \underline{\underline{H}}}{\partial x_r} \cdot \underline{\underline{O}}_p \cdot \frac{\partial \underline{\underline{H}}}{\partial x_q} \\ &= \frac{\partial}{\partial x_q}\left((\underline{\underline{\Lambda}}^p + \underline{\underline{H}} \cdot \underline{\underline{O}}_p) \cdot \frac{\partial \underline{\underline{H}}}{\partial x_r} \right) \\ &\quad + \frac{\partial \underline{\underline{H}}}{\partial x_r} \cdot \underline{\underline{O}}_p \cdot \frac{\partial \underline{\underline{H}}}{\partial x_q} - \frac{\partial \underline{\underline{H}}}{\partial x_q} \cdot \underline{\underline{O}}_p \cdot \frac{\partial \underline{\underline{H}}}{\partial x_r} \\ &= \frac{\partial}{\partial x_q}\left(\frac{\partial \underline{\underline{H}}}{\partial x_{p+r}} \right) + \frac{\partial \underline{\underline{H}}}{\partial x_r} \cdot \underline{\underline{O}}_p \cdot \frac{\partial \underline{\underline{H}}}{\partial x_q} - \frac{\partial \underline{\underline{H}}}{\partial x_q} \cdot \underline{\underline{O}}_p \cdot \frac{\partial \underline{\underline{H}}}{\partial x_r} \, . \end{aligned} \tag{17}$$

Eq. (17) is a general recursion relation from which various types of nonlinear PDE's can be derived[4,5]. Specializing to the case p=1, in which case the matrix $\underline{\underline{O}}_p$ reduces to the projection matrix $\underline{\underline{O}}$, we find the following equation for the matrix $H=H_{0,0}$

$$\frac{\partial^2 H}{\partial x_r \partial x_{q+1}} - \frac{\partial^2 H}{\partial x_q \partial x_{r+1}} = \left[\frac{\partial H}{\partial x_r} , \frac{\partial H}{\partial x_q}\right] \quad . \tag{18}$$

Eq. (18) gives the basic equations for explicitely building a hierarchy of N×N matrix PDE's. For r=0, $\partial H/\partial x_0 = [J,H]$ this has been considered e.g. in refs. 18,22,23,46 and 48, and the case q,r≠0 has been also considered in ref. 22. So, I will not bother to give a complete discussion of all PDE's that can be obtained from (18), but only mention that a great many of the well-known soliton equations falls into this class. The equations arise also in another way, namely as the compatibility conditions of the overdetermined system (15), i.e. from the Zakharov-Shabat equations[22-24]

$$\frac{\partial Q_k^{(p)}}{\partial x_q} - \frac{\partial Q_k^{(q)}}{\partial x_p} + \left[Q_k^{(p)} , Q_k^{(q)}\right] = 0 \quad . \tag{19}$$

Of course eq. (18) is still in an unreduced form: it contains many auxiliary, nondynamical, fields and many coupled differentiations, whereas the 'physical' dimensionality is 1+1 (one may pick out any two different coordinates x_p and x_q and reduce the system to a set of nonlinear PDE's in terms of only a resticted set of fields and their derivatives with respect to these two coordinates). There is, however,some virtue in keeping eq. (18) as it is, namely that in this form it is very easy to write down a system of Lagrange densities from which eqs. (18) are obtained as the Euler-Lagrange equations. The Lagrangians are given by

$$L_{r,q} = \frac{1}{2}\mathrm{tr}\left(\frac{\partial H}{\partial x_q}\frac{\partial H}{\partial x_{r+1}} - \frac{\partial H}{\partial x_r}\frac{\partial H}{\partial x_{q+1}} \right) + \frac{1}{3}\mathrm{tr}\left(H \left[\frac{\partial H}{\partial x_q},\frac{\partial H}{\partial x_r}\right]\right), \tag{20}$$

which in principle should be integrated to an action over all the coordinates. This might give some problems, but I shall show that these can be circumvented in section 5. Anyway, we see here the advantage of having a Lagrange formulation rather than a Hamiltonian one. In the latter we have to get rid of the auxiliary fields beforehand, whereas in the former one we may keep them and have a much simpler formulation. The bridge between the two is provided by the Dirac's constraint formalism, which enables us in principle to recover the Poisson structure from the Lagrangians.

Remark

Up to now we have been mainly concerned with the element $H_{0,0}$ of the $\mathbb{Z}\times\mathbb{Z}$ matrix $\underline{\underline{H}}$. There are, however also some other elements of $\underline{\underline{H}}$ which give rise to interesting PDE's. The different elements of $\underline{\underline{H}}$ are connected via Miura transformations, whereas the different components of the vector $\underline{\Phi}_k$ are related via gauge transformations[4 5]. As an example consider eq. (11) with p=1 for $H_{-1,-1}$. This gives

$$(\mathbb{1} + H_{-1,0})(\mathbb{1} + H_{0,-1}) = \mathbb{1} \quad ,$$

which in combination with (16) with p=1 for $H_{-1,0}$ yields

$$Q_{q+1} = -\frac{\partial H}{\partial x_q} = -(\frac{\partial g}{\partial x_{q+1}})g^{-1} \quad , \; g \equiv \mathbf{1} - H_{0,-1} \; . \qquad (21)$$

Furthermore, from (7a) taking p=1 we find for $\Phi_k^{(-1)}$

$$\Phi_k = \Phi_k^{(0)} = k\, g\, \Phi_k^{(-1)} \quad , \qquad (22)$$

which in combination with (14) for p=1 yields

$$\frac{\partial \Phi_k^{(-1)}}{\partial x_{q+1}} = k\left\{\frac{\partial \Phi_k^{(-1)}}{\partial x_q} + S_q \, \Phi_k^{(-1)}\right\} \quad , \qquad (23)$$

in which $S_q \equiv g^{-1}\partial g/\partial x_q = -\partial H_{-1,-1}/\partial x_{q+1}$, cf. eq. (16) with p=1 for $H_{-1,-1}$. Eq. (23) leads to a hierarchy in terms of the 'spin matrix' $S=S_0=g^{-1}Jg$, related to a generalized Takhtadzhyan problem[49,50].

4. NON-COMMUTING FLOWS

In the previous sections I have argued that with every matrix J there corresponds a hierarchy of PDE's, which is given implicitely in eqs. (7)-(11). Here J was taken to be fixed, but we need not restrict ourselves to PDE's arising from hierarchies with a single matrix J. It is a legitimate question to ask whether it is possible to combine hierarchies associated with different matrices J, J and $\mathfrak{J}$ say. Here the following statement is essential[45].

Theorem: Two hierarchies of differential relations such as given in (7)-(11) above, one associated with a matrix J and another associated with a matrix $\tilde{J}$, can be superimposed on one and the same $\underline{\Phi}_k$ and $\underline{\underline{H}}$, such that these equations are preserved under the DLT, provided that $[J,\tilde{J}]=0$.

This theorem on 'commuting hierarchies' is implicit in many considerations in the literature, and is essential to combine hierarchies associated with different matrices J. Usually one restricts oneself to diagonal matrices, where the condition is trivially satisfied, and the most obvious example of a PDE arising from the combination of two different J is the N-wave interaction system[3]

$$\left[\frac{\partial H}{\partial x_1}, \tilde{J}\right] - \left[\frac{\partial H}{\partial \tilde{x}_1}, J\right] + \left[[H,\tilde{J}], [H,J]\right] = 0 \quad , \qquad (24)$$

where $\tilde{x}_p$ ($\forall p \in \mathbb{Z}$) are the coordinates corresponding to differential relations (9),(10) with the matrix $\tilde{J}$ instead of J. So for J and $\tilde{J}$ commuting we can simply take the compatibility relations $\left[\frac{\partial}{\partial x_p}, \frac{\partial}{\partial \tilde{x}_q}\right]\Phi_k=0$ for obtaining PDE's. This is no longer true if the matrices do not commute. Nevertheless, in the derivation of (7)-(10) no use has been made of a requirement that J should live in an Abelian subalgebra of $Mat_N(\mathbb{C})$. Hence, it is natural to expect that there should be a way to combine hierarchies even if $[J,\tilde{J}]\neq 0$. In that case one has to adapt the compatibility condition for derivatives with respect to coordinates associated with different matrices J and $\tilde{J}$ in a way that I am now about to derive. In fact, let us re-do the calculation in (12):

$$\frac{\partial}{\partial \tilde{x}_q}\frac{\partial}{\partial x_p}\underline{\Phi}_k = \frac{\partial}{\partial \tilde{x}_q}\left(\underline{\underline{\Lambda}}^p + \underline{\underline{H}}\cdot\underline{\underline{O}}_p \right)\cdot J\ \underline{\Phi}_k$$

$$= (\underline{\underline{\Lambda}}^p + \underline{\underline{H}}\cdot\underline{\underline{O}}_p) J \cdot (\underline{\underline{\Lambda}}^q + \underline{\underline{H}}\cdot\underline{\underline{O}}_q) \tilde{J} \cdot \underline{\Phi}_k + \frac{\partial \underline{\underline{H}}}{\partial \tilde{x}_q} \cdot \underline{\underline{O}}_p \cdot J \underline{\Phi}_k$$

$$= (\underline{\underline{\Lambda}}^{p+q} + \underline{\underline{H}}\cdot\underline{\underline{O}}_{p+q}) \cdot J\tilde{J}\underline{\Phi}_k + \frac{\partial \underline{\underline{H}}}{\partial \tilde{x}_q} \cdot \underline{\underline{O}}_p \cdot J\underline{\Phi}_k$$

$$+ (\underline{\underline{\Lambda}}^p \cdot J \, \underline{\underline{H}} - \underline{\underline{H}} \cdot {}^t\underline{\underline{\Lambda}}^p \, J + \underline{\underline{H}}\cdot\underline{\underline{O}}_p \cdot J \, \underline{\underline{H}}) \cdot \underline{\underline{O}}_q \cdot \tilde{J} \, \underline{\Phi}_k \;, \quad (25)$$

where I have used (13). From (10) we note that the third term on the right-hand side of (25) is the same as the previous one if one interchanges p and q, x_p and $\tilde{x}_q$ and J and $\tilde{J}$. Hence, we find from (25)

$$\left[\frac{\partial}{\partial \tilde{x}_q}, \frac{\partial}{\partial x_p}\right]\underline{\Phi}_k = (\underline{\underline{\Lambda}}^{p+q} + \underline{\underline{H}}\cdot\underline{\underline{O}}_{p+q}) \cdot [J, \tilde{J}]\underline{\Phi}_k = \frac{\partial}{\partial \hat{x}_{p+q}} \underline{\Phi}_k \quad (26)$$

where $\hat{x}_p$ are the coordinates for a hierarchy of differential relations associated with a matrix $\hat{J} = J, \tilde{J}$. Eq. (26) tells us that differentiations with respect to the generalized coordinates $x_p = x_p[J]$, depending functionally on the matrix J, do no longer commute when acting on the vector $\underline{\Phi}_k$. In fact, they build a representation of a loop algebra, this being precisely the algebraic interpretation of eq. (26). Note that here the loop algebra is obtained in terms of the *flows*, rather than as a symmetry algebra, nor did I have to introduce τ-functions and vertex operators. The draw-back of course is that we have to live with these strange differentiations that do not commute with each other. It is interesting that these kind of objects have recently attracted much attention within the context of string theory and non-commutative differential geometry[51-53]. Furthermore, the idea of extending the notion of integrability to non-commuting flows is already present in the work of Mishchenko and Fomenko[54], cf. also ref. 55. My idea is that eq. (26) is the right way to implement this on the level of a linear problem.

Let us now turn to the PDE's. A similar calculation, starting from (10), reveals that the algebraic property of the flows is the same when acting on H, i.e.

$$\left[\frac{\partial}{\partial \tilde{x}_q}, \frac{\partial}{\partial x_p}\right] \underline{\underline{H}} = \frac{\partial}{\partial \hat{x}_{p+q}} \underline{\underline{H}} \quad . \tag{27}$$

Furthermore, noting that eq. (16) is valid independent of the choice of J, we can re-do the calculation in (17), but now for coordinates corresponding to different choices of J. The result is

$$\frac{\partial}{\partial \tilde{x}_r}\left(\frac{\partial \underline{\underline{H}}}{\partial x_{p+q}}\right) - \frac{\partial}{\partial x_q}\left(\frac{\partial \underline{\underline{H}}}{\partial \tilde{x}_{p+r}}\right) + \frac{\partial \underline{\underline{H}}}{\partial x_q} \cdot \underline{\underline{O}}_p \cdot \frac{\partial \underline{\underline{H}}}{\partial \tilde{x}_r} - \frac{\partial \underline{\underline{H}}}{\partial \tilde{x}_r} \cdot \underline{\underline{O}}_p \cdot \frac{\partial \underline{\underline{H}}}{\partial x_q}$$

$$= \frac{\partial \underline{\underline{H}}}{\partial \hat{x}_{p+q+r}} \quad . \tag{28}$$

Note that due to the non-commutativity of the flows there is a new term which is absent in (17). Eq. (28), e.g. for p=1 in which case (28) reduces to an equation for $H=H_{0,0}$, defines in principle a hierarchy of nonlinear PDE's in terms of differentiations which obey the algebraic property (27). I do not know, however, at the moment a systematic way for reducing the system in order to obtain explicit evolution equations in terms of as little as possible variables (e.g. via a kind of generating operator). These are questions for future research.

5. COMPOUNDING THE HIERARCHIES

In this section I want to make a connection between the hierarchies as described in the previous sections and a class of nonlinear σ models which have attracted a great deal of interest over the last two years. The aim is to show that a full Kac-Moody algebra arises naturally if one looks at the Lagrangian strucure of these hierarchies, but in a special way[56]. To obtain this connection let us introduce a new kind of derivatives, that are obtained by a formal summation of differentiations with respect to the coordinates x_p with a weight ℓ ($\ell \in \mathbb{C}$), ℓ being a parameter

$$\partial_\ell \equiv \sum_{j=0}^{\infty} \frac{1}{\ell^{j+1}} \frac{\partial}{\partial x_j} \quad . \tag{29}$$

(A sum over differentiations with respect to coordinates x_j bearing a negative label could also have been considered). We would now like to write down the equations for the hierarchy in terms of these differentiations ∂_ℓ. From (9) we have

$$\partial_\ell \underline{\Phi}_k = (\underline{\underline{1}} + \underline{\underline{H}} \cdot \frac{1}{\ell - {}^t\underline{\underline{\Lambda}}} \cdot \underline{\underline{O}}) \cdot J \frac{1}{\ell - \underline{\underline{\Lambda}}} \cdot \underline{\Phi}_k \quad , \tag{30}$$

using (8). The factors $(\ell - \underline{\underline{\Lambda}})^{-1}$ and $(\ell - {}^t\underline{\underline{\Lambda}})^{-1}$ in (30) should be understood as formal power series in the index-raising operators $\underline{\underline{\Lambda}}$ and ${}^t\underline{\underline{\Lambda}}$. Let us call the expression between brackets on the right-hand side of (30) $\underline{\underline{G}}_\ell$, then we have from (30) and a formula that is obtained in a similar way from (7), i.e. by summation with powers ℓ^{-p-1},

$$\partial_\ell \underline{\Phi}_k = \underline{\underline{G}}_\ell \cdot J \frac{1}{\ell - \underline{\underline{\Lambda}}} \cdot \underline{\Phi}_k \qquad \frac{1}{\ell - k} \underline{\Phi}_k = \underline{\underline{G}}_\ell \cdot \frac{1}{\ell - \underline{\underline{\Lambda}}} \cdot \underline{\Phi}_k \quad . \tag{31}$$

Eliminating the term $(\ell-\underline{\underline{\Lambda}})^{-1}\cdot\underline{\Phi}_k$ and projecting down on the element $G_\ell \equiv (\underline{\underline{G}}_\ell)_{0,0}$ we arrive at

$$\partial_\ell \Phi_k = \frac{R_\ell}{\ell-k}\Phi_k \quad , \quad R_\ell \equiv G_\ell J (G_\ell)^{-1} \quad . \tag{32}$$

Furthermore, from (10) and (11) we find

$$(\underline{\underline{\mathbb{1}}} + \underline{\underline{H}}\cdot\frac{1}{\ell-{}^t\underline{\underline{\Lambda}}}\cdot\underline{\underline{O}})\cdot(\underline{\underline{\mathbb{1}}} - \frac{1}{\ell-\underline{\underline{\Lambda}}}\cdot\underline{\underline{H}}\cdot\underline{\underline{O}}) = \underline{\underline{\mathbb{1}}} \tag{33}$$

and

$$J-\partial_\ell\underline{\underline{H}}\cdot\underline{\underline{O}} = (\underline{\underline{\mathbb{1}}} + \underline{\underline{H}}\cdot\frac{1}{\ell-{}^t\underline{\underline{\Lambda}}}\cdot\underline{\underline{O}})\cdot J \, (\underline{\underline{\mathbb{1}}} - \frac{1}{\ell-\underline{\underline{\Lambda}}}\cdot\underline{\underline{H}}\cdot\underline{\underline{O}}) . \tag{34}$$

Eqs. (33) and (34) give us an expression for R_ℓ in terms of H. Another expression is obtained from (32), taking k a fixed value, e.g. $k\to 0$, in which case $\Phi_k \to g$, cf. eq. (21). Hence, we have for R_ℓ the following expressions

$$R_\ell = J - \partial_\ell H = \ell(\partial_\ell g)g^{-1} \quad . \tag{35}$$

Now eq. (32) can be thought of as a linear spectral problem containing a differentiation with respect to a certain variable ξ_ℓ, and with a potential R_ℓ. We can of course also consider another spectral problem of this form but now with another parameter ℓ' say. For commuting flows we can just take the compatibility relations of these two equations, remembering that ℓ and ℓ' are fixed, which lead to

$$\partial_{\ell'} R_\ell = \partial_\ell R_{\ell'} \quad , \quad \ell\partial_\ell R_{\ell'} - \ell'\partial_{\ell'} R_\ell + [R_{\ell'}, R_\ell] = 0 \quad . \tag{36a,b}$$

It is easily noted that the two expressions for R in (35) solve (36a) and (36b) respectively. Substituting these expressions in the other equation in (36) we obtain respectively an equation for H and for g, namely

$$\partial_\ell \partial_{\ell'} H = \frac{[J-\partial_\ell H, J-\partial_{\ell'} H]}{\ell' - \ell} \tag{37}$$

and

$$\ell \partial_{\ell'}(\partial_\ell g g^{-1}) = \ell' \partial_\ell (\partial_{\ell'} g\, g^{-1}) \quad . \tag{38}$$

Both equations can be obtained from an action principle. For (37) we can write down the following Lagrangian density

$$L = \frac{1}{2}\,\mathrm{tr}(\,\partial_\ell H\, \partial_{\ell'} H\,) + \frac{1}{2}\,\mathrm{tr}\left\{ J \left[H, \frac{\partial_\ell H - \partial_{\ell'} H}{\ell - \ell'} \right] \right\} - \frac{1}{3}\,\mathrm{tr}\left\{ H \frac{[\partial_\ell H, \partial_{\ell'} H]}{\ell - \ell'} \right\}, \tag{39}$$

the Euler-Lagrange equation for which gives the desired equation of motion. Eq. (38) is not so easily cast into a Lagrangian form, except when $\ell' = -\ell$. In that case (38) is the equation of motion for the chiral field model. When $\ell' \neq -\ell$ (38) becomes the equation of motion of the Wess-Zumino-Witten σ model[41,42], the chiral field model with an anomaly term. The action for (38) falls into the class of multivalued actions[43], and is

$$L' = \mathrm{tr}(\partial_\ell g \partial_{\ell'} g^{-1}) + \frac{\ell' + \ell}{\ell' - \ell} \int_0^1 dt\ \mathrm{tr}(\,[\partial_\ell g g^{-1}, \partial_{\ell'} g g^{-1}]\, \dot{g} g^{-1}\,). \tag{40}$$

Here we must introduce an extra parameter t , where $\dot{g}=dg/dt$, and $g(t=0)=\mathbf{1}$, $g(t=1)=g$. The nonlinear terms in (39) and (40) are both topological in nature, cf. ref.44. Eq. (39) could also have been obtained by taking a formal summation of the Lagrangians (20) over p and q with powers $\ell^{-p-1}, \ell'^{-q-1}$.

Starting from (39) and (40) one can now calculate the Poisson structure associated with the hierarchies via the weighed summation (29), using Dirac's method of constraints. Thus, (following Witten[41]), the limit $\ell'\to\infty$, i.e. $\ell'\partial_{\ell'}\to\partial/\partial x_0$, (which does not imply a loss of generality, because (40) is in fact a 1-parameter commuting family of Lagrangians), gives rise to a Kac-Moody algebra. In fact, if the conserved currents are given by $J_A(\xi_\ell)=\mathrm{tr}(\ A(\xi_\ell)g^{-1}\partial_\ell g\ )$, in which A is an arbitrary matrix independent of x_0, in this limit, then the Poisson algebra of currents is

$$\{J_A(\xi_\ell),J_B(\eta_\ell)\}=J_{[A,B]}(\xi_\ell)\delta(\xi_\ell-\eta_\ell)+\mathrm{tr}(AB)\delta'(\xi_\ell-\eta_\ell). \tag{41}$$

Eq. (41) is of Kac-Moody type, and the second term on the right-hand side (the socalled Schwinger term) corresponds to the central extension.

What happens if we extend the above discussion to the case of non-commuting differentiations? In the previous section I have derived an algebraic rule (26) for interchanging differentiations. This rule can be translated to the case of weighed derivatives (29) as follows

$$[\partial_\ell,\tilde{\partial}_{\ell'}]\Phi_k = \frac{\hat{\partial}_\ell-\hat{\partial}_{\ell'}}{\ell-\ell'}\Phi_k \ , \tag{42}$$

where $\tilde{\partial}_\ell$ and $\hat{\partial}_\ell$ denote weighed derivatives associated with the matrices J and $\hat{J}$ respectively. Now we retain (32) but only take care of (42) if we calculate the compatibility conditions. These become in this case

$$\tilde{\partial}_{\ell'} R_\ell - \partial_\ell \tilde{R}_{\ell'} = - \frac{\hat{R}_\ell - \hat{R}_{\ell'}}{\ell - \ell'} \tag{43a}$$

$$\ell' \tilde{\partial}_{\ell'} R_\ell - \ell \partial_\ell \tilde{R}_{\ell'} + [R_\ell, \tilde{R}_{\ell'}] + \frac{\ell' \hat{R}_\ell - \ell \hat{R}_{\ell'}}{\ell - \ell'} = 0, \tag{43b}$$

in which obviously $\tilde{R}_{\ell'}$ and $\hat{R}_\ell$ are the potentials in (32) associated with the derivatives $\tilde{\partial}_{\ell'}$ and $\hat{\partial}_\ell$ respectively. Using again the algebraic property for the differentiations it is easily checked that both expressions in (35) remain to be solutions of the equations ((43a) and (43b) respectively). Substituting these expressions in the other equation yields the following PDE's

$$\ell \partial_\ell \tilde{\partial}_{\ell'} H - \ell' \tilde{\partial}_{\ell'} \partial_\ell H + [J - \partial_\ell H, \tilde{J} - \tilde{\partial}_{\ell'} H] = \hat{J} + \frac{\ell' \hat{\partial}_\ell H - \ell \hat{\partial}_{\ell'} H}{\ell - \ell'} \tag{44}$$

and

$$\ell \tilde{\partial}_{\ell'} (\partial_\ell g\, g^{-1}) - \ell' \partial_\ell (\tilde{\partial}_{\ell'} g\, g^{-1}) = - \frac{\ell' \hat{\partial}_{\ell'} g - \ell \hat{\partial}_\ell g}{\ell' - \ell} g^{-1}. \tag{45}$$

Eqs. (44) and (45) form the non-commuting counterparts of eqs. (37) and (38), but it is not so easy to give a counterpart for the Lagrangians (39) and (40) in this case. This would require a noncommutative version of variational calculus, which to my knowledge does not yet exist.

Remark

The method of compounding the hierarchy by means of the weighed derivatives (29) is reminiscent of the approach for finding an infinite sequence of symmetry generators for certain field-theoretic models by means of infinitesimal Riemann-Hilbert transformations, cf. e.g. ref. 57 and references therein. It was already noted by Veselov and Takhtadzhyan[58] that the equations of motion for the Wess-Zumino-Witten model admit a Lax pair.

6. HIERARCHIES IN 2+1 DIMENSIONS

Up to now I have been concerned with the construction of hierarchies of PDE's in 1+1 dimensions. In this section I shall indicate what changes occur in the transition from 1+1 to 2+1 dimensions.

In 1+1 dimensions the basic equations for the hierarchies consisted of differential equations (9) and (10) as well as of algebraic relations (7) and (11). The main difference with the 2+1-dimensional case is now, that the latter relations are no longer valid and are replaced by another set of differential equations associated with another constant matrix, preferably the unit matrix, in terms of a new sequence of coordinates y_p ($p \in \mathbb{Z}$). So, we retain eqs. (9) and (10) and replace (7) and (11) by

$$\frac{\partial}{\partial y_p} \underline{\Phi}_k = (\underline{\underline{\Lambda}}^p + \underline{\underline{H}} \cdot \underline{\underline{O}}_p) \cdot \underline{\Phi}_k \ , \quad \frac{\partial}{\partial y_p} {}^t\underline{\Phi}_k = - {}^t\underline{\Phi}_k \cdot ({}^t\underline{\underline{\Lambda}}^p - \underline{\underline{O}}_p \cdot \underline{\underline{H}}) \qquad (46)$$

$$\frac{\partial}{\partial y_p}\underline{\underline{H}} = \underline{\underline{\Lambda}}^p\cdot\underline{\underline{H}} - \underline{\underline{H}}\cdot{}^t\underline{\underline{\Lambda}}^p + \underline{\underline{H}}\cdot\underline{\underline{O}}_p\cdot\underline{\underline{H}} \quad , (\forall p\in\mathbb{Z}) \ . \qquad (47)$$

In order to have an obstruction to the validity of eqs. (7) and (11), we have to introduce some new structure in the DLT. This can be done by making the integral transformations non-local, cf. also refs. 33,35,37. Consider the following non-local versions of eqs. (5)

$$\underline{\Phi}_k = \underline{\Phi}_k^0 + \iint_D \underline{\Phi}_\ell \, d\zeta(\ell,\ell') \, G^0_{k\ell'} \qquad (48a)$$

$${}^t\underline{\Phi}_{k'} = {}^t\underline{\Phi}_{k'} + \iint_D G_{\ell k'} \, d\zeta(\ell,\ell') \, {}^t\underline{\Phi}^0_{\ell'} \quad , \qquad (48b)$$

and the DLT for $\underline{\underline{H}}$

$$\underline{\underline{H}} = \underline{\underline{H}}^0 + \iint_D \underline{\Phi}_\ell \, d\zeta(\ell,\ell') \, {}^t\underline{\Phi}^0_{\ell'} \quad . \qquad (49)$$

It is clear that now there cannot be in general a relation of the form (7), because there is no direct coupling in (48) between the labels k and ℓ , viz. k' and ℓ', of the solutions Φ, ${}^t\Phi$. The measure in (48) is now a general function of the two variables ℓ and ℓ' and the integration is performed over an arbitrary area D in the complex hyperplane of ℓ and ℓ'. We have to assume again that the solution of (48) is unique for given choice of $d\zeta$ and D. The kernel in (48) is an integral over a path in the space of all the variables x_p, y_p starting from an arbitrary point P up to the point $X=(\{x_p, y_p\}_{p\ \mathbb{Z}})$ over a closed 1-form, i.e.

$$G_{k\ell'} = \int_P^X \Omega_{k\ell'} \; , \; \Omega_{k\ell'} = \sum_p \left({}^t\underline{\Phi}_{\ell'} \cdot \underline{\underline{O}}_p \cdot \underline{\Phi}_k \, dy_p + {}^t\underline{\Phi}_{\ell'} \cdot \underline{\underline{O}}_p \cdot J\underline{\Phi}_k \, dx_p \right) . \quad (50)$$

Due to the following 'Wronskian' relation the integral in (50) is independent of the path chosen :

$$\frac{\partial}{\partial y_q} \left({}^t\underline{\Phi}_{\ell'} \cdot \underline{\underline{O}}_p \cdot J\underline{\Phi}_k \right) = \frac{\partial}{\partial x_p} \left({}^t\underline{\Phi}_{\ell'} \cdot \underline{\underline{O}}_q \cdot \underline{\Phi}_k \right) \quad . \qquad (51)$$

Eq. (51) and similar ones involving two derivatives with respect to only x-coordinates or only y-coordinates follow from a simple calculation starting from (46). Using (50) it is not difficult to take derivatives of $G_{k\ell'}$ with respect to x_p and y_p, and from this it is easily derived that (9), (46), are indeed preserved by the DLT (48).

From eqs. (10) and (47), which are preserved under the DLT as follows from (9),(46) through (49), one can in principle derive a lattice of nonlinear PDE's by varying the order not only of the x-variable (chosen as 'time'), but also of the y-variable. In a similar way as in section 3 one can derive again recursion type of equations, but they are more involved due to the absence of the algebraic relation (11). (In order to derive 'closed' equations, i.e. removing the operators $\underline{\underline{\Lambda}}$ and ${}^t\underline{\underline{\Lambda}}$, we have of course to introduce extra derivations as compared to the 1+1-dimensional case). In fact, one of the problems here is that in introducing higher-order x_p and y_p one is not always able to project down on a single element of the matrix $\underline{\underline{H}}$, because we seem to keep stuck with some matrix

$\underline{\underline{0}}_p$, which seems to imply that we rather obtain a system of coupled equations rather than a single matrix equation. I shall not go into these details here, but rather give a few examples of equations associated with first-order x and y, i.e. $x=x_1$, $y=y_1$, where these problems do not occur. For these variables one can derive by combining (9) and (46) the classical Ablowitz-Haberman problem[26], i.e.

$$\frac{\partial}{\partial x_1}\Phi_k = J\frac{\partial}{\partial y_1}\Phi_k + [H,J]\,\Phi_k \tag{52}$$

for $\Phi_k=\Phi_k^{(0)}$ and $H=H_{0,0}$. The nonlinear equations that one can derive for H for the flows with respect to $t=x_2$ and $z=y_2$ are

$$[J,H_t] = H_{xx} - J\,H_{yy}\,J + [[J,H],H_x] + [J,H]H_y\,J - J\,H_y[J,H] \tag{53}$$

$$[J,H_z] = 2\,H_{xy} - \{J,H_{yy}\} + 2\,[[J,H],H_y]\ , \tag{54}$$

which can be considered as matrix versions of the Davey-Stewartson equations. The gauge transformation between Φ_k and $\Phi_k^{(-1)}$ also goes through here, but now $\mathbb{1}+H_{-1,0}\neq(\mathbb{1}-H_{0,-1})^{-1}$ as was the case in 1+1 dimensions. We now have

$$\frac{\partial}{\partial x_1}\Phi_k^{(-1)} = g^{-1}J\,\Phi_k \quad , \quad \frac{\partial}{\partial y_1}\Phi_k^{(-1)} = g^{-1}\,\Phi_k \quad , \tag{55}$$

in which $g=(\mathbb{1}+H_{-1,0})^{-1}$, $g^{-1}\neq(\mathbb{1}-H_{0,-1})^{-1}$. Eliminating Φ_k we obtain a 2+1-dimensional version of the spectral problem (23), for q=0, namely

$$\frac{\partial}{\partial x_1}\Phi_k^{(-1)} = g^{-1}Jg\,\frac{\partial}{\partial y_1}\Phi_k^{(-1)} \quad , \tag{56}$$

where the spin matrix $S=g^{-1}Jg$ can be expressed as

$$S = \left(J - \frac{\partial}{\partial x_1} H_{-1,-1} \right) \left(\mathbb{1} - \frac{\partial}{\partial y_1} H_{-1,-1} \right)^{-1} \tag{57}$$

using (10) and (47) with p=1 for $H_{-1,-1}$. An especially interesting case of a nonlinear PDE associated with (56) is the isotropic Heisenberg ferromagnet in 2+1 dimensions[35,59]

$$\begin{aligned} &\vec{S}_t + \frac{1}{2}\vec{S}\times(\vec{S}_{xx} \mp \vec{S}_{yy}) = \phi_x \vec{S}_y + \phi_y \vec{S}_x \\ &\phi_{xx} \mp \phi_{yy} = \pm \vec{S}\cdot(\vec{S}_x \times \vec{S}_y) \end{aligned} \tag{58}$$

which is obtained in the case that we work in su(2), i.e. in which H is (anti-) hermitean $J=\sigma_3, gg^{\dagger}=\mathbb{1}$, $S=i\vec{S}\cdot\vec{\sigma}$, $\vec{S}^2=1$. The application of the DLT to the case of 2+1-dimensional lattice equations associated with eqs. (52)-(58) has been considered in ref. 37.

Acknowledgement

The author would like to to thank Profs. M. Boiti and F. Pempinelli for their hospitality during the authors visit to the University of Lecce, where this work was completed.

REFERENCES

1. M.J. Ablowitz, D.J. Kaup, A.C. Newell and H. Segur: Stud. Appl. Math. 53 (1974) 249.
2. V.E. Zakharov and S.V. Manakov: Sov. Sci. Revs. A2(1981).

3. S.P.Novikov, S.V.Manakov L.P.Pitaevskii and V.E.Zakharov *Theory of Solitons*.(Plenum Press, New York, 1984).

4. A.S.Fokas and M.J.Ablowitz: in *Nonlinear Phenomena*, ed. B.Wolf, Lect. Notes Phys. 189 (Springer Verlag,1983) 167.

5. V.G.Dubrovin: Funct. Anal. Appl. 11 (1977) 265, Journ. Sov. Math. 28 (1985) 20.

6. I.M.Krichever: Journ. Sov. Math. 28 (1985) 51.

7. M.Jimbo and T.Miwa: Publ.RIMS 19 (1983) 943.

8. E.Date, M.Kashiwara, M.Jimbo and T.Miwa: in *Nonlinear Integrable Systems-Classical Theory and Quantum Theory*, eds. M.Jimbo and T.Miwa,(World Sci. Publ. Co., Signapore, 1983), 39.

9. G.Segal and G.Wilson: Publ. Math. IHES 16 (1985) 5.

10. K.Ueno: in *Infinite Dimensional Groups with Applications*, ed. V.G.Kac (Springer Verlag, New York, 1985),

11. Yu.I.Manin: Journ. Sov. Math. 11(1979) 1.

12. L.A. Dickey: Commun. Math. Phys. 82 (1981) 345,361; ibid. 88 (1983) 27.

13. P.P.Kulish and A.G.Reiman: Journ. Sov. Math. 28 (1985) 505.

14. A.G.Reiman: Journ. Sov. Math. 30 (1985) 2319.

15. V.G.Kac: *Infinite-dimensional Lie Algebras*, (Birkhäuser Verlag, Boston, 1984).

16. V.G.Drinfel'd and V.V.Sokolov: Journ. Sov. Math. 24 (1984) 1975.

17. K.Ueno and K.Takesaki: in *Group Representations and Systems of Differential Equations*, Adv. Stud. Appl. Math. 4 (1984) 1.

18. G.Wilson: Ergod. Theory Dyn. Syst. 1 (1981) 361.

19. V.G.Drinfel'd and V.V.Sokolov: Sov. Math. Dokl. 23 (1981) (1981) 475.

20. A.G.Reiman and M.A.Semenov-Tian-Shanskii: Inv. Math. 54 (1979) 81; ibid. 63 (1981) 423.

21. M.Adler and P.van Moerbeke: Adv. Math. 38 (1980) 267,318.

22. H.Flaschka, A.C.Newell and T.Ratiu: Physica 9D (1983) 300,324.

23. A.C.Newell: *Solitons in Mathematics and Physics*, (SIAM, Philadelphia, 1985).

24. A.G.Reiman and M.A.Semenov-Tian-Shanskii: Sov. Math. Dokl. 21 (1980) 630; Funct. Anal. Appl. 14 (1980) 146.

25. D.Olive and N.Turok: Nucl. Phys. B265 [FS15] (1986) 469.

26. M.J.Ablowitz and R.Haberman: Phys. Rev. Lett. 35(1975)1185.

27. A.S.Fokas and M.J. Ablowitz: Phys. Lett. 94A(1983) 67.

28. A.S.Fokas and M.J.Ablowitz : Journ. Math. Phys. 25 41984) 2494; Phys. Rev. Lett. 51 (1983) 7.

29. R.Beals and R.R.Coifman: Proc. Symp. Pure Math. 43(1985)45.

30. B.G.Konopelchenko: Commun. Math. Phys. 87 (1982) 105; ibid. 88 (1983) 531.

31. M.Boiti,B.G.Konopelchenko and F.Pempinelli: Inv.Probl. 1 (1985) 33.

32. V.E.Zakharov and S.V.Manakov: Funct. Anal. Appl. 19 (1985) 89; Journ. Sov. Math. 31 (1985) 3307.

33. M.J.Ablowitz, R.Beals and K.Tenenblat: Preprint Clarkson, (1985).

34. P.M.Santini, M.J.Ablowitz and A.S.Fokas: Journ. Math. Phys. 25 (1984) 2614.

35. F.W.Nijhoff, H.W.Capel and G.L.Wiersma: in *Geometric Aspects of the Einstein Equations and Integrable Systems*, ed. R.Martini, Lect. Notes Phys. 239 (1985) 263.

36. F.W.Nijhoff, G.R.W.Quispel, H.W.Capel and J.van der Linden, Physica 119A (1983) 101.

37. F.W.Nijhoff: Lett. Math. Phys. 9 (1985) 217.

38. J.Lepowsky and R.L.Wilson: Commun. Math. Phys. 62 (1978) 43.

39. V.G.Kac, D.A.Kazhdan,J.Lepowsky and R.L.Wilson: Adv. Math. 42 (1981) 83.

40. I.Frenkel and V.G.Kac: Inv. Math. 62 (1980) 23

41. E.Witten: Commun. Math. Phys. 92 (1984) 455.

42. A.M.Polyakov and P.B.Wiegmann: Phys. Lett. 131B (1983) 121; ibid. 141B (1984) 223.

43. S.P.Novikov: Russ. Math. Surv.37 (1982) 1 .

44. I.M.Krichever, M.A.Olshanetsky and A.M.perelomov: Nucl. Phys.B264 (1986) 415.

45. F.W.Nijhoff: *Linear Integral Transformations and Hierarchies of Nonlinear Evolution Equations*, Preprint University of Utrecht (1986).

46. R.Beals and R.R.Coifman: Commun. Pure Appl. Math. 37 (1984) 39 ; ibid. 38 (1985) 29.

47. K.Ueno: in *Vertex Operators in Mathematics and Physics*, eds. J.Lepowsky and S.Mandelstam (Springer Verlag, New York,1985), 191.

48. T.Sattinger: Stud. Appl. Math. 72 (1985) 65.

49. A.N.Leznov and M.V.Saveliev: Commun. Math. Phys. 89 (1983) 59.

50. L.A. Takhtadzhyan: Phys. Lett. 64A (1977) 235.

51. E.Witten: Nucl. Phys. B268 (1986) 79.

52. J.-L. Gervais: Nucl. Phys. B276 (1986) 349.

53. A.Connes: Publ. Math. IHES (1986).

54. A.S. Mishchenko and A.T. Fomenko: Funct. Anal. Appl. 12 (1978) 113.

55. C.M.Marle: Lett. Math. Phys. 7 (1983) 55.

56. F.W.Nijhoff and D.J.Smit: *Hierarchies of Integrable Non-linear Evolution Equations and Wess-Zumino Terms*, Preprint University of Utrecht,(1986).

57. Wu Yong-Shi: Commun. Math. Phys. 90 (1983) 461.

58. A.P.Veselov and L.A.Takhtadzhyan: Sov. Phys. Dokl. 29 (1984) 994.

59. Y.Ishimori: Progr. Theor. Phys. 72 (1984) 33.

RELATIVISTIC CALOGERO-MOSER SYSTEMS AND SOLITONS

S.N.M. Ruijsenaars

CWI, Kruislaan 413
Amsterdam, Netherlands

1. INTRODUCTION

In the following we survey a number of results pertaining to a novel class of completely integrable N-particle systems, which generalize the well-known Calogero-Moser (CM) systems, and which were discovered in collaboration with H. Schneider [1)]. The new systems are relativistically invariant, and the CM systems result upon taking a parameter to ∞ which may be regarded as the speed of light. They can be quantized in such a fashion that integrability is preserved [2)]. However, we shall restrict ourselves here to the classical level, and present results from Refs. [1)–4)] with a bias towards relations with soliton PDE, in keeping with the spirit of the conference. Related results concerning external fields and Toda systems, and a Lax pair formulation for the elliptic case can be found in Refs. [5)–7)], resp. (See also F. Calogero's contribution to these proceedings with regard to the last-mentioned topic.)

In Section 2 we describe our systems and their relation to the CM systems, and discuss their relativistic invariance and complete integrability. Section 3 is concerned with a matrix whose symmetric functions yield the N independent commuting Hamiltonians. This matrix is an essential tool in the construction of the action-angle map, which is discussed in Section 4. This section also prepares the ground for Section 5, which deals with the relation of the particle systems to various well-known soliton equations. It turns out that the solutions consisting of solitons, antisolitons, breathers,... may all be viewed as manifestations of underlying point particle dynamics associated with our systems. The relation leads in particular to a natural concept of soliton space-time trajectory, which is presented and discussed in Section 6.

2. A DESCRIPTION OF THE SYSTEMS

Recall that the Hamiltonian of the CM system reads

$$H_{nr} \equiv \sum_{i=1}^{N} \theta_i^2/2 + g^2 \sum_{1 \le i < j \le N} V(q_i - q_j). \tag{1}$$

Here, the particle mass equals 1, and the pair potential V equals the Weierstrass $\wp$-function (elliptic case) or one of its degenerate cases, viz., $1/sh^2$ (hyperbolic case), $1/\sin^2$ (trigonometric case), $1/q^2$ (rational case). (More precisely, this defines the CM system associated with the Lie algebra A_{N-1}, cf. the review [8].) Supplementing H_{nr} with the functions

$$P_{nr} \equiv \sum_{i=1}^{N} \theta_i \tag{2}$$

(space translation generator) and

$$B_{nr} \equiv -\sum_{i=1}^{N} q_i \tag{3}$$

(boost generator) leads to a representation of the nonrelativistic space-time symmetry group (the Galilei group) in one space dimension.

The integrable systems to be discussed here lead to a representation of the relativistic space-time symmetry group (the Poincaré group). The generators are given by

$$H \equiv c^2 \sum_{i=1}^{N} ch(\theta_i/c)\prod_{j\neq i} f(q_i - q_j) \tag{4}$$

$$P \equiv c \sum_{i=1}^{N} sh(\theta_i/c)\prod_{j\neq i} f(q_i - q_j) \tag{5}$$

$$B \equiv -\sum_{i=1}^{N} q_i. \tag{6}$$

Here, c is the speed of light and f satisfies

$$f^2(q) = a + b\wp(q), \tag{7}$$

where a,b are arbitrary constants. Picking

$$a = 1,\ b = g^2/c^2, \tag{8}$$

it is obvious that one obtains the CM systems in the nonrelativistic limit. That is, one has

$$H_{nr} = \lim_{c\to\infty} (H - Nc^2) \tag{9}$$

$$P_{nr} = \lim_{c\to\infty} P \qquad B_{nr} = \lim_{c\to\infty} B \tag{10}$$

It is not supposed to be obvious that H,P,B satisfy the Poincaré Lie algebra

$$\{H,P\} = 0,\ \{H,B\} = P,\ \{P,B\} = H/c^2. \tag{11}$$

More precisely, the vanishing of the first bracket (translation invariance) is not obvious.. One readily verifies that for even f its vanishing is equivalent to the functional equation

$$\sum_{i=1}^{N} \partial_i \prod_{j\neq i} f^2(q_i - q_j) = 0. \tag{12}$$

This equation is trivially satisfied for $N = 2$, but for $N = 3$ it amounts to the well-known functional equation of the $\wp$-function. Thus, (7) results.

The fact that the $\wp$-function satisfies (12) for arbitrary N appears to be new. A direct proof is given in Appendix A of Ref. [1]. It is also shown there that (12) entails that the functions

$$S_k \equiv \sum_{\substack{|I|=k \\ I\subset\{1,\dots,N\}}} \exp\Big(\sum_{i\in I}\theta_i\Big) \prod_{\substack{i\in I \\ j\notin I}} f(q_i - q_j), \quad k = 1,\dots,N \tag{13}$$

commute with H. (Here and henceforth we set $c \equiv 1$.) Moreover, it is proved that these integrals are in involution if and only if f^2 satisfies

$$\sum_{\substack{|I|=k\\ I\subset\{1,\ldots,N\}}} \sum_{i\in I} \partial_i \prod_{\substack{i\in I\\ j\notin I}} f^2(q_i-q_j) = 0,\ \forall k\in\{1,\ldots,N\},\ \ \forall N>1 \tag{14}$$

The validity of these more general functional equations in the hyperbolic case has been established in Ref. [1] by invoking scattering theory. The fact that they still hold in the elliptic case is proved in Ref. [2]. They follow by taking a limit in new functional equations for the Weierstrass σ-function, which express the fact that a suitably chosen quantization preserves integrability. The limit involved here may be viewed as taking Planck's constant to 0 [2].

3. LAX MATRICES AND EXPLICIT SOLUTIONS

In the hyperbolic case it is convenient to use the parametrization

$$f(q) = [1-sh^2\frac{\mu}{2}/sh^2\frac{q}{2}]^{\frac{1}{2}}. \tag{15}$$

Here and in the sequel we shall take

$$\mu \in i(0,\pi]. \tag{16}$$

The commuting functions $S_1,\ldots,S_N$ are then equal to the symmetric functions of the $N\times N$ matrix with elements

$$L_{ij}(q,\theta) \equiv d_i(q,\theta)C_{ij}(q)d_j(q,\theta) \tag{17}$$

where

$$d_j \equiv e^{\theta_j/2}\prod_{j\neq i} f(q_i-q_j)^{\frac{1}{2}} \tag{18}$$

$$C_{ij} \equiv sh\frac{\mu}{2}/sh\frac{1}{2}(\mu+q_i-q_j). \tag{19}$$

This assertion can be verified by making suitable use of Cauchy's identity,

$$|(\frac{1}{x_i-y_j})| = \prod_i \frac{1}{x_i-y_i}\prod_{i<j}\frac{(x_i-x_j)(y_i-y_j)}{(x_i-y_j)(y_i-x_j)} \tag{20}$$

In the elliptic case there seems to be no matrix generalizing (17) whose symmetric functions are equal to $S_1,\ldots,S_N$. However, a matrix whose symmetric functions are proportional to $S_1,\ldots,S_N$ does exist; the proportionality follows from a generalization of Cauchy's identity. This matrix depends on an extra parameter λ and leads to Krichever's [9] Lax matrix in the nonrelativistic limit, so that the integrability of the CM systems follows as a corollary. Also, the hyperbolic Lax matrix (17) results by taking one period of $\wp$ to ∞, and then sending the spectral parameter λ to ∞ [2]. Bruschi and Calogero independently arrived at the same elliptic Lax matrix, and also obtained a matrix M such that $\{S_1,L\} = [L,M]$ [7)-10)].

It is not yet clear whether the elliptic Lax matrix can be used to obtain explicit solutions, as a generalization of Krichever's results in the nonrelativistic case [2)-9)]. However, in the hyperbolic case this problem has been solved. For instance, the solutions $q_1(t),\ldots,q_N(t)$ to Hamilton's equations for S_1 are equal to the logarithms of the eigenvalues of the positive matrix $e^{Q(0)/2}e^{tL(0)}e^{Q(0)/2}$, where

$$Q \equiv diag(q_1,\ldots,q_N). \tag{21}$$

This is proved in Ref. [1] by exploiting Rayleigh-Schrödinger perturbation theory. This result generalizes the results of Olshanetski and Perelomov in the nonrelativistic case [8].

More generally, the Lax matrix (17) can be used to construct the action-angle transformation. This transformation is crucial for making contact with soliton solutions. We proceed by describing it, without entering into the technicalities associated with its construction and with the proof of its canonicity.

4. THE HYPERBOLIC ACTION-ANGLE MAP

The key to the construction of the action-angle map is the commutation relation

$$\frac{1}{2}cth\frac{\mu}{2}[e^{Q},L] = e^{Q/2}d\otimes e^{Q/2}d-\frac{1}{2}(e^{Q}L+Le^{Q}) \tag{22}$$

which is readily verified, cf. (17-19), (21). From this one can infer that L has positive and simple spectrum for any (q,θ) in the phase space

$$\Omega \equiv\{(q,\theta)\in\mathbf{R}^{2N}|q_1<...<q_N\}, \tag{23}$$

and that there exists a uniquely determined unitary U satisfying

$$UL(q,\theta)U^{-1} = diag(e^{\hat{\theta}_1},...,e^{\hat{\theta}_N}) \tag{24}$$

$$Ue^{Q(q)}U^{-1} = \bar{L}(\hat{\theta},\hat{q}), \tag{25}$$

where $(\hat{q},\hat{\theta})$ belongs to the action-angle phase space

$$\hat{\Omega} \equiv \{(\hat{q},\hat{\theta})\in\mathbf{R}^{2N}|\hat{\theta}_1<...<\hat{\theta}_N\}. \tag{26}$$

(This generalizes results obtained for the nonrelativistic $1/q^2$-system by Airault, McKean and Moser [11].)

It is a matter of simple linear algebra to show that one obtains a bijection $\Phi:\Omega\to\hat{\Omega}$ in this way, but to show that Φ is a symplectic diffeomorphism and hence may be regarded as the action-angle map involves a lot more work [3]. Once this is proved, it is obvious from (24-25) that Φ linearizes the flows generated by Hamiltonians of the form

$$H_h = Tr\ h(lnL),\ \ h\in C^{\infty}_{\mathbf{R}}(\mathbf{R}). \tag{27}$$

Indeed, the transform of H_h to $\hat{\Omega}$ reads

$$(H_h\circ\Phi^{-1})(\hat{q},\hat{\theta}) \equiv \hat{H}_h(\hat{q},\hat{\theta}) = \sum_{i=1}^{N}h(\hat{\theta}_i), \tag{28}$$

so that the nonlinear flow

$$(q,\theta)\to(q(t),\theta(t)) \equiv \exp(tH_h)(q,\theta) \tag{29}$$

maps into the linear flow

$$(\hat{q},\hat{\theta})\to\exp(t\hat{H}_h)(\hat{q},\hat{\theta}) = (\hat{q}_1+th'(\hat{\theta}_1),...,\hat{q}_N+th'(\hat{\theta}_N),\hat{\theta}). \tag{30}$$

More generally, consider the multi-parameter flow

$$(q,\theta)\to(q(t_0,...,t_m),\theta(t_0,...,t_m)) \equiv exp(\sum_{k=0}^{m}t_kH_{h_k})(q,\theta). \tag{31}$$

Let us introduce

$$Q(t_0,...,t_m) \equiv diag(q_1(t_0,...,t_m),...,q_N(t_0,...,t_m)) \tag{32}$$

and

$$A(t_0,...,t_m) \equiv \bar{L}(\hat{\theta},\hat{q}_1 + \sum_{k=0}^{m} t_k h_k'(\hat{\theta}_1),...,\hat{q}_N + \sum_{k=0}^{m} t_k h_k'(\hat{\theta}_N)), \tag{33}$$

where $(\hat{q},\hat{\theta}) \equiv \Phi(q,\theta)$. Then it readily follows from the above that the matrices $A(\vec{t})$ and $\exp(Q(\vec{t}))$ are similar, so that

$$ln(\det(1+A(\vec{t}))) = \sum_{i=1}^{N} ln(1+\exp(q_i(\vec{t}))) \tag{34}$$

$$Tr(Arctg A(\vec{t})) = \sum_{i=1}^{N} Arctg(exp(q_i(\vec{t}))). \tag{35}$$

As will be explained below, for $m = 1$ and suitable μ these equalities can be used to arrive at a picture of N-soliton solutions as being deformations of an elastic medium, which hides an underlying motion of N point particles. The arbitrary m case encodes the relation of our systems to the commuting flows of various soliton hierarchies.

The scattering theory for the special Hamiltonian S_1 (for which $h(u)=e^u$) is a crucial ingredient in our proof that Φ is canonical. In fact, Φ^{-1} is very closely related to the $t \to \infty$ Møller transformation. For the case at hand there is only one scattering channel; physically speaking, the interparticle forces are repulsive, so that all particles move freely for asymptotic times. (Cf. also (42) below, with $h_0(u)=e^u$, $x \equiv 0$.

However, let us now make the substitutions $q_j \to q_j + i\pi$, $1 \leq n < j \leq N$, in the Hamiltonians $H,P,S_1,...,S_{N-1}$ of Section 2. Then we get again integrable N-particle systems, since the functional equations expressing integrability can be analytically continued, and since the resulting functions are real-valued (recall (15-16)). In the nonrelativistic case it is obvious that the substitutions lead to attraction between the particles $1,...,n$ (solitons) and $n+1,...,N$ (antisolitons), since (e.g.) the repulsive potential $1/sh^2\frac{1}{2}(q_1 - q_N)$ turns into the attractive potential $-1/ch^2\frac{1}{2}(q_1 - q_N)$ (as observed first by Calogero [12]).

It is not obvious, but true, that the substitutions have the same physical consequence for our more general systems. Correspondingly, the phase space of initial conditions splits up into a number of components which are invariant under S_1 and which correspond to various scattering channels, and an exceptional set which is nowhere dense and has measure zero. These channels correspond to asymptotics of freely evolving solitons, antisolitons and k soliton-antisoliton pairs (breathers); one has $k \in \{0,...,\min(n,N-n)\}$ and all possibilities occur. The exceptional set contains channels with many-body bound states, and also states that do not admit a scattering interpretation.

As suggested by our terminology, the systems just discussed can be used to describe solutions to soliton PDE with similar characteristics. For instance, the various particle-like solutions of the sine-Gordon equation can be obtained via (the generalization of) (35) with $m=1$ and $\mu = i\pi$.

Unfortunately, a more precise description of the latter relation cannot be given without detailing the action-angle map in the general case, something which is well beyond the scope of our survey. Here, we only mention that (the analytic continuation of) the commutation relation (22) is the starting-point in the general case, too. Again, it can be exploited to derive spectral properties of L. In contrast to the pure soliton case, these properties depend on the region of phase space involved, and this gives rise to the splitting mentioned above. For instance, the no-bound-state channel corresponds to L having simple and real spectrum, whereas the states without scattering interpretation correspond to points in phase space where L is not diagonalizable. For the full story we refer to our forthcoming paper [4].

5. THE RELATION TO SOLITON EQUATIONS

We shall begin by describing the relation of our N-particle systems to the KdV equation

$$\dot{u} - 6uu' + u''' = 0. \tag{36}$$

Set

$$\tau(\vec{t}) \equiv \det(1 + A(\vec{t})), \tag{37}$$

where $A(\vec{t})$ is given by (33), and specialize to

$$\mu = i\pi,\ m = 1,\ t_0 = t,\ t_1 = -x. \tag{38}$$

Now choose

$$h_0(u) = e^u,\ \ h_1(u) = \frac{1}{3}e^{3u}. \tag{39}$$

Then

$$u(t,x) \equiv -2\partial_x^2(\ln\tau(t,x)) \tag{40}$$

is an N-soliton solution to (36). This can be seen from Hirota's pioneering paper [13] by invoking Cauchy's identity (20) [1]. Moreover, the multi-parameter KdV τ-function of the Kyoto school can be obtained analogously by taking $\mu = i\pi$ and suitable Hamiltonians $h_1,...,h_m$ in (37), cf. pp. 960, 963 in the review [14].

More generally, picking

$$\mu_n = 2i\pi/n,\ \ n = 2,3,4,... \tag{41}$$

one obtains the τ-function of the $A_n^{(1)}$-reduction of the KP-hierarchy. In particular, the case $n = 3$ corresponds to the Boussinesq hierarchy [14].

The N-soliton solutions of the Toda lattice can also be obtained via (37,38): One needs only take $x \in \mathbf{Z}$ and make another choice of h_0, h_1. Similarly, the solitons of the modified KdV and sine-Gordon equations can be written in terms of $A(t,x)$ by picking suitable h_0, h_1. For these cases the function $Tr\ ArctgA$ plays the role of $\ln(det(1+A))$. We refer to [1] for the details.

Relations between the above-mentioned soliton PDE and motions of point particles have been known for over a decade. The idea that there might be such connections was first put forward by Kruskal, who proposed to study the pole motions of soliton solutions [15]. Subsequent work in this area includes Refs. [16)–20]. We mention in particular the work by the Choodnovsky's [17], Airault, McKean and Moser [11], and Krichever [9], who discovered relations with the nonrelativistic CM systems.

It is likely that some of the latter results may be viewed as degenerate cases of the relations detailed in [1,3,4]. However, for function-theoretic reasons one cannot obtain the known results concerning elliptic solutions and the elliptic nonrelativistic CM systems from the hyperbolic context considered by us. Rather, the latter relations suggest that the action-angle map for the relativistic elliptic systems (which has not been constructed yet) can be used to parametrize the quasi-periodic solutions to the above soliton equations, as an analogue of the situation in the hyperbolic case. More generally, there are several hints which lead us to conjecture that the KP hierarchy can be tied in with integrable particle systems generalizing the ones at issue here.

6. SOLITON SPACE-TIME TRAJECTORIES

Let us now return to the context of N-soliton solutions in a two-dimensional space-time. As we have seen in the previous section, the r.h.s. of (34) and (35), with $m=1$, $t_0=t$, $t_1=-x$, gives rise to various well-known soliton functions, provided we choose appropriate μ and space-time translation generators $Tr\ h_\nu(lnL)$, $\nu = 0,1$. For the moment, let us only assume

$$\mu \in i(0,\pi],\ h_0''>0. \tag{42}$$

This assumption is satisfied in all cases of interest, up to flipping the sign of h_0, which amounts to time reversal.

The point of this assumption is, that it already suffices to determine the long-time asymptotics of the functions $q_i(t,x)$. In view of the fact that $A(t,x)$ has eigenvalues $exp(q_i(t,x))$, this boils down to a determination of spectral asymptotics. The result of this reads [3]

$$\underset{N-i+1}{q_i}\ (t,x) \sim \hat{q}_i + th_0'(\hat{\theta}_i) - xh_1'(\hat{\theta}_i) \mp \frac{1}{2}\Delta_i(\hat{\theta}),\ t\to\pm\infty \tag{43}$$

Here, the phase shift Δ_i is given by

$$\Delta_i = -\sum_{j<i}\delta(\hat{\theta}_i-\hat{\theta}_j)+\sum_{j>i}\delta(\hat{\theta}_i-\hat{\theta}_j) \tag{44}$$

and δ is the pair phase shift,

$$\delta(x) = ln(1-sh^2\frac{\mu}{2}/sh^2\frac{x}{2}). \tag{45}$$

This result amounts to a unified derivation of soliton scattering, and is crucial in justifying a notion of soliton space-time trajectory we are about to describe [1]. For concreteness we shall specialize from now on to the sine-Gordon equation

$$\phi''-\ddot{\phi} = \sin\phi. \tag{46}$$

Its N-soliton solutions can be written

$$\phi(t,x) = 4Tr(ArctgA(t,x)) = 4\sum_{i=1}^{N} Arctg(exp(q_i(t,x))), \tag{47}$$

provided one sets

$$\mu = i\pi,\ h_0(u) = chu,\ h_1(u) = shu. \tag{48}$$

This entails $H_{h_0} = H$, $H_{h_1} = P$, so that

$$\dot{q}_i = sh\theta_i\prod_{j\neq i}f(q_i-q_j) \tag{49}$$

$$q_i' = -ch\theta_i\prod_{j\neq i}f(q_i-q_j) \tag{50}$$

by virtue of Hamilton's equations.

Let us now fix t and require

$$q_i(t,x_i(t)) = 0,\ i = 1,...,N. \tag{51}$$

Since $q_i'<-1$, this yields uniquely determined space-time trajectories $x_1(t),...,x_N(t)$. It is not hard to see that these trajectories are Poincaré invariant, and (49-51) entail

$$\dot{x}_i(t) = th(\theta_i(t,x_i(t))). \tag{52}$$

Hence, the speeds $|\dot{x}_i|$ are always smaller than the speed of light. (Recall $c\equiv 1$.) Finally, for $|t|\to\infty$ the particle trajectories coincide with the N maxima of $\phi'(t,x)$, as a consequence of (43). We conclude with a picture which illustrates this state of affairs.

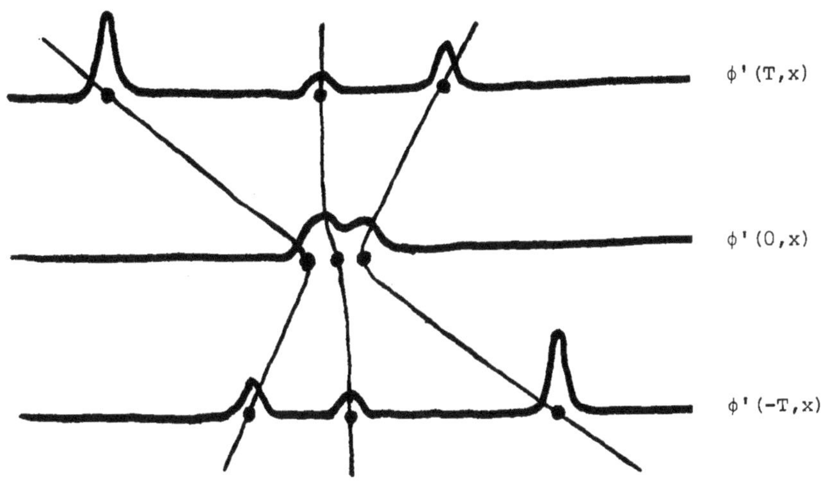

Fig.1. Particle trajectories for a 3-soliton solution to the sine-Gordon equation $\phi'' - \ddot{\phi} = \sin\phi$.

REFERENCES

1. Ruijsenaars, S.N.M., Schneider, H., "A new class of integrable systems and its relation to solitons", Ann. Phys. (NY) 170, 370-405 (1986).
2. Ruijsenaars, S.N.M., "Complete integrability of relativistic Calogero-Moser systems and elliptic function identities", to appear in Comm. Math. Phys.
3. Ruijsenaars, S.N.M., "Action-angle maps and scattering theory for some finite-dimensional integrable systems I The pure soliton case", to appear.
4. Ruijsenaars, S.N.M., "Action-angle maps and scattering theory for some finite-dimensional integrable systems II Solitons, antisolitons, and their bound states", to appear.

5. Schneider, H., "Integrable relativistic *N*-particle systems in an external potential", to appear in Physica D.

6. Ruijsenaars, S.N.M., "Relativistic Toda systems", to appear.

7. Bruschi, M., Calogero, F., "The Lax representation for an integrable class of relativistic dynamical systems", to appear.

8. Olshanetsky, M.A., Perelomov, A.M., "Classical integrable finite-dimensional systems related to Lie algebras", Phys. Reps. 71, 313-400 (1981).

9. Krichever, I.M., "Elliptic solutions of the Kadomtsev-Petviashvili equation and integrable systems of particles", Func. Anal. Appl. 14, 282-290 (1980).

10. Calogero, F., Contribution to these proceedings.

11. Airault, H., McKean, H.P., Moser, J., "Rational and elliptic solutions of the Korteweg- de Vries equation and a related many-body problem", Comm. Pure Appl. Math. 30, 95-148 (1977).

12. Calogero, F., "Exactly solvable one-dimensional many-body problems", Lett. Nuovo Cim. 13, 411-416 (1975).

13. Hirota, R., "Exact solution of the Korteweg- de Vries equation for multiple collisions of solitons", Phys. Rev. Lett. 27, 1192-1194 (1971).

14. Jimbo, M., Miwa, T., "Solitons and infinite dimensional Lie algebras", Publ. RIMS 19, 943-1001 (1983).

15. Kruskal, M.D., "The Korteweg- de Vries equation and related evolution equations", Lect. Appl. Math. 15, 61-83 (1974).

16. Thickstun, W.R., "A system of particles equivalent to solitons", J. Math. Anal. Appl. 55, 335-346 (1976).

17. Choodnovsky, D.V., Choodnovsky, G.V., "Pole expansions of nonlinear partial differential equations", Nuovo Cim. B40, 339-353 (1977).

18. Calogero, F., Degasperis, A., "Special solutions of coupled nonlinear evolution equations with bumps that behave as interacting particles", Lett. Nuovo Cim. 19, 525-533 (1977).

19. Bowtell, G., Stuart, A.E.G., "Interacting sine-Gordon solitons and classical particles: a dynamic equivalence", Phys. Rev. D15, 3580-3591 (1977).

20. Bowtell, G., Stuart, A.E.G., "A particle representation for Korteweg- de Vries solitons", J. Math. Phys. 24, 969-981 (1983).

Poisson-Drinfeld Groups

Yvette Kosmann-Schwarzbach
UFR de Mathématiques
Université de Lille I
59655 Villeneuve d'Ascq

ABSTRACT

A Poisson-Drinfeld group is a Lie group G with a Poisson structure such that the group multiplication is a Poisson morphism from $G \times G$, with the product Poisson structure, to G. When G is a Poisson-Drinfeld group, $\mathcal{g}^*$ is a Lie algebra, and the Lie brackets on $\mathcal{g}$ and $\mathcal{g}^*$ satisfy a distributivity property. Poisson-Drinfeld structures on Lie groups can be constructed from solutions of the generalized Yang-Baxter equation. They give rise to bihamiltonian structures and also permit an interpretation of the group of dressing transformations.

RESUME

Un groupe de Poisson-Drinfeld est un groupe de Lie G muni d'une structure de variété de Poisson telle que la loi de multiplication de groupe soit un morphisme de Poisson de la variété de Poisson produit $G \times G$ dans G. Lorsque G est un groupe de Poisson-Drinfeld, $\mathcal{g}^*$ est une algèbre de Lie, et les crochets de Lie sur $\mathcal{g}$ et sur $\mathcal{g}^*$ possèdent une propriété de distributivité. On peut construire des structures de Poisson-Drinfeld sur des groupes de Lie à partir de solutions de l'équation de Yang-Baxter généralisée. Les structures de Poisson-Drinfeld définissent des structures bihamiltoniennes ; elles permettent aussi une interprétation des propriétés du groupe des transformations d'habillage.

Poisson-Drinfeld Groups

Introduction.

There are two reasons for studying the Poisson-Drinfeld groups to be defined below. The first is the observation that the Poisson-Drinfeld structures are basic to a geometric understanding of the factorization method and of the properties of the group of dressing transformations (see Semenov-Tian-Shansky [11]), thus they play an important rôle in the mathematical theories of quantization . The basic idea is that the group of dressing transformations is a Poisson-Drinfeld group which acts on the phase space, a Poisson manifold, by a Poisson-Drinfeld action. In view of this, anyone working on nonlinear integrable equations should start thinking in terms of Poisson-Drinfeld actions in addition to the usual Hamiltonian actions.

The second motivation is the theory's close relationship with that of the bihamiltonian structures whose importance in the theory of integrable systems is now generally recognized. (A good reference which relates both topics is an unpublished booklet by Magri and Morosi [8]). Further work is still required to make this relationship more precise and it is in progress.

In section 1 we define the Poisson-Drinfeld groups, and in section 2 we show that there is a Poisson-Drinfeld structure associated with each solution of the "Yang-Baxter equation", i.e., with each "Poisson cocycle" on the Lie algebra of the Lie group. The most useful examples of Poisson-Drinfeld structures may be defined on infinite-dimensional groups (though they are not really Lie groups) of smooth mappings from $\mathbb{R}$ or S^1 or $\mathbb{C}$ into a finite-dimensional Lie group such as $GL(2,\mathbb{R})$. In section 3 we study the Lie-Drinfeld algebras, which are the Lie algebras of Poisson-Drinfeld groups . In section 4 we study the Poisson-Drinfeld groups, and their Lie-Drinfeld algebras, defined by the solutions of the "generalized Yang-Baxter equation", which we propose to call the Poisson-Sklyanin groups and the Lie-Sklyanin algebras, respectively. In section 5 we briefly define Poisson-Drinfeld actions and their infinitesimal counterpart.

The basic reference for what we propose to call Poisson-Drinfeld groups is the concise article by Drinfeld [3]. We have supplied a more intrinsic formulation, the proofs of theorems 1 and 2, and that of theorem 5, which is due to F. Magri. We have generalized some of the properties of the solutions of the Yang-Baxter equation to those of the generalized Yang-Baxter equation. The ideas in Drinfeld's paper have been greatly developed by Semenov-Tian-Shansky in two important articles [10] [11], and in a set of as yet unpublished lecture notes [12]. As related bibliography let us cite articles by Chudnovsky and Chudnovsky [2], Reiman [9], Gelfand and Dorfman [5], Gelfand and Cherednik [4] and Sklyanin [13].

1. Poisson-Drinfeld groups.

By definition, a Poisson-Drinfeld group is a Lie group G with a Poisson structure such that the group multiplication is a Poisson morphism from $G\times G$, equipped with the product Poisson structure, to G. We shall require the following basic definitions which we now state, omitting all the obvious quantifiers here and in the rest of this paper.

1) A Poisson manifold is a manifold M with a Poisson bracket, i.e., a Lie-algebra structure $\{\ ,\ \}_M$ on the vector space $C^\infty(M)$ of smooth, real functions over M such that

$$\{f_1,f_2f_3\}_M = f_2\{f_1,f_3\}_M + f_3\{f_1,f_2\}_M .$$

2) The product of two Poisson manifolds $(M,\{\ ,\ \}_M)$ and $(N,\{\ ,\ \}_N)$ is the Poisson manifold $M\times N$ equipped with the Poisson bracket $\{\ \ \}_{M\times N}$ defined by

$$\{f_1,f_2\}_{M\times N}(x,y) = \{f_1^x,f_2^x\}_N(y) + \{f_1^y,f_2^y\}_M(x),$$

where f_1 and f_2 are smooth functions of $(x,y) \in M\times N$. For each $x \in M$, $y \in N$ we have denoted by f^x (resp., f^y) the smooth function of $y \in N$ (resp., $x \in M$) obtained from a smooth function f by keeping x (resp., y) constant.

3) A <u>Poisson morphism</u> from a Poisson manifold $(M,\{\ ,\ \}_M)$ to a Poisson manifold $(N,\{\ ,\ \}_N)$ is a smooth mapping ϕ from M into N such that

$$\{f_1,f_2\}_N \circ \phi = \{f_1 \circ \phi, f_2 \circ \phi\}_M .$$

We can now formulate precisely and make explicit the definition we stated above.

<u>Definition.</u> <u>A Poisson-Drinfeld group is a Lie group</u> G <u>with a Poisson structure</u> $\{\ ,\ \}_G$ <u>such that the group multiplication</u> $\pi : (x,y) \in G\times G \to \pi(x,y) = x\cdot y \in G$ <u>is a Poisson morphism from</u> $(G\times G,\{\ ,\ \}_{G\times G})$ <u>to</u> $(G,\{\ ,\ \}_G)$.

By definition, π is a Poisson morphism if and only if

$$\{f_1,f_2\}_G \circ \pi = \{f_1 \circ \pi, f_2 \circ \pi\}_{G\times G} ,$$

or, equivalently, if

$$\{f_1,f_2\}_G(\pi(x,y)) = \{f_1 \circ \pi, f_2 \circ \pi\}_{G\times G}(x,y),$$

for all x and y in G, <u>i.e.</u>, if

$$\{f_1,f_2\}_G(x\cdot y) = \{f_1(x\cdot y),f_2(x\cdot y)\}_{G\ \substack{\\ x=\text{constant}}} + \{f_1(x\cdot y),f_2(x\cdot y)\}_{G\ \substack{\\ y=\text{constant}}}$$

Here and below we denote the <u>left and right translation</u> on G by an element x of G by λ_x and ρ_x, respectively. In this notation it is clear that π is a Poisson morphism if and only if

$$(1)\qquad \{f_1,f_2\}_G(x\cdot y) = \{f_1 \circ \lambda_x, f_2 \circ \lambda_x\}_G(y) + \{f_1 \circ \rho_y, f_2 \circ \rho_y\}_G\ (x).$$

Let us introduce the <u>Poisson bivector</u> Λ, a twice contravariant, antisymmetric tensor, defined by

$$\Lambda(df_1,df_2) = \{f_1,f_2\}_G .$$

Then formula (1) is equivalent to

$$\Lambda_{x\cdot y}((df_1)_{x\cdot y},(df_2)_{x\cdot y}) = \Lambda_y(d(f_1\circ\lambda_x)_y,d(f_2\circ\lambda_x)_y) + \Lambda_x(d(f_1\circ\rho_y)_x,d(f_2\circ\rho_y)_x).$$

We denote the tangent map at a point z in G to a smooth map ϕ from G to G by $(T\phi)_z$. Thus

$$\Lambda_z(d(f_1\circ\phi)_z,d(f_2\circ\phi)_z) = \Lambda_z((df_1)_{\phi z}\circ(T\phi)_z,(df_2)_{\phi z}\circ(T\phi)_z)$$

$$= ((T\phi)_z \otimes (T\phi)_z)(\Lambda_z)((df_1)_{\phi z},(df_2)_{\phi z}).$$

Applying this formula with $z = x$ and $\phi = \rho_y$, and with $z = y$ and $\phi = \lambda_x$ to the right-hand side of (1) shows that formula (1) is equivalent to

$$\Lambda_{x\cdot y} = ((T\lambda_x)_y \otimes (T\lambda_x)_y)(\Lambda_y) + ((T\rho_y)_x \otimes (T\rho_y)_x)(\Lambda_x). \tag{2}$$

This condition expresses the fact that the <u>value of</u> Λ <u>at</u> $x\cdot y$ <u>is obtained as the sum of the right translate by</u> y <u>of the value of</u> Λ <u>at</u> x <u>and the left translate by</u> x <u>of the value of</u> Λ <u>at</u> y.

If we denote the direct image of the contravariant vector Λ under the map ϕ by $\phi_*(\Lambda)$, then we may also write (2) as

$$\Lambda_{x\cdot y} = ((\lambda_x)_*\Lambda)_{x\cdot y} + ((\rho_y)_*\Lambda)_{x\cdot y}. \tag{3}$$

We observe that the bivector Λ is neither left- nor right-invariant.

Setting both x and y equal to e, the identity of G, in formula (2) yields $\Lambda_e = 2\Lambda_e$, whence

$$\Lambda_e = 0.$$

Therefore a Poisson structure on G such that G is a Poisson-Drinfeld group is always <u>of rank</u> 0 <u>at</u> e. (Consequently, it may never be a symplectic structure.) If we set $y = x^{-1}$ we obtain

$$\Lambda_x = -((T(\rho_x\circ\lambda_x))_{x^{-1}} \otimes (T(\rho_x\circ\lambda_x))_{x^{-1}})(\Lambda_{x^{-1}}).$$

2. Construction of Poisson-Drinfeld groups from solutions of the generalized Yang-Baxter equation.

Let g be a real or complex Lie algebra and g^* be its dual. We denote elements in g by Roman letters X, Y,... and elements in g^* by Greek letters $\xi,\eta,\ldots$.

If g is finite-dimensional, a linear map r from g^* to g can be considered as an element in $g \otimes g$ which we shall also denote by r. (r may also be regarded as a linear form on $g^* \otimes g^*$.)

Let $r : g^* \to g$ be antisymmetric, i.e., r satisfies

$$<\xi,r(\eta)> = -<\eta,r(\xi)>.$$

We consider the trilinear antisymmetric form $\langle\![r,r]\!\rangle$ on g^* defined by

$$\langle\![r,r]\!\rangle(\xi_1,\xi_2,\xi_3) = \oint<\xi_3,[r(\xi_1),r(\xi_2)]>$$

where $\oint$ denotes cyclic permutation over the indices 1, 2, 3. ($\langle\![r,r]\!\rangle$ is an algebraic version of the Schouten bracket, as we shall see shortly.) We shall consider the antisymmetric maps r that satisfy the condition

(YB) $$\langle\![r,r]\!\rangle = 0.$$

It is easy to see that if r is one-to-one, (YB) is the condition for the inverse mapping $r^{-1} : g \to g^*$ to be a symplectic cocycle in the sense of Souriau [14].

Drinfeld [3] called the condition (YB), which he wrote in a basis, the classical Yang-Baxter equation. (Here "classical" is opposed to "quantum" , and the reference is to the Yang-Baxter equation which arises in statistical mechanics.) We shall simply call it the Yang-Baxter equation. Drinfeld essentially proved that this condition is equivalent to the Jacobi identity for a right- or left-invariant Poisson bracket on a Lie group G with Lie algebra g, while in [5] Gelfand and Dorfman introduced (YB) for kernels depending upon two complex variables in the form

$$\oint [r^{12}(u_1,u_2), r^{13}(u_1,u_3)] = 0,$$

calling it, too, the classical Yang-Baxter equation, and they showed that it was the expression in coordinates of a Hamiltonian condition, i.e., that it was equivalent to the vanishing of the Schouten bracket of a Hamiltonian operator, in a suitable algebraic sense. They called the solutions of the Yang-Baxter equation nondegenerate, generalized 2-cocycles. Magri and Morosi in their unpublished book [8], independently introduced condition (YB) in its present form and called its solutions Poisson cocycles, and developed applications. They also gave the following example.

Let g be the smooth maps from $\mathbb{R}$ to $\mathfrak{gl}(2,\mathbb{R})$ which are rapidly decreasing at infinity. Then

$$r(\xi)(t) = \int_{-\infty}^{t} \xi(\tau)d\tau$$

is a Poisson cocycle, i.e., a solution of the Yang-Baxter equation. In [1] Belavin and Drinfeld give results concerning the determination of all Poisson-Drinfeld structures on the group of smooth mappings from $\mathbb{C}$ into a simple Lie group. Other examples are discussed by Kulish and Sklyanin [7].

Let G be a Lie group with the Lie algebra g. We now show that each solution r of the Yang-Baxter equation on g defines a Poisson-Drinfeld structure on G, i.e., the difference of the left- and right-invariant Poisson structures generated by r.

Let r be a solution of the Yang-Baxter equation, i.e., an antisymmetric linear map from g^* to g satisfying the Yang-Baxter condition (YB). From r we can construct a left-invariant Poisson structure whose Poisson bivector Λ_r^λ is defined by

$$\Lambda^\lambda_{r,x} = ((T\lambda_x)_e \otimes (T\lambda_x)_e)(r).$$

In other words, the left-invariant Poisson bracket $\{\ ,\ \}_r^\lambda$ associated with r is defined by

$$\{f_1,f_2\}_r^\lambda(x) = \Lambda_{r,x}^\lambda((df_1)_x,(df_2)_x) = r((df_1)_x \circ (T\lambda_x)_e,(df_2)_x \circ (T\lambda_x)_e)$$

or, equivalently, by

$$\{f_1,f_2\}_r^\lambda(x) = r(d(f_1 \circ \lambda_x)_e, d(f_2 \circ \lambda_x)_e).$$

For smooth functions f_1 and f_2 on G, $d(f_1 \circ \lambda_x)_e$ and $d(f_2 \circ \lambda_x)_e$ are elements of $\mathcal{g}^*$, and the above is well defined. Moreover, this bracket is antisymmetric since r is. The fact that it satisfies the Jacobi identity is precisely equivalent to the Yang-Baxter condition for r. So $\{\ ,\ \}_r^\lambda$ is indeed a Poisson bracket on G. (See the appendix.)

In exactly the same fashion, we construct from r the right-invariant Poisson structure with Poisson bivector Λ_r^ρ defined by

$$\Lambda_{r,x}^\rho = ((T\rho_x)_e \otimes (T\rho_x)_e)(r)$$

and Poisson bracket

$$\{f_1,f_2\}_r^\rho(x) = r(d(f_1 \circ \rho_x)_e, d(f_2 \circ \rho_x)_e).$$

Moreover, these left- and right-invariant Poisson brackets are indeed compatible, i.e., their sum (or any linear combination of them) is still a Poisson bracket.

More generally, for any solutions r and r' of the Yang-Baxter equation, $\{\ ,\ \}_r^\lambda$ and $\{\ ,\ \}_{r'}^\rho$ are compatible Poisson brackets. When one of the Poisson structures is invertible this result follows from proposition 2.1 of Magri and Morosi [8]. Compare also theorem 1 of Semenov-Tian-Shansky [11]. A proof is given in the appendix.

This discussion is summarized in

Theorem 1.

(i) For any solution r of the Yang-Baxter equation on $\mathcal{g}$, $\{\ ,\ \}_r^\lambda$ and $\{\ ,\ \}_r^\rho$ are Poisson structures on G.

(ii) For any solutions r and r' of the Yang-Baxter equation on $\mathcal{g}$, the Poisson structures $\{\ ,\ \}_r^\lambda$ and $\{\ ,\ \}_{r'}^\rho$ on G are compatible.

By this theorem we know that, for any solution r of the Yang-Baxter equation, the bivector

$$\Lambda_r = \Lambda_r^{\lambda} - \Lambda_r^{\rho}$$

is a Poisson bivector or, equivalently, that

$$\{f_1,f_2\}_r = \{f_1,f_2\}_r^{\lambda} - \{f_1,f_2\}_r^{\rho}$$

defines a Poisson structure $\{\ ,\ \}_r$ on G. Moreover, this Poisson structure on G is a Poisson-Drinfeld structure as we shall prove below.

In section 4 we shall consider antisymmetric linear maps r from $\mathcal{g}^*$ to $\mathcal{g}$ such that $\langle\!\{r,r\}\!\rangle$ is invariant under the adjoint representation of $\mathcal{g}$. Such a mapping is called a solution of the generalized Yang-Baxter equation (GYB). While it is clear that the assertions of theorem 1 are not valid for solutions of the generalized Yang-Baxter equation , we shall prove that if we construct the bracket

$$\{\ ,\ \}_r = \{\ ,\ \}_r^{\lambda} - \{\ ,\ \}_r^{\rho}$$

from any solution r of the generalized Yang-Baxter equation, it is a Poisson-Drinfeld bracket. This generalization is important because the solutions to the modified Yang-Baxter equation, which define the Lie-Baxter algebras [11], satisfy the generalized Yang-Baxter equation but not the Yang-Baxter equation. These Lie-Baxter algebras permit a simple algebraic interpretation of the infinitesimal version of the Riemann-Hilbert factorization problem [10]. For the relationships between Lie-Drinfeld algebras (to be defined below), Lie-Baxter algebras [11] and double Lie algebras [10], see [6].

Theorem 2. *Let r be an antisymmetric map from $\mathcal{g}^*$ to $\mathcal{g}$ such that $\langle\!\{r,r\}\!\rangle$ is invariant under the adjoint action of G. Then $\{\ ,\ \}_r$ is a Poisson-Drinfeld structure on G.*

Proof : It follows from the proof of theorem 1 given in the appendix that for any antisymmetric map r from $\mathcal{g}^*$ to $\mathcal{g}$

$$\langle[\Lambda_r^\lambda-\Lambda_r^\rho,\Lambda_r^\lambda-\Lambda_r^\rho]\rangle_x = -((T\rho_x)_e\otimes(T\rho_x)_e\otimes(T\rho_x)_e)((Ad_x\otimes Ad_x\otimes Ad_x)\langle[r,r]\rangle - \langle[r,r]\rangle).$$

Thus $\Lambda_r=\Lambda_r^\lambda-\Lambda_r^\rho$ is a Poisson bivector if and only if $\langle[r,r]\rangle$ is invariant under the adjoint action of G. Moreover, equality (1) for $\{\ ,\ \}_r$ is proved by a short calculation using $\lambda_x\circ\rho_y = \rho_y\circ\lambda_x$ and $\lambda_{x\cdot y} = \lambda_x\circ\lambda_y$, $\rho_{x\cdot y} = \rho_y\circ\rho_x$ in order to evaluate $\{f_1,f_2\}_r(x\cdot y)$ and $\{f_1\circ\lambda_x,f_2\circ\lambda_x\}_r(y) + \{f_1\circ\rho_y,f_2\circ\rho_y\}_r(x)$.

<u>In particular, $\{\ ,\ \}_r$ is a Poisson-Drinfeld structure on G when r is a solution of the Yang-Baxter equation or, if G is connected, when r is a solution of the generalized Yang-Baxter equation.</u>

3. Lie-Drinfeld algebras.

Whenever G is a Poisson-Drinfeld group with Poisson bracket $\{\ ,\ \}_G$, the vector space $\mathfrak{g}^*$ has a Lie-algebra structure $[\ ,\ ]_{\mathfrak{g}^*}$ defined in the following way. Let ξ_1 and ξ_2 be in $\mathfrak{g}^*$. We can choose smooth functions f_1 and f_2 on G such that $(df_1)_e = \xi_1$, $(df_2)_e = \xi_2$. Then we set

$$[\xi_1,\xi_2]_{\mathfrak{g}^*} = (d(\{f_1,f_2\}_G))_e.$$

The result is independent of the choice of f_1 and f_2 —this follows from the next lemma—, and the bracket thus defined, which is obviously antisymmetric, satisfies the Jacobi identity since $\{\ ,\ \}_G$ does.

<u>Lemma. Let Λ be the Poisson bivector of the Poisson-Drinfeld structure $\{\ ,\ \}_G$. Let ℓ be the smooth mapping from G to $\mathfrak{g}\otimes\mathfrak{g}$ defined by</u>

$$\ell(x) = ((T\rho_x^{-1})_x \otimes (T\rho_x^{-1})_x)\Lambda_x.$$

<u>We set $\varepsilon = (T\ell)_e$ and we denote by $\gamma = \varepsilon^*$ the transpose of ε. Then</u>

$$[\xi_1,\xi_2]_{\mathfrak{g}^*} = \gamma(\xi_1 \otimes \xi_2).$$

Proof : For x in G, Λ_x is in $T_xG \otimes T_xG$, so $\ell(x)$ is indeed in $T_eG \otimes T_eG = \mathfrak{g} \otimes \mathfrak{g}$. The tangent mapping at e, $\varepsilon = (T\ell)_e$, of the smooth mapping ℓ from G to $\mathfrak{g} \otimes \mathfrak{g}$ maps $\mathfrak{g}$ into $\mathfrak{g} \otimes \mathfrak{g}$. Thus for X in $\mathfrak{g}$, and ξ_1 and ξ_2 in $\mathfrak{g}^*$, $\varepsilon(X)(\xi_1 \otimes \xi_2)$ is a number which, by the definition of the transpose $\gamma = \varepsilon^*$ of ε, is equal to $\langle\gamma(\xi_1 \otimes \xi_2), X\rangle$.

Let ξ_1 and ξ_2 be in $\mathfrak{g}^*$, and let f_1 and f_2 be smooth functions on G such that $(df_1)_e = \xi_1$ and $(df_2)_e = \xi_2$. By the definition of ℓ,

$$\{f_1, f_2\}_G(x) = \Lambda_x((df_1)_x, (df_2)_x) = ((T\rho_x)_e \otimes (T\rho_x)_e)(\ell(x))((df_1)_x, (df_2)_x)$$
$$= \ell(x)((T\rho_x)_e^*(df_1)_x \otimes (T\rho_x)_e^*(df_2)_x).$$

When X is an arbitrary element of $\mathfrak{g}$, we may evaluate $\langle[\xi_1, \xi_2]_{\mathfrak{g}^*}, X\rangle$ by taking the derivative of the function $\{f_1, f_2\}_G$ in the direction of X. Taking into account the fact that $\ell(e) = 0$ we obtain

$$\langle[\xi_1, \xi_2]_{\mathfrak{g}^*}, X\rangle = \langle(d\{f_1, f_2\}_G)_e, X\rangle = (T\ell)_e(X)((df_1)_e \otimes (df_2)_e)$$
$$= \varepsilon(X)(\xi_1 \otimes \xi_2) = \langle\gamma(\xi_1 \otimes \xi_2), X\rangle,$$

thus proving the lemma.

The next result is the key to the relationship between Poisson-Drinfeld groups and Lie-Drinfeld algebras.

Theorem 3. <u>Let G be a Poisson-Drinfeld group with Lie algebra $\mathfrak{g}$, and let $[\ ,\ ]_{\mathfrak{g}^*}$ be the Lie algebra bracket on $\mathfrak{g}^*$ defined above. The Lie algebra structures $[\ ,\ ]$ on $\mathfrak{g}$ and $[\ ,\ ]_{\mathfrak{g}^*}$ on $\mathfrak{g}^*$ satisfy the distributivity condition</u>

$$(*) \quad \langle[\xi_1, \xi_2]_{\mathfrak{g}^*}, [X_1, X_2]\rangle = -\langle[ad^*_{X_1}\xi_1, \xi_2]_{\mathfrak{g}^*}, X_2\rangle - \langle[\xi_1, ad^*_{X_1}\xi_2]_{\mathfrak{g}^*}, X_2\rangle$$
$$+ \langle[ad^*_{X_2}\xi_1, \xi_2]_{\mathfrak{g}^*}, X_1\rangle + \langle[\xi_1, ad^*_{X_2}\xi_2]_{\mathfrak{g}^*}, X_1\rangle.$$

Proof : The bivector Λ satisfies the relation (2) if and only if ℓ satisfies

$$\ell(x\cdot y) = (Ad_x \otimes Ad_x)(\ell(y)) + \ell(x),$$

i.e., if ℓ is a 1-cocycle of G with values in $g \otimes g$ relative to the adjoint action of G. The mapping $\varepsilon = (T\ell)_e : g \to g \otimes g$ is then a 1-cocycle of g with values in $g \otimes g$ relative to the adjoint action of g (and conversely, if G is connected and simply connected), i.e.,

$$\varepsilon([X,Y]) = (ad_X \otimes 1 + 1 \otimes ad_X)(\varepsilon(Y)) - (ad_Y \otimes 1 + 1 \otimes ad_Y)(\varepsilon(X)).$$

By the lemma we know that

$$<[\xi_1,\xi_2]_{g^*}, [X_1,X_2]> = <\gamma(\xi_1 \otimes \xi_2), [X_1,X_2]> = <\xi_1 \otimes \xi_2, \varepsilon([X_1,X_2])>.$$

When we express the fact that $\varepsilon = \gamma^*$ is a 1-cocycle of g with values in $g \otimes g$ we obtain the distributivity property (*) as follows :

$$\begin{aligned}
<[\xi_1,\xi_2]_{g^*}, [X_1,X_2]> &= <\xi_1 \otimes \xi_2, ad_{X_1}\gamma^* X_2> - <\xi_1 \otimes \xi_2, ad_{X_2}\gamma^* X_1> \\
&= -<ad^*_{X_1}(\xi_1 \otimes \xi_2), \gamma^* X_2> + <ad^*_{X_2}(\xi_1 \otimes \xi_2), \gamma^* X_1> \\
&= -<ad^*_{X_1}\xi_1 \otimes \xi_2, \gamma^* X_2> - <\xi_1 \otimes ad^*_{X_1}\xi_2, \gamma^* X_2> \\
&\quad + <ad^*_{X_2}\xi_1 \otimes \xi_2, \gamma^* X_1> + <\xi_1 \otimes ad^*_{X_2}\xi_2, \gamma^* X_1> \\
&= -<\gamma(ad^*_{X_1}\xi_1 \otimes \xi_2), X_2> - <\gamma(\xi_1 \otimes ad^*_{X_1}\xi_2), X_2> \\
&\quad + <\gamma(ad^*_{X_2}\xi_1 \otimes \xi_2), X_1> + <\gamma(\xi_1 \otimes ad^*_{X_2}\xi_2), X_1> \\
&= -<[ad^*_{X_1}\xi_1, \xi_2]_{g^*}, X_2> - <[\xi_1, ad^*_{X_1}\xi_2]_{g^*}, X_2> \\
&\quad + <[ad^*_{X_2}\xi_1, \xi_2]_{g^*}, X_1> + <[\xi_1, ad^*_{X_2}\xi_2]_{g^*}, X_1>.
\end{aligned}$$

We observe that (*) is indeed an intrinsic form of the "compatibility condition" (2), p. 69, of Drinfeld [3]. Magri has remarked that this condition (*) expresses the fact that $\gamma \in \Lambda^2(g) \otimes g^*$ is closed in a suitably defined sense.

We shall now state and prove the converse of theorem 3 for connected and simply connected groups. Let us first formally define Lie bialgebras or Lie-Drinfeld structures on Lie algebras.

Definition. A Lie bialgebra is a Lie algebra $(g,[\ ,\ ])$ together with a Lie-algebra structure $[\ ,\ ]_{g^*}$ on g^* that satisfies the distributivity property (*).

Alternatively, we propose to define a Lie-Drinfeld algebra as a Lie algebra $(g,[\ ,\ ])$ with an antisymmetric 1-cocycle ε of g with values in $g \otimes g$ relative to the adjoint action of g such that $\gamma = \varepsilon^*$ satisfies

$$\oint \gamma(\xi_3 \otimes \gamma(\xi_1 \otimes \xi_2)) = 0,$$

where $\oint$ denotes the sum over the cyclic permutations of 1, 2, 3.

It is easy to see [6] that g is a Lie bialgebra if and only if it is a Lie-Drinfeld algebra.

Theorem 4. Each Poisson-Drinfeld structure on a Lie group G defines a Lie-Drinfeld structure on g. Conversely, if G is connected and simply connected, each Lie-Drinfeld structure $(g,[\ ,\ ]_{g^*})$ on g defines a Poisson-Drinfeld structure $\{\ ,\ \}_G$ on G in such a fashion that the associated Lie-Drinfeld structure on g is $(g,[\ ,\ ]_{g^*})$.

Proof : The first part of the theorem is just a reformulation of theorem 3. To prove the converse we denote the given Lie algebra structure on g^* by $[\ ,\ ]_{g^*}$, and we define $\gamma : g^* \otimes g^* \to g^*$ by $[\xi_1,\xi_2]_{g^*} = \gamma(\xi_1 \otimes \xi_2)$. We note that, by the calculation in the proof of theorem 3, asserting property (*) is equivalent to requiring that $\varepsilon = \gamma^*$ be a 1-cocycle of g with values in $g \otimes g$ relative to the adjoint action of g. But if G is connected and simply connected, each such 1-cocycle ε of g defines a unique 1-cocycle ℓ of G with values in $g \otimes g$ relative to the adjoint action of G, satisfying $(T\ell)_e = \varepsilon$. Indeed $(T\ell)_x = (Ad_x \otimes Ad_x)(\varepsilon \circ (T\lambda_x^{-1})_x)$. We then define a bivector Λ by

$$\Lambda_x = ((T\rho_x)_e \otimes (T\rho_x)_e)(\ell(x)),$$

and an antisymmetric bracket $\{\ ,\ \}_G$ on the smooth functions on G by

$$\{f_1,f_2\}_G = \Lambda(df_1,df_2).$$

As we remarked in the proof of theorem 3, since ℓ is a cocycle of G, Λ satisfies relation (2).

It remains to be proved, in view of the Jacobi identity for the Lie bracket $[\ ,\ ]_{g^*}$ on g^* and the cocycle property of ε, that the bracket $\{\ ,\ \}_G$ satisfies the Jacobi identity, i.e., that the Schouten bracket $[\![\Lambda,\Lambda]\!]$ vanishes. We first observe that it vanishes at e since Λ does. The result will follow if we prove that $[\![\Lambda,\Lambda]\!]$ is invariant under any right-invariant vector field.

Let ξ_1,ξ_2 and ξ_3 be elements in g^*. The Jacobi identity for $[\ ,\ ]_{g^*}$ is

$$\oint \gamma(\xi_3 \otimes \gamma(\xi_1 \otimes \xi_2)) = 0$$

or

$$\oint \langle \xi_3 \otimes \gamma(\xi_1 \otimes \xi_2), \varepsilon(X)\rangle = 0 \quad \text{for each } X \text{ in } g,$$

or, when we regard ε as a bilinear map from $g\times g^*$ to g,

$$\oint \langle \xi_2, \varepsilon(\varepsilon(X;\xi_3);\xi_1)\rangle = 0, \text{ for each } X \text{ in } g.$$

Therefore theorem 4 is a consequence of the following result which is due to Magri (unpublished).

Theorem 5. *Let ℓ be an antisymmetric 1-cocycle of G with values in $g \otimes g$ relative to the adjoint action of G on $g \otimes g$ and let Λ be the bivector obtained from ℓ by right translations. For any ξ in g^* we denote the corresponding right-invariant 1-form by ξ^ρ. For any X in g we denote the corresponding right-invariant vector field by X^ρ. Then*

$$X^\rho\cdot([\![\Lambda,\Lambda]\!](\xi_1^\rho,\xi_2^\rho,\xi_3^\rho))(x) = -\oint \langle \bar\xi_2(x), \varepsilon(\varepsilon(\bar X(x);\bar\xi_3(x));\bar\xi_1(x))\rangle$$

where $\bar\xi_i(x) = \mathrm{Ad}^*_{x^{-1}}\,\xi_i$ $(i = 1,2,3)$ and $\bar X(x) = \mathrm{Ad}_{x^{-1}} X$.

The proof of this theorem is given in the appendix.

4. Poisson-Sklyanin groups and Lie-Sklyanin algebras.

In this section, let r be a solution of the generalized Yang-Baxter equation (see the end of section 2) on g. Let us consider

$$\Lambda_r = \Lambda_r^\lambda - \Lambda_r^\rho$$

as in section 2, and we set

$$\{f_1,f_2\}_r = \Lambda_r(df_1,df_2).$$

If r is a solution of the generalized Yang-Baxter equation on $\mathcal{g}$, and if G is connected, then by theorem 2, $\{\ ,\ \}_r$ is a Poisson-Drinfeld structure on G. (To prove that $\{\ ,\ \}_r$ is a Poisson-Drinfeld structure on G, we could also have used the fact that the bracket defined below is a Lie-algebra bracket on $\mathcal{g}^*$ which satisfies the distributivity condition (*) [6] and the calculation in the proofs of theorems 4 and 5.)

<u>Theorem</u> 6 : <u>The Lie-Drinfeld structure on</u> $\mathcal{g}$ <u>associated with the Poisson-Drinfeld structure</u> $\{\ ,\ \}_r$ <u>on</u> G <u>is defined by</u>

$$[\xi_1,\xi_2]_{\mathcal{g}^*} = ad^*_{r\xi_1}\xi_2 - ad^*_{r\xi_2}\xi_1 .$$

<u>Proof</u> : Λ_r is the bivector on G defined by

$$\Lambda_x = \Lambda_{r,x} = \Lambda^\lambda_{r,x} - \Lambda^\rho_{r,x} = ((T\lambda_x)_e \otimes (T\lambda_x)_e)r - ((T\rho_x)_e \otimes (T\rho_x)_e)r,$$

whence $\ell(x) = (Ad_x \otimes Ad_x)r - r$.
Computing $(T\ell)_e = \varepsilon$, and setting $\gamma = \varepsilon^*$, we find that

$$\langle\gamma(\xi_1 \otimes \xi_2),X\rangle = \frac{d}{dt}\langle\xi_1 \otimes \xi_2,(Ad_{\exp(tX)} \otimes Ad_{\exp(tX)})r - r\rangle|_{t=0}$$

$$= \frac{d}{dt}(r(Ad^*_{\exp(-tX)}\xi_1 \otimes Ad^*_{\exp(-tX)}\xi_2))|_{t=0}$$

$$= -r(ad^*_X\xi_1 \otimes \xi_2) - r(\xi_1 \otimes ad^*_X\xi_2) = \langle ad^*_X\xi_1,r\xi_2\rangle - \langle ad^*_X\xi_2,r\xi_1\rangle$$

$$= \langle ad^*_{r\xi_1}\xi_2 - ad^*_{r\xi_2}\xi_1,X\rangle.$$

We observe that

$$<[\xi_1,\xi_2]_{g^*},X> = <\xi_1,[r\xi_2,X]> - <\xi_2,[r\xi_1,X]>.$$

Let us denote a basis of g by (e_i), and the structure constants of g in this basis by C^k_{ij}. Let (ε^i) be the dual basis, and let (r^{ij}) be the components of r in this basis. The structure constants of the Lie-algebra structure $[\ ,\ ]_{g^*}$ on g^* defined by r are given in the basis (ε^i) by

$$f^{ij}_k = <[\varepsilon^i,\varepsilon^j]_{g^*},e_k> = <\varepsilon^i,[r\varepsilon^j,e_k]> - <\varepsilon^j,[r\varepsilon^i,e_k]>$$

$$= r^{jm}<\varepsilon^i,[e_m,e_k]> - r^{im}<\varepsilon^j,[e_m,e_k]> = r^{jm}C^i_{mk} - r^{im}C^j_{mk}.$$

This local formula coincides with that of Drinfeld [3], § 6. Therefore theorem 6 is an invariant formulation of Drinfeld's statement.

We propose to call those Poisson-Drinfeld groups (resp. Lie-Drinfeld algebras) which are defined by solutions of the generalized Yang-Baxter equation Poisson-Sklyanin groups (resp. Lie-Sklyanin algebras). In fact, g is a Lie-Sklyanin algebra if and only if it is an exact Lie-Drinfeld algebra, i.e., the 1-cocycle ε is exact.

5. Poisson-Drinfeld actions.

Once Poisson-Drinfeld groups are defined, one can study the Poisson-Drinfeld actions of such a group on a Poisson manifold in the following sense.

Definition. A Poisson-Drinfeld action of a Poisson-Drinfeld group G on a Poisson manifold M is a Lie-group action $\Phi : G\times M \to M$ such that Φ is a Poisson morphism from $G\times M$ (equipped with the product Poisson structure) to M.

For (g,x) in $G\times M$, we set $\Phi(g,x) = g^M(x) = x^G(g)$. Thus for each g in G, g^M is a diffeomorphism of M, and for each x in M, x^G is a smooth map from G to M. The Lie-group action Φ is a Poisson-Drinfeld action if and only if

$$g^{M^*}\{f_1,f_2\}_M(x) = \{x^{G^*}f_1,x^{G^*}f_2\}_G(g) + \{g^{M^*}f_1,g^{M^*}f_2\}_M(x).$$

In fact, $g^{M^*}\{f_1,f_2\}_M(x) = \{f_1,f_2\}_M(g^M(x)) = \{f_1,f_2\}_M(\Phi(g,x))$, while by the definition of the product Poisson structure on $G\times M$, the right-hand side is $\{f_1 \circ \Phi, f_2 \circ \Phi\}_{G\times M}(g,x)$.

Differentiating this formula yields a characterization of the infinitesimal generators of the Poisson-Drinfeld actions.

<u>Theorem</u> 7. <u>Let</u> X <u>be in</u> $\mathfrak{g}$, <u>and let</u> X^M <u>be the corresponding infinitesimal generator of the Poisson-Drinfeld action</u> Φ. <u>Then</u>

$$X^M\cdot\{f_1,f_2\}_M = \{X^M\cdot f_1,f_2\}_M + \{f_1,X^M\cdot f_2\}_M - \langle[\xi_{f_1},\xi_{f_2}]_{\mathfrak{g}^*},X\rangle ,$$

<u>where we have set</u>

$$\xi_f(x) = (d(f\circ x^G))_e$$

<u>for a smooth function</u> f <u>on</u> M.

<u>Proof</u> : The result follows from

$$\frac{d}{dt}\{f_1\circ x^G,f_2\circ x^G\}_G(\exp(-tX))\Big|_{t=0} = -\langle(d\{f_1\circ x^G,f_2\circ x^G\})_e,X\rangle$$

$$= -\langle[\xi_{f_1}(x),\xi_{f_2}(x)]_{\mathfrak{g}^*},X\rangle.$$

In turn, the formula in theorem 7 can be used to define an infinitesimal Poisson-Drinfeld action of a Lie-Drinfeld algebra on a Poisson manifold.

It is clear from the above formulae that in the nontrivial case, i.e., when G is not equipped with the trivial Poisson-Drinfeld structure defined by $r = 0$,

1) Poisson-Drinfeld actions are not the same as Poisson actions (often called Hamiltonian actions). In fact, when Φ is a Poisson action, each mapping g^M is assumed to be a Poisson diffeomorphism, i.e.,

$$\{f_1,f_2\}_M \circ g^M = \{f_1\circ g^M, f_2\circ g^M\}_M,$$

while, for a Poisson-Drinfeld action, $\{f_1,f_2\}_M \circ g^M$ contains an additional term, and

2) the infinitesimal Poisson-Drinfeld actions are not the same as the infinitesimal Poisson actions. In fact, each infinitesimal generator X^M of a Poisson action is a derivation of the Poisson bracket (i.e., in the symplectic case, a locally Hamiltonian vector field) satisfying

$$X^M\cdot\{f_1,f_2\}_M = \{X^M\cdot f_1,f_2\}_M + \{f_1,X^M\cdot f_2\}_M,$$

while, for any infinitesimal generator M of a Poisson-Drinfeld action, $X^M\cdot\{f_1,f_2\}_M$ contains an additional term.

Thus new types of vector fields on Poisson manifolds, in addition to the familiar locally Hamiltonian vector fields, arise when Poisson-Drinfeld actions are considered.

Semenov-Tian-Shansky has shown that the action of the group of dressing transformations on phase-space is a Poisson-Drinfeld action in the discrete [11] as well as the continuous [12] cases.

The preceding exposition suggests that the geometry of the Poisson-Drinfeld groups is a new and rich domain for investigation and that it has applications to problems in the theory of nonlinear evolution equations.

Appendix

Proofs of theorems 1 and 5.

Notations. Let us denote $(T\rho_x)_e X$ by X_x^ρ and $(T\lambda_x)_e X$ by X_x^λ. Let us set

$$\xi_x^\rho = (T\rho_x)_e^{*-1}\,\xi = \xi\circ(T\rho_x)_x^{-1}$$

and

$$\xi_x^\lambda = (T\lambda_x)_e^{*-1}\,\xi = \xi\circ(T\lambda_x)_x^{-1}.$$

By convention, $(\mathrm{Ad}_x)^* = \mathrm{Ad}^*_{x^{-1}}$ and $(\mathrm{ad}_X)^* = -\mathrm{ad}^*_X$.

Proof of theorem 1.

Let r be an antisymmetric linear map from $\mathcal{g}^*$ to $\mathcal{g}$, and let $\Lambda^\rho_{r,x} = ((T\rho_x)_e \otimes (T\rho_x)_e)r$. For each ξ in $\mathcal{g}^*$, the vector field $X_\xi = \#_{\Lambda_r^\rho}\xi^\rho$ associated with the right-invariant 1-form ξ^ρ is right-invariant. In fact, $X_\xi = (r\xi)^\rho$. The Lie derivative of X_η with respect to X_ξ is therefore

$$L_{X_\xi}X_\eta = [X_\xi,X_\eta] = [(r\xi)^\rho,(r\eta)^\rho] = -[r\xi,r\eta]^\rho.$$

We now compute the Schouten bracket $[\![\Lambda_r^\rho,\Lambda_r^\rho]\!]$ on three right-invariant 1-forms, ξ_1^ρ, ξ_2^ρ and ξ_3^ρ. By definition,

$$[\![\Lambda_r^\rho,\Lambda_r^\rho]\!](\xi_1^\rho,\xi_2^\rho,\xi_3^\rho) = -\oint \langle L_{X_{\xi_1}}\xi_2^\rho, X_{\xi_3}\rangle.$$

Since $\langle\xi_2^\rho, X_{\xi_3}\rangle = \langle\xi_2^\rho,(r\xi_3)^\rho\rangle$ is a constant its Lie derivative vanishes. Therefore

$$\begin{aligned}[\![\Lambda_r^\rho,\Lambda_r^\rho]\!](\xi_1^\rho,\xi_2^\rho,\xi_3^\rho) &= \oint \langle\xi_2^\rho, L_{X_{\xi_1}}X_{\xi_3}\rangle = \oint \langle\xi_2^\rho,[r\xi_3,r\xi_1]^\rho\rangle \\ &= \oint \langle\xi_2,[r\xi_3,r\xi_1]\rangle \\ &= [\![r,r]\!](\xi_1,\xi_2,\xi_3),\end{aligned}$$

and

$$[\Lambda_r^\rho,\Lambda_r^\rho]_x = ((T\rho_x)_e \otimes (T\rho_x)_e \otimes (T\rho_x)_e)\,[r,r].$$

In the same fashion we prove that

$$[\Lambda_r^\lambda,\Lambda_r^\lambda]_x = -((T\lambda_x)_e \otimes (T\lambda_x)_e \otimes (T\lambda_x)_e)\,[r,r].$$

Therefore Λ_r^ρ (resp. Λ_r^λ) is a Poisson tensor if and only if r is a solution of the Yang-Baxter equation, and we have proved (i) of theorem 1.

We shall now prove that if r and r' are two antisymmetric maps from g^* to g, then

$$[\Lambda_r^\rho,\Lambda_{r'}^\lambda] = 0\ .$$

For each ξ in g^*, we set

$$Z_\xi = \#_{\Lambda_r^\lambda}\,\xi^\rho, \qquad \text{and} \qquad Z'_\xi = \#_{\Lambda_{r'}^\lambda}\,\xi^\rho\ .$$

Then

$$Z_\xi(x) = (r(\mathrm{Ad}^*_{x^{-1}}\,\xi))^\lambda_x\ .$$

We evaluate

$$\begin{aligned} X^\rho\cdot\langle\xi^\rho,Z_\eta\rangle(x) &= X^\rho\cdot\langle\xi,\mathrm{Ad}_x r\,\mathrm{Ad}^*_{x^{-1}}\,\eta\rangle \\ &= \langle\xi,\mathrm{ad}_X\,\mathrm{Ad}_x r\,\mathrm{Ad}^*_{x^{-1}}\eta - \mathrm{Ad}_x r\,\mathrm{Ad}^*_{x^{-1}}\,\mathrm{ad}^*_X\eta\rangle \\ &= -\langle(\mathrm{ad}^*_X\xi)^\rho_x,Z_\eta(x)\rangle + \langle(\mathrm{ad}^*_X\eta)^\rho_x,Z_\xi(x)\rangle, \end{aligned}$$

where we have used the antisymmetry of r. Thus

$$X^\rho\cdot\langle\xi^\rho,Z_\eta\rangle = -\langle(\mathrm{ad}^*_X\xi)^\rho,Z_\eta\rangle + \langle(\mathrm{ad}^*_X\eta)^\rho,Z_\xi\rangle.$$

Using this formula, the fact that

$$L_{X^\rho}\,\xi^\rho = -(\mathrm{ad}^*_X\xi)^\rho$$

and that $\langle\xi^\rho,X^\rho\rangle$ is a constant, we have proven that

$$<\xi^\rho, [X^\rho, Z_\eta]> = <(ad^*_X\eta)^\rho, Z_\xi>.$$

We now compute the Schouten bracket $<[\Lambda^\rho_r, \Lambda^\lambda_{r'}]>$ on three right-invariant 1-forms. Using the last two formulae, we obtain

$$2<[\Lambda^\rho_r, \Lambda^\lambda_{r'}]>(\xi^\rho_1, \xi^\rho_2, \xi^\rho_3)$$

$$= -\oint(<L_{X_{\xi_1}}\xi^\rho_2, Z'_{\xi_3}> + <L_{Z'_{\xi_1}}\xi^\rho_2, X_{\xi_3}>)$$

$$= -\oint(<L_{(r\xi_1)^\rho}\xi^\rho_2, Z'_{\xi_3}> + <L_{Z'_{\xi_1}}\xi^\rho_2, (r\xi_3)^\rho>)$$

$$= -\oint(<L_{(r\xi_1)^\rho}\xi^\rho_2, Z'_{\xi_3}> + <\xi^\rho_2, [(r\xi_3)^\rho, Z'_{\xi_1}]>)$$

$$= \oint(<(ad^*_{r\xi_1}\xi_2)^\rho, Z'_{\xi_3}> - <(ad^*_{r\xi_3}\xi_1)^\rho, Z'_{\xi_2}>) = 0,$$

which proves (ii) of theorem 1 since, for any constants α and β,

$$<[\alpha\Lambda^\rho_r + \beta\Lambda^\lambda_{r'}, \alpha\Lambda^\rho_r + \beta\Lambda^\lambda_{r'}]> = \alpha^2<[\Lambda^\rho_r, \Lambda^\rho_r]> + 2\alpha\beta<[\Lambda^\rho_r, \Lambda^\lambda_{r'}]> + \beta^2<[\Lambda^\lambda_{r'}, \Lambda^\lambda_{r'}]> = 0$$

when r and r' are solutions of the Yang-Baxter equation.

Proof of theorem 5 (F. Magri).

Let ℓ be an antisymmetric 1-cocycle of G with values in $g \otimes g$ relative to the adjoint action of G. From the cocycle condition $\ell(x\cdot y) = Ad_x\,\ell(y)\,Ad^*_{x^{-1}} + \ell(x)$, we obtain

$$(T\ell)_x(X^\lambda_x) = Ad_x\,\varepsilon(X)\,Ad^*_{x^{-1}}$$

whence

(1)
$$(T\ell)_x(X^\rho_x) = Ad_x\,\varepsilon(Ad_{x^{-1}}X)\,Ad^*_{x^{-1}}$$

and

(2)
$$(T\ell)_x(X^\rho_x) = ad_X\,\ell(x) - \ell(x)\,ad^*_X + \varepsilon(X).$$

The condition that ε be a 1-cocycle of g with values in $g \otimes g$ relative to the adjoint action is written in terms of the mapping

$\varepsilon(\ ;\ ) : g \times g^* \to g$ as

(3) $\varepsilon([X,Y];\xi) = \mathrm{ad}_X \varepsilon(Y;\xi) - \varepsilon(Y;\mathrm{ad}_X^*\xi) - \mathrm{ad}_Y \varepsilon(X;\xi) + \varepsilon(X;\mathrm{ad}_Y^*\xi)$.

For each ξ in g^*, we define $\#_\ell \xi$ as the vector field

$$(\#_\ell \xi)(x) = (T\rho_x)_e (\ell(x)\xi).$$

The vector field $\#_\ell \xi$ is the Hamiltonian vector field $\#_\Lambda \xi^\rho$ associated with the right-invariant 1-form ξ^ρ, defined by

$$\xi^\rho_x = (T\rho_x)_e^{*-1} \xi = \xi \circ (T\rho_x)_x^{-1} ,$$

with respect to the bivector Λ defined by

$$\Lambda_x = ((T\rho_x)_e \otimes (T\rho_x)_e)(\ell(x)).$$

The following formulae are valid :

$$L_{X^\rho} \xi^\rho = -(\mathrm{ad}_X^*\xi)^\rho ,$$

$$L_{X^\rho}(\#_\ell \xi) = \varepsilon(X;\xi)^\rho - \#_\ell(\mathrm{ad}_X^*\xi),$$

and

(4) $\langle L_{\#_\ell \xi_1} \xi_2^\rho, \#_\ell \xi_3 \rangle(x) = \langle \xi_2, (T\ell)_x (\#_\ell \xi_1)_x (\xi_3) + [\ell(x)\xi_3, \ell(x)\xi_1] \rangle$.

In addition, we shall use the following results :

(5) $Y^\rho \cdot \mathrm{Ad}_x X = \mathrm{ad}_Y\, \mathrm{Ad}_x X$,

(6) $Y^\rho \cdot \mathrm{Ad}_{x^{-1}} X = -\mathrm{Ad}_{x^{-1}}\, \mathrm{ad}_Y X$,

(7) $Y^\rho \cdot \mathrm{Ad}^*_{x^{-1}} \xi = -\mathrm{Ad}^*_{x^{-1}}\, \mathrm{ad}^*_Y \xi$,

and the commutation formulae

(8) $\mathrm{Ad}_x\, \mathrm{ad}_X = \mathrm{ad}_{\mathrm{Ad}_x X}\, \mathrm{Ad}_x$,

(9) $\mathrm{Ad}^*_x\, \mathrm{ad}^*_X = \mathrm{ad}^*_{\mathrm{Ad}_x X}\, \mathrm{Ad}^*_x$.

We now compute the Schouten bracket $\langle[\Lambda,\Lambda]\rangle$ on three right-invariant 1-forms $\xi_1^\rho, \xi_2^\rho, \xi_3^\rho$. By definition

$$J(x) = \langle[\Lambda,\Lambda]\rangle\ (\xi_1^\rho,\xi_2^\rho,\xi_3^\rho)(x) = -\oint \langle L_{\#_\Lambda \xi_1^\rho}\ \xi_2^\rho, \#_\Lambda\ \xi_3^\rho\rangle(x)$$

$$= -\oint \langle L_{\#_\ell \xi_1}\ \xi_2^\rho, \#_\ell\ \xi_3\rangle(x)$$

By applying formulae (4) and (1), we obtain

$$J(x) = -\oint \langle \xi_2, \mathrm{Ad}_x \varepsilon(\mathrm{Ad}_{x^{-1}}(\ell(x)\xi_1)\ ;\ \mathrm{Ad}^*_{x^{-1}}\xi_3) + [\ell(x)\xi_3, \ell(x)\xi_1]\rangle .$$

We now take the derivative of J in the direction of an arbitrary right-invariant vector field X^ρ. We obtain a sum of six terms, applying formulae (5), (6) and (7), as follows :

$$L_{X^\rho} J(x) = \oint \langle \xi_2, -\mathrm{ad}_X\ \mathrm{Ad}_x\ \varepsilon(\mathrm{Ad}_{x^{-1}}(\ell(x)\xi_1)\ ;\ \mathrm{Ad}^*_{x^{-1}}\xi_3)$$

$$+ \mathrm{Ad}_x\ \varepsilon(\mathrm{Ad}_{x^{-1}}\ \mathrm{ad}_X(\ell(x)\xi_1)\ ;\ \mathrm{Ad}^*_{x^{-1}}\xi_3)$$

$$- \mathrm{Ad}_x\ \varepsilon(\mathrm{Ad}_{x^{-1}}(T\ell)_x(X^\rho_x)(\xi_1)\ ;\ \mathrm{Ad}^*_{x^{-1}}\xi_3)$$

$$+ \mathrm{Ad}_x\ \varepsilon(\mathrm{Ad}_{x^{-1}}(\ell(x)\xi_1)\ ;\ \mathrm{Ad}^*_{x^{-1}}\ \mathrm{ad}^*_X\xi_3)$$

$$+ [(T\ell)_x(X^\rho_x)(\xi_1), \ell(x)\xi_3] + [\ell(x)\xi_1, (T\ell)_x(X^\rho_x)(\xi_3)]\rangle .$$

We shall replace $(T\ell)_x(X^\rho_{x_*})$ by its value and we shall set $\bar{X}(x) = \mathrm{Ad}_{x^{-1}}X$, $\bar{\xi}_i(x) = \mathrm{Ad}^*_{x^{-1}}\xi_i$ $(i = 1,2,3)$. We obtain

$$L_{X^\rho} J(x) = -\oint \langle \bar{\xi}_2, \varepsilon(\varepsilon(\bar{X};\bar{\xi}_1);\bar{\xi}_3)\rangle(x) + K(x),$$

and we shall prove that the remaining term $K(x)$ vanishes because of the cocycle condition (3).

Using formulae (1), (8) and (9), we compute

$$<\xi_2, \mathrm{ad}_{\ell(x)\xi_1} (T\ell)_x (X_x^\rho)(\xi_3)>$$

$$= <\xi_2, \mathrm{ad}_{\ell(x)\xi_1} \mathrm{Ad}_x\, \varepsilon(\mathrm{Ad}_{x^{-1}} X\ ;\ \mathrm{Ad}^*_{x^{-1}}\ \xi_3)>$$

$$= <\mathrm{ad}^*_{\overline{\ell(x)\xi_1}}\ \bar{\xi}_2,\ \varepsilon(\bar{X};\bar{\xi}_3)> = -<\bar{\xi}_2,\ \mathrm{ad}_{\overline{\ell(x)\xi_1}}\ \varepsilon(\bar{X};\bar{\xi}_3)>,$$

where we have set $\overline{\ell(x)\xi_i}(x) = \mathrm{Ad}_{x^{-1}}(\ell(x)\xi_i)$.

The expression for $K(x)$ is obtained by repeated use of (8) and (9) :

$$K(x) = \oint <\bar{\xi}_2, \mathrm{ad}_{\bar{X}} \varepsilon(\overline{\ell(x)\xi_1};\bar{\xi}_3) - \varepsilon(\mathrm{ad}_{\bar{X}} \overline{\ell(x)\xi_1};\bar{\xi}_3)$$

$$- \varepsilon(\overline{\ell(x)\xi_1};\mathrm{ad}^*_{\bar{X}}\ \bar{\xi}_3)$$

$$- \mathrm{ad}_{\overline{\ell(x)\xi_1}}\ \varepsilon(\bar{X};\bar{\xi}_3) + \mathrm{ad}_{\overline{\ell(x)\xi_3}}\ \varepsilon(\bar{X};\bar{\xi}_1)>.$$

Taking the cocycle condition (3) into account, we obtain

$$K(x) = \oint <\bar{\xi}_2,\ \mathrm{ad}_{\overline{\ell(x)\xi_3}}\ \varepsilon(\bar{X};\bar{\xi}_1) - \varepsilon(\bar{X};\mathrm{ad}^*_{\overline{\ell(x)\xi_1}}\ \bar{\xi}_3)>$$

$$= \oint (<\bar{\xi}_1, \varepsilon(\bar{X};\mathrm{ad}^*_{\overline{\ell(x)\xi_3}}\ \bar{\xi}_2)> - <\bar{\xi}_2, \varepsilon(\bar{X};\mathrm{ad}^*_{\overline{\ell(x)\xi_1}} \bar{\xi}_3)>),$$

where we have used the antisymmetry of ε. We conclude that $K(x) = 0$, proving the theorem.

REFERENCES

1. Belavin, A.A. and Drinfeld, V. G., "On the solutions of the classical Yang-Baxter equation for simple Lie algebras", Funct. Anal. Appl. 16 (3), 159-180 [Russian : 1-29] (1982).

2. Chudnovsky, D. V. and Chudnovsky, G. V., "Hamiltonian structure of isospectral deformation equation and semi-classical approximation to factorized S-matrices", Lett. Math. Phys. 4 , 485-493 (1980).

3. Drinfeld, V. G., "Hamiltonian structures on Lie groups, Lie bialgebras and the geometric meaning of the classical Yang-Baxter equations", Soviet Math. Dokl. 27(1), 68-71 (1983). [Russian : Dokl. Akad. Nauk. SSSR 268(2), 285-287 (1983)].

4. Gelfand, I. M and Cherednik I.V., "The abstract Hamiltonian formalism for the classical Yang-Baxter bundles", Russian Math. Surveys 38(3), 1-22 (1983). [Russian : Uspekhi Mat. Nauk 38(3), 3-21 (1983)].

5. Gelfand, I. M. and Dorfman, I. Ya., "Hamiltonian operators and the classical Yang-Baxter equation", Funct. Anal. Appl. 16(4), 241-248 [Russian 1-9] (1982).

6. Kosmann-Schwarzbach, Y., "Algèbres de Lie-Drinfeld", preprint, Université de Lille I, 1986.

7. Kulish, P.P. and Sklyanin, E. K., "Solutions of the Yang-Baxter equation", Journal Soviet Math. 19, 1596-1620 (1982). [Russian : Zap. Nauchn. Sem. LOMI 95, 129-160 (1980)].

8. Magri, F. and Morosi, C., "A geometrical characterization of integrable Hamiltonian systems through the theory of Poisson-Nijenhuis manifolds", Quaderno S 19/1984, Università di Milano.

9. Reiman, A. G., "Integrable Hamiltonian systems connected with graded Lie algebras", Journal Soviet Math. 19, 1507-1545 (1982). [Russian : Zap. Nauchn. Sem. LOMI, 95, 3-54 (1980)].

10. Semenov-Tian-Shansky, M. A., "What is a classical r-matrix ?", Funct. Anal. Appl. 17(4), 259-272 [Russian : 17-33] (1983).

11. Semenov-Tian-Shansky, M. A., "Dressing transformation and Poisson group actions", Publ. RIMS, Kyoto University 21, 1237-1260 (1985).

12. Semenov-Tian-Shansky, M. A., "Classical r-matrices, Lax equations, Poisson Lie groups and dressing transformations", Lecture notes of a course given at the Ecole Normale Supérieure (Paris) in 1986.

13. Sklyanin, E. K., "Quantum version of the method of inverse scattering problem", Journal Soviet Math. 19, 1546-1596 (1982). [Russian : Zap. Nauchn. Sem. LOMI 95, 55-128 (1980)].

14. Souriau, J.-M., Structure des systèmes dynamiques, Dunod. Paris 1970.

SYMMETRIES AND CONSTANTS OF MOTION FOR SOME NONLINEAR EVOLUTION EQUATIONS

LI, Yi−shen† and CHENG, Yi≠ *

† Department of Mathematics,
University of Science and Technology of China (USTC).
Hefei, Anhui, People's Republic of China.

≠ Department of Mathematics,
UMIST, P.O. Box 88,
Manchester M60 1QD., U.K.

ABSTRACT

This is a brief survey of some recent results about the symmetries and constants of motion for some nonlinear evolution equations (NEEs) including 1+1 and 2+1 dimensional and lattice systems. In general, these NEEs are non−local (i.e. are the integro−differential equations) and depend explicitly on the time and space variables.

1. Introduction

It is well known that many integrable nonlinear evolution equations (NEEs), like the KdV equation, in general possess an infinite number of constants of motion (e.g. Ref.1). These constants of motion are related to the symmetries of Langrangian (called K−symmetries) through Noether's theorem.[2,3,4] However, some recent papers.[5−10] have shown that besides the K−symmetries, there exist, in many cases, an infinite number of additional symmetries (called τ−symmetries). The discovery of these additional symmetries for integrable NEEs has much more significance[5−10] As an example of this, the τ−symmetries together with K−symmetries of a system form a graded infinite dimensional Lie algebra.

The features of the τ−symmetries are quite different from those of K−symmetries. They are not the symmetries of a Lagrangian (except in the 2+1 dimensional KP case)[11]. Therefore, Noether's theorem is not applicable and no corresponding constants of motion can be found. For

* On leave of absence from USTC, Hefei, Anhui, P.R. China

the KP equation, although the corresponding constants exist, they are not in involution.[11] On the other hand, the τ-symmetries are in general non-local and explicitly depend on the time and space variables. Thus the classes of NEEs generated by τ-symmetries are also non-local and depend explicitly on the variables of time and space. In the 1+1 dimensional case, these NEEs were indicated to be nothing but the NEEs associated with non-isospectral deformations of the spectral problems.[9,10] An example is for the AKNS system[9], where the NEEs generated by τ-symmetries can be put into the Lax-formulation with the same AKNS spectral problem, but such that the spectrum is no more invariant in time. This property is also valid for the Toda lattice[12] and we think it should be valid for the KP equation too.[13] Thus we call these NEEs generated by τ-symmetries a "new hierarchy of equations".

Examples of these so called new hierarchies of equations in 1+1 dimensional case have been studied for several years.[14-18] They can be solved by the inverse scattering transform with a time varying spectrum[15,16] and can be used to describe nonlinear waves in an inhomogeneous medium.[14,18] Therefore, it is worthwhile investigating the symmetries and constants of motion for these new hierarchies of equations.

The present paper is a short summary of our recent results in dealing with this question. We first give a typical example and then establish a more general theory, which can be applied to other NEEs. In the last section, we discuss the KP system and use another approach, which avoids finding the usual recursion operator.

Some basic notations used in this paper can be found in the Refs.3,4.

2. A Typical Example − The AKNS System

Let us begin with an example to present our main idea and also to summarize the results in Refs.9,19.

It is known that associated with the isospectral deformation of the AKNS spectral problem

$$\varphi_x = M\varphi, \quad \varphi = \begin{bmatrix}\varphi_1\\ \varphi_2\end{bmatrix}, M = \begin{bmatrix}-i\xi/_2 & q(x,t)\\ r(x,t) & i\xi/_2\end{bmatrix} \tag{2.1}$$

is the hierarchy of NEEs

$$u_t = K_\ell(u), \ell = 0,1,2, \cdots \tag{2.2}$$

where $u = (q,r)^T$, $K_o = (-iq, ir)^T$, $K_\ell = \varphi^\ell K_o$ and φ is an operator

$$\varphi = \frac{1}{2}\begin{bmatrix} -D+2qD^{-1}r & 2qD^{-1}q \\ -2rD^{-1}r & D-2D^{-1}q \end{bmatrix}, \quad D = \partial/\partial_x \tag{2.3}$$

The NEEs (2.2) are completely integrable and have a common set of an infinite number of constants of motion. I_n which are in involution[2])

$$I_{nt} = 0, \quad \{I_m, I_n\} = 0 \tag{2.4}$$

The Poisson bracket for any two functionals is defined through a symplectic operator θ

$$\{F,G\} = \langle \text{grad } F, \theta \text{ grad } G\rangle, \quad \theta = \begin{bmatrix} 0 & 1 \\ -1 & 0 \end{bmatrix} \tag{2.5}$$

in which $\langle \; , \; \rangle$ is the standard scalar product and grad $= (\delta/\delta q, \delta/\delta r)$

The symmetry for a specific NEE $u_t = K(u)$ is an algebraic expression $\sigma(u)$ that satisfies the linear variation equation

$$\sigma_t = K'[\sigma], \quad K'[\sigma] = \lim_{\epsilon \to 0} \frac{\partial}{\partial \epsilon} K(u+\epsilon\sigma) \tag{2.6}$$

which is the Gateaux derivative of K(u)in the direction of σ. In the present AKNS case $K = K_\ell$, $\ell = 0,1,2,\cdots$ and the two sets of symmetries of (2.2) are[9])

$$\begin{aligned} &\text{K-symmetries:} \quad K_n = \varphi^n K_o \\ &\tau\text{-symmetries:} \quad \tau_n^\ell = \varphi^n \tau_o^\ell, \; \tau_o^\ell = \ell t \, K_{\ell-1} + \sigma_o, \; \sigma_o = xK_o \end{aligned} \tag{2.7}$$

The proofs that these are symmetries are based on the properties of the operator φ in (2.3), which is a hereditary and strong symmetry[4,9]), namely, φ is the Nijenhuis operator mapping a tangent space of a manifold into itself and satisfies the invariance condition.[3]

$$\varphi'[K_\ell] = K_\ell'\varphi - \varphi K_\ell' \tag{2.8}$$

The constants of motion I_n then can be related to the K−symmetries by θ [3])

$$K_n = \theta \text{ grad } I_n, \quad n = 0,1,2 \cdots \tag{2.9}$$

and this of course means that the hierarchy of equations (2.2) are Hamiltonian systems.

If we introduce the Lie bracket between any two symmetries as

$$[A,B] = A'[B] - B'[A] \tag{2.10}$$

then we find that there exists a graded infinite dimensional Lie algebra involving both the K− and τ−symmetries[9], namely

$$[K_m, K_n] = 0, \quad [K_m, \tau_n^{\ell}] = m\, K_{m+n-1} \;, \qquad [\tau_m^{\ell}, \tau_n^{\ell}] = (m-n)\, \tau_{m+n-1}^{\ell}, \quad m,\ n = 0,1,2, \cdots . \tag{2.11}$$

Our "new hierarchy of equations" in the AKNS case is then generated by the τ−symmetries as

$$u_t = \tau \equiv \tau_{N+1}^{\ell} \tag{2.12}$$

Here, for simplicity, we have changed the labelling and also put $\tau \equiv \tau_{N+1}^{\ell}$. Equations (2.12) correspond to the non-isospectral deformation of the spectral problem (2.1) with the spectrum ξ allowed to be time varying, i.e. $\xi_t = \xi^{N+1}$. Our purpose, now, is to discuss the symmetries and constants of motion for the new NEEs (2.12).

In general, one simple way of finding the symmetries of an equation is by looking for a strong symmetry, which can generate the hierarchy of symmetries. For Equations (2.12) if the strong symmetry exist, say $\bar{\Phi}$, it maybe explicitly dependent on t and must satisfy[19]

$$\partial\bar{\Phi}/\partial t + \bar{\Phi}'[\tau] = \tau'\bar{\Phi} - \bar{\Phi}\tau' \tag{2.13}$$

In the cases when $N = -1$ and $N = 0$ (2.12) such operator can be easily found; they are $\bar{\Phi} = \varphi - t$, $\bar{\Phi} = e^{-t}\varphi$, respectively. The corresponding spectra satisfy $\xi_t = 1$ and $\xi_t = \xi$, from which one finds that the new parameter $\bar{\xi} = \xi - t$, $\bar{\xi} = e^{-t}\xi$ in each case is time invariant. The point is that as functionals of φ and ξ, respectively, $\bar{\Phi}$ and $\bar{\xi}$ are coincident! Thus in the general case of (2.12), it is preferable to find a time invariant parameter $\bar{\xi} = f(\xi,t) = \xi(1+Nt\xi^N)^{-1/N}$ ($\bar{\xi}_t = 0$ when $\xi_t = \xi^{N+1}$) and consider the operator $\bar{\Phi} = f(\varphi,t)$ as a strong symmetry. This idea has been developed in Ref.19, as the theorem

Theorem 2.1 let $\bar{\Phi} = f(\varphi,t)$, $f(\xi,t) = \xi(1+Nt\xi^{n})^{\frac{-1}{N}}$ (2.14)

then $\bar{\Phi}$ satisfy (2.13) and so is the strong symmetry of Eqs. (2.12).

By means of a strong symmetry, once we find a symmetry, we then find a hierarchy of symmetries. In this way we can check that $\bar{K} = K_0$, $\bar{\tau}_0 = \tau - K_\ell$ are symmetries, thus we have two sets of symmetries of (2.12) they are[19] $\bar{K}-$ and $\bar{\tau}-$symmetries

$$\bar{K}_n = \bar{\Phi}^n \bar{K}_0, \quad \bar{\tau}_n = \bar{\Phi}^n \bar{\tau}_0 \tag{2.15}$$

An infinite dimensional Lie algebra also exists for these symmetries[19]

$$\begin{aligned} &[\bar{K}_m, \bar{K}_n] = 0, \quad [\bar{K}_m, \bar{\tau}_n] = m\, \bar{K}_{m+n+N} \\ &[\bar{\tau}_m, \bar{\tau}_n] = (m-n)\, \bar{\tau}_{m+n+N}, \quad m,n = 0,1,2, \cdots \end{aligned} \tag{2.16}$$

Let us consider the constants of motion of Eqs. (2.12). A functional $\bar{I}(u)$ is a constant of the motion of (2.12) when $\bar{I}_t = 0$ if u satisfies (2.12). Note that for (2.12) the $\bar{K}-$ symmetries no longer yield the constants of motion,[19] nevertheless, we can still construct them in terms of the quantities I_n in (2.4), namely[19]

Theorem 2.2. Let

$$\bar{I}_n = \begin{cases} \sum_{k=o}^{n} C_n^k (-t)^{n-k} I_k, \quad C_n^k = \dfrac{n!}{k!(n-k)!}, & N = -1 \\ \exp(-(n+1)t) I_n & N = 0 \\ \sum_{k=o}^{\infty} g_{n+kN}\, I_{n+kN} & N \geq 1, \quad n = 0,1,2, \cdots \end{cases} \tag{2.17}$$

where g_{n+kN} are coefficients of $g(\xi,t) = f^n . \partial f / \partial \xi = \sum g_{n+kN}\, \xi^{n+kN}$. Then the $\bar{I}_n$ are constants of motion of (2.12) and obviously in involution with the Poisson bracket defined in (2.5).

The proofs of all these results can be found in Ref.19 and will be demonstrated briefly in the next section for a more general case.

According to the inverse scattering theories of (2.1), the quantities I_n can be expressed in terms of the scattering data. Here we assume $r = -q^*$ for a simplified example. In this case, we have

$$I_n = \sum_{k=o}^{m} \frac{\xi_k^{n+1} - \xi_k^{*n+1}}{n} + \frac{1}{\pi i} \int_{-\infty}^{+\infty} \xi^n \ell n\, |a|\, d\xi, \quad n = o,1,2, \cdots \tag{2.18}$$

where ξ_k, $k = 0,1,2, \cdots m$ and $a(\xi)$ are scattering data of the spectral problem (2.1).[2] By means of the definition of the functions $f(\xi,t)$ and $g(\xi,t)$, our constants of motion $\bar{I}_n$ of Equations (2.12) can be also expressed in a similar way

$$\bar{I}_n = \sum_{k=o}^{m} \frac{f^{n+1}(\xi_k, t) - f^{n+1}(\xi_k^*, t)}{n} + \frac{1}{\pi i} \int_{-\infty}^{+\infty} f^n \, \ln|a| \, df \qquad (2.19)$$

Thus, we find that the expression of $\bar{I}_n$ in (2.19) conforms to our previous replacement, namely, (2.19) can be obtained by using $f(\xi,t)$ to replace ξ, etc. in (2.18)

3. The Theory.

In this section, we shall show that tbe above structures suit not only for the AKNS systems, but also for a more general systems, which covers many equations of physical interest.

Given the fields $K_o(u)$ and $\sigma_o(u)$ (σ_o may be a constant) and an operator φ, then if they satisfy the conditions

a) hereditary: $\varphi'[\varphi f]g - \varphi'[\varphi g]f = \varphi(\varphi'[f]q - \varphi'[g]f)$, $\forall f,g \epsilon T$ (3.1)
T is a proper defined space;

b) for a strong symmetry of K_o: $\varphi'[K_o] = K_o'\varphi - \varphi K_o'$ (3.2)

c) for α a parameter: $\varphi'[\sigma_o] = \sigma_o'\varphi - \varphi\sigma_o' + \alpha$ (3.3)

d) for β a parameter: $[K_o, \sigma_o] = 0$, $[K_1, \sigma_o] = \beta K_o, K_1 = \varphi K_o$ (3.4)

we have[9,10,20]

Theorem 3.1. For the hierarchy of equations

$$u_t = K_\ell(u), \; K_\ell(u) = \varphi^\ell K_o, \; \ell = 0,1,2, \cdots \qquad (3.5)$$

there are two sets of symmetries, i.e. K- and τ-symmetries

$$K_n = \varphi^n K_o, \; \tau_n^\ell = \varphi^n \tau_o^\ell, \; \tau_o^\ell = (\alpha\ell+\nu)tK_{\ell-1} + \sigma_o, n = 0,1,2, \cdots \qquad (3.6)$$

where $\gamma = \beta - \alpha$. An infinite dimensional Lie algebra exists for these symmetries, namely

$$[K_m, K_n] = 0, \; [K_m, \tau_n^\ell] = (\alpha m + \gamma) \, K_{m+n-1}$$
$$[\tau_m^\ell, T_n^\ell] = \alpha(m-n)\tau_{m+n-1}^\ell, \; m,n = 0,1,2, \cdots \qquad (3.7)$$

The proofs can be found in Refs. 9,10,20.

Now, let us consider the new hierarchy of equations

$$u_t = \tau \equiv \tau_{N+1}^\ell \; \ell \geqslant 0, \; N \geqslant -1 \qquad (3.8)$$

For these equations, we can find the following relations from conditions a,b,c) and (3.7)

$$(\varphi^k)'[K_n] = K_n'\varphi^k - \varphi^k K_n'$$
$$(\varphi^k)'[\tau] = \tau'\varphi^k - \varphi^k\tau' + \alpha k\varphi^{k+N}$$

and $K_{nt} = \tau'[K_n] + (\alpha n + \gamma)K_{n+N}$, $\tau_t = \tau'[\tau] + (\alpha\ell + \gamma)K_{\ell+N}$ (3.9)

Thus, we have the following theorems

Theorem 3.2. The strong symmetry of Eqs. (3.8) is

$$\bar{\Phi} = f(\varphi,t), \quad f(\xi,t) = \xi(1+\alpha Nt\xi^N)^{-\frac{1}{N}} \tag{3.10}$$

Theorem 3.3. The two sets of symmetries of Eqs. (3.8) are

$$\bar{K}\text{-symmetries: } \bar{K}_n = \bar{\Phi}^n\bar{K}_o, \quad \bar{K}_o = \begin{cases} K_o, & N = -1 \\ \exp(-\gamma_t)K_o, & N = 0 \\ \bar{\Phi}_1 K_o, & N > 0 \end{cases}$$

$$\bar{\tau}\text{-symmetries: } \bar{\tau}_n = \bar{\Phi}^n\bar{\tau}_o, \quad \bar{\tau}_o = \tau - K_\ell, \tag{3.11}$$

for $n = 0,1,2,\cdots$ where $\bar{\Phi}_1 = f_1(\varphi,t)$, $f_1,(\xi,t) = (1+\alpha Nt\xi^N)^{-\gamma/\alpha N}$. Note that when $\gamma = 0$, $\bar{K}_o = K_o$ for all $N \geqslant -1$. A graded infinite dimensional Lie algebra is also valid for these $\bar{K}$- and $\bar{\tau}$-symmetries

$$[\bar{K}_m,\bar{K}_n] = 0, \quad [\bar{K}_m,\bar{\tau}_n] = (\alpha m+\gamma)\bar{K}_{m+n+N}$$
$$[\bar{\tau}_m,\bar{\tau}_n] = \alpha(m-n)\bar{\tau}_{m+n+N}, \quad m,n = 0,1,2,\cdots \tag{3.12}$$

The proofs of theorems 3.2 and 3.3 are as follows. Firstly expanding $\bar{\Phi}$ in (3.10) in terms of powers of φ and using (3.9), one can check that $\bar{\Phi}$ satisfies

$$\bar{\Phi}_t = \partial\bar{\Phi}/\partial t + \bar{\Phi}'[\tau] = \tau'\bar{\Phi} - \bar{\Phi}\tau'$$

Thus Φ is a strong symmetry. From (3.9), we find both $\bar{K}_o$ and $\bar{\tau}_o$ are symmetries. So $\bar{K}_n$, $\bar{\tau}_n$ in (3.11) are also symmetries of Eqs. (3.8). To prove the Lie bracket relations in (3.12), we note that $\bar{K}_n$ and $\bar{\tau}_n$ are combinations of the K's, and the K's and τ's, respectively. Thus by using the Lie brackets (3.7), we can demonstrate the Lie brackets of (3.14).[20]

If for Eqs.(3.5), the K−symmetries can yield the constants of motion I_n in involution through a symplectic operator θ

$$K_n = \theta \text{ grad } I_n, \quad I_{nt} = 0, \quad \{I_m,I_n\} = 0 \tag{3.13}$$

Then, for the new hierarchy of equations (3.8), we have[19,20]

$$I_{nt} = I_n'[\tau] = (\alpha n + \alpha N + \beta)\, I_{n+N} \tag{3.14}$$

and from this equation, we can obtain the constants of motion $\bar{I}_n$ for Eqs.(3.8), i.e.

theorem 3.4. Let

$$\bar{I}_n = \begin{cases} \sum_{k=o}^{n} C_n^k(-t)^{n-k} I_k, & N = -1, \text{ if } \alpha = 1, \nu = 0 \\ \exp(-(\alpha n+\beta)t) I_n, & N = 0 \\ \sum_{n=\infty}^{\infty} g_{n+kN}\, I_{n+kN}, & N > 0, \ n = 0,1,2,\cdots \end{cases} \tag{3.15}$$

where $g(\xi,t) = f^{n+\gamma/\alpha}\, \partial f/\partial\xi = \xi^{\gamma/\alpha} \sum_{k=o-}^{\infty} g_{n+kN}\, \xi^{n+kN}$. The coefficients g_{n+kN} are the functions of t. Then, $\bar{I}_n$ are constants of motion of Eqs. (3.8) and in involution under the same Poisson bracket defined in (3.13).

The equation (3.14) has been provded in Refs. 19,20), and the theorem 3.4 can be proved by means of (3.14) and properties of the function $g(\xi,t)$.

4. Some Examples

For brevity, here we only show four examples and list the main results for them. The first one is the familiar KdV equation: it is certainly the most popular equation in the soliton theory. The second example is the Toda lattice, and we choose it to show that our theory is also valid for a lattice system. The third one is the classical shallow water wave equation[21]) It is completely integrable up the advent of a shock, but there is no analogous soliton solution in this case. The last one is the Burger's equation. Its hierarchy of equations (the higher–order Burger's equation) are not Hamiltonian and no constants of motion exist for them. However we shall see that part of our theory is still valid in this case. Some details can be found in Ref. 20.

4.1 The KdV equation

The KdV equation is $= u_t = u_{xxx} + 6uu_x$ (4.1)

The operator is $\varphi = D^2 + uu + 2u_xD^{-1}$ (4.2)

and the fields are $K_0 = u_x$, $\sigma_0 = \frac{1}{2}$. With the parameters $\alpha = 2$, $\beta = 3$, $\gamma = 1$, the operator φ and the fields satisfy the conditions a–d) in Sec. 3. So all the corresponding results are valid for the KdV system.

The symplectic operator θ in (3.13) of the present example is

$\theta = D = \partial/\partial x$. By means of the inverse scattering theories, we know that the constants I_n for the KdV system take the form

$$\bar{I}_n = \sum_{j=1}^{m} \frac{k_j^{2n+3}}{2n+3} + \frac{1}{\pi i}\int_{-\infty}^{+\infty} (ik)^{2n+2}\, \ell n|a|d(ik) \tag{4.3}$$

where k_j and $a(k)$ are scattering data of the Schrödinger spectral problem[2]. The corresponding $\bar{I}_n$ in the present case then also take a similar form

$$\bar{I}_n = \sum_{j=1}^{m} \frac{\bar{f}^{2n+3}(k_j,t)}{2n+3} + \frac{1}{\pi i}\int_{-\infty}^{+\infty} \bar{f}^{2n+2}(ik,t)\, \ell n|a|d\, \bar{f}(ik,t) \tag{4.4}$$

where $\bar{f}(k,t) = (f(k^2,t))^{\frac{1}{2}}$, $f(k,t)$ is in (3.10) with $\alpha = 2$.

4.2 The Toda Lattice[12,23]

$$\text{Equation} \begin{cases} g_{nt} = \gamma_{n-1} - \gamma_n \\ \gamma_{nt} = -\gamma_n(g_{n+1} - g_n) \end{cases} \tag{4.5}$$

$$\text{Operator} = \varphi = \begin{bmatrix} g_n & 1-\Delta^+\gamma\Delta^{-1}\gamma^{-1} \\ \gamma_n(\Delta+2) & \gamma_n\Delta g\Delta^{-1}\gamma^{-1} \end{bmatrix} \tag{4.6}$$

where $\Delta^+\chi = \chi_{n-1} - \chi_n$, $\Delta\chi = \chi_{n+1} - \chi_n$ for any χ_n. The operator φ in (4.6) has been proved to satisfy (3.1) on a well defined space and is strong symmetry of the Toda lattice[23]. If we request that $\Delta^{-1}0 = 1$ [23] and define the fields as $K_0 = (0,0)^T$, $\sigma_0 = (1,0)^T$ then we have $K_1 = \varphi K_0 = (\Delta^+\gamma, -\gamma_n\Delta g)^T$, $\sigma_1 = \varphi\sigma_0 = (g_n, 2\gamma_n)^T$. The parameters in this case can be determined and they are $\alpha = 1$, $\beta = 1$ and $\gamma = 0$. Therefore, we obtain all the corresponding results for the Toda lattice system.

The symplectic operator of the Toda lattice is[23]

$$\theta = \begin{bmatrix} 0 & \Delta^+\gamma \\ -\gamma_n\Delta & 0 \end{bmatrix}$$

Substituting it into (3.13),we can get the constants of motion I_n for the hierarchy of Toda lattices and then obtain $\bar{I}_n$ for the new hierarchy of Toda lattices, which are in the forms of (3.15). According to the inverse scattering theories of the lattice version of the Schrödinger spectral problem, the I_n take the form[24].

$$I_n = \frac{1}{2\pi i}\int_0^{2\pi} \ell n|a| e^{i(n-1)\varphi} de^{i\varphi} - \frac{1}{2\pi i}\int_0^{2\pi} \ell n|a| e^{-i(n-1)\varphi} de^{-i\varphi} + \sum_{j=1}^{m} \frac{z_j^{\,n} - z_j^{\,-n}}{n}$$

Thus, the $\bar{I}_n$ have the form[12])

$$\bar{I}_n = \frac{1}{2\pi i}\int_0^{2\pi} \ell n|a| f^{n-1}(e^{i\varphi},t)df(e^{i\varphi},t) - \frac{1}{2\pi i}\int_0^{2\pi} \ell n|a| f^{n-1}(e^{-i\varphi},t)df(e^{-i},t$$

$$+ \sum_{j=0}^{m} n^{-1}(f^n(z_j,t) - f^n(z_j^{-1}, t))$$

4.3 Classical Shallow Water Wave Equation

The equation is: $\begin{cases} u_t = uu_x + h_x \\ h_t = (hu)_x \end{cases}$

The operator is: $\varphi = \begin{bmatrix} \frac{1}{2}v + \frac{1}{2}v_x D^{-1} & 1 \\ h + \frac{1}{2}h_x D^{-1} & \frac{1}{2}v \end{bmatrix}$

The fields are: $K_0 = (v_x, h_x)^T$ $\sigma_0 = (2,0)^T$

The parameters are: $\alpha = 1,\ \beta = 2,\ \gamma = 1$

The Symplectic operator is: $\theta = \begin{bmatrix} 0 & D \\ D & 0 \end{bmatrix}$

With these definitions, the whole theory is valid in this case.

4.4 Burger's Equation

The equation is $u_t = u_{xx} + 2uu_x$

The operator is $\varphi = D + u + u_x D^{-1}$

and the fields are $K_0 = u_x$, $\sigma_0 = 1$. With the parameter $\alpha = 1$, $\beta = 2$, $\gamma = 1$, the theorems 3.2, 3.2, 3.3 are valid in this case. Since there are no constants of motion for the Burger's and hierarchy of Burger's equations, the theory is only partially valid in this case.

5. The KP Equation

We take the KP equation in the form

$$u_t = K(u) = D_x^{-1}u_{yy} - 6uu_x - u_{xxx} \tag{5.1}$$

Chen et al.[6]) have derived its two sets of symmetries and a graded Lie algebra for the symmetries. Soon after, Case[25]) generalized their results to the KP hierarchies (the higher–order KP equations). These results can be put into a general form[13])

The KP hierarchies: $u_t = K_\ell(u)$, $K_2 = K$ (5.2)

The K- and τ-symmetries: K_n, τ_n^ℓ, $n = 0,1,2,\cdots$ (5.3)

The Lie algebra: $[K_m,K_n] = 0$, $[K_m,\tau_n^\ell] \frac{1}{3} - (m+1)K_{m+n-2}$

$[\tau_m^\ell,\tau_n^\ell] = \frac{1}{3}(m-n)\tau_{m+n-2}^\ell$, $m,n = 0,1,2,\cdots$ (5.4)

For $\ell = 2$ i.e. for the KP equation, the K– and τ–symmetries were constructed· not by an recursion operator (strong symmetry) in the usual way, but through the Lie algebra (5.4), namely, by direct calculation, one can obtain K_0 and $\tau_0{}^2, \tau_3{}^2, \tau_4{}^2$ as the "seed" elements and then obtain the hierarchies of symmetries from (5.4)[6]. For other members of Eqs. (5.2), the τ-symmetries are as follows

$$\tau_n{}^\ell = \frac{\ell+1}{3}\, t\, K_{n+\ell-2} + \sigma_n,\quad \sigma_n = \tau_n{}^2 - tK_n \tag{5.5}$$

The Lie brackets among $K_n, \tau_n{}^\ell$ then can be derived in terms of the Lie brackets among K_n and $\tau_n{}^2$.[13] The new hierarchy of KP equations is

$$u_t = \tau \equiv \tau_{N+2}^{\ell},\quad \ell \geqslant 0,\quad N \geqslant 0 \tag{5.6}$$

In order to construct their symmetries, we introduce the following functions

$$F_n(\xi, t) = \xi^n \left(1 + \tfrac{1}{3}Nt\xi^N\right)^{-\frac{n+1}{N}} = \sum_{k=o}^{\infty} f_{n,k}(t)\xi^{n+kN}$$

$$G_n(\xi, t) = \xi^n \left(1 + \tfrac{1}{3}Nt\xi^N\right)^{-\frac{n}{N}} = \sum_{k=o}^{\infty} g_{n,k}(t)\xi^{n+kN} \tag{5.7}$$

For n = 0,1,2, $\cdots$. The coefficients $f_{n,k}$, $g_{n,k}$ are well defined functions of t and have some internal relations among them[13]. According to F_n and G_n we can obtain the symmetries for (5.6), the are $\bar{K}$- and $\bar{\tau}$-symmetries

$$\bar{K}_n = F_n{*}K_0,\quad \bar{\tau}_n = G_n{*}(\tau_{N+2}^{\ell} - K_\ell),\quad n = 0,1,2,\cdots \tag{5.8}$$

where for any sequence χ_n we define

$$F_n{*}\chi \equiv \sum_{k=o}^{\infty} f_{n,k}\chi_{m+n+kN} \text{ for any m, etc.} \tag{5.9}$$

The $\bar{K}$- and $\bar{\tau}$-symmetries in (5.8) also form a graded Lie algebra

$$[\bar{K}_m, \bar{K}_n] = 0,\quad [\bar{K}_m, \bar{\tau}_n] = \tfrac{1}{3}(m-1)\bar{K}_{m+n+N}$$

$$[\bar{\tau}_m, \bar{\tau}_n] = \tfrac{1}{3}(m-n)\,\bar{\tau}_{m+n+N},\quad m,n = 0,1,2,\cdots \tag{5.10}$$

Follow Chen et al.[6,11] both K_n and $\tau_n{}^2$ for (5.1) are symmetries of a Lagrangian; thus our $\tau_n{}^\ell$, $\ell \neq 2$ are also the symmetries of a Lagrangian and the corresponding constants of motion $I_n, J_n{}^\ell$ can be found which satisfy[11,13]

$$K_n = D_x\, \delta I_n/\delta u,\quad \tau_n{}^\ell = D_x\, \delta J_n{}^\ell/\delta u \tag{5.11}$$

the Poisson bracket between the constants of motion is defined as

$$\{F,G\} = \iint (\delta F/\delta u)\, D_x(\delta G/\delta u)\, dxdy \tag{5.12}$$

It is related to the Lie bracket between the corresponding symmetries in the following

$$[A,B] = D_x(\delta\{F,G\}/\delta u) \tag{5.13}$$

Hence, both Eqs. (5.2) and Eqs. (5.6) are Hamiltonian systems. According to (5.11) and (5.13), we find that there exists a Poisson algebra for I_n and J_n^{ℓ}, which is naturally isomorphic to the Lie algebra for K_n and τ_n^{ℓ} in (5.4). This implies that the I_n are in involution, while the J_n^{ℓ} are not.

The constants of motion of the new hierarchy of KP equations are

$$\bar{I}_n = F_n * I_0, \quad \bar{J}_n = G_n * (J_{N+2}^{\ell} - I_{\ell}) \tag{5.14}$$

here, we used the notation (5.9) again. $\bar{I}_n$, and $\bar{J}_n$ are corresponding to the $\bar{K}_n$ and $\bar{\tau}_n$ in the same way of (5.11) and the Poisson algebra for them are also valid, which is isomorphic to the Lie algebra (5.10). Thus the $\bar{I}_n$ are in involution. According to the inverse scattering theories of the "time dependent" Schrödinger spectral problem,[26], the I_n take the form[13]

$$I_n = \frac{1}{6^3} \iint_{\xi^+>\xi^-} (\xi^{+n+1} - \xi^{-n+1}) p(\xi^+,\xi^-) d\xi^+ d\xi^- - \frac{2\pi i}{9} \sum_{j=1}^{m} (\xi_j^{+n+1} - \xi_j^{-n+1}) \tag{5.15}$$

So our $\bar{I}_n$ have the form[13]

$$\bar{I}_n = \frac{1}{6^3} \iint_{\xi^+>\xi^-} (G^{n+1}(\xi^+,t) - G^{n+1}(\xi^-,t)) p(\xi^+,\xi^-) d\xi^+ d\xi^- - \frac{2\pi i}{9} \sum_{j=1}^{m} (G^{n+1}(\xi_j^+,t) - G^{n+1}(\xi_j^-,t)) \tag{5.16}$$

where $\xi_j^{\pm}$, $j = 1, \cdots m$, $p(\xi^+,\xi^-)$, $\xi^+ > \xi^-$ constitute the spectrum.

As we have pointed out, in the 1+1 dimensional and lattice case, the new hierarchies of equations are generally related to the non– isospectral deformations of spectral problems. For the KP system, it is not clear whether the above property is also valid. However, we can still calculate the evolution equations about $\xi_j^{\pm}$ and $P(\xi^+,\xi^-)$ for Eqs. (5.6). From the Poisson algebra for I_n and J_n^{ℓ} we find that if u satisfies Eqs. (5.6), then.

$$I'_n[\tau] = I_{nt} = \{I_n, J^{\ell}_{n+2}\} = \frac{1}{3}(n+1)I_{n+N} \quad (5.17)$$

Substituting the I_n of (5.15) into this equations, we find[13])

$$(\xi_j^{\pm})_t = (\xi_j^{\pm})'[\tau] = \frac{1}{3}(\xi_j^{\pm})^{N+1}, \quad j = 1,2, \cdots m. \quad (5.18)$$

$$P'[\tau] = -\frac{1}{3}(N+1)(\xi^{+N} + \xi^{-N})P - \frac{1}{3}(\xi^{+N+1}\,\partial P/\partial\xi^{+} + \xi^{-N+1}\partial P/\partial\xi^{-}) \quad (5.19)$$

Hence, if $\xi_t = \frac{1}{3}\xi^{N+1}$, then (5.19) becomes

$$P_t(\xi^{+},\xi^{-}) = -\frac{1}{3}(N+1)\ (\xi^{+N} + \xi^{-N})p(\xi^{+},\xi^{-})$$

The conclusion is that if the new hierarchy of KP equations (5.6) can be solved by the inverse scattering transform, the spectrum should be time varying. It is possible that it satisfied $\xi_t = {}^1/_3\ \xi^{N+1}$, and we shall give a further discussion of this elsewhere.

Acknowledgements

Both authors would like to thank R.K. Bullough for his help and critical reading the manuscript. Li Yi–shen also thanks him for the invitation to visit UMIST and Cheng Yi is grateful to the British Council for their Academic Link with China Scheme which currently supports him. The work was partially supported by the Science Fund of the P.R. of China.

References

1. Ablowitz, M.J. and Segur, H., Solitons and the Inverse Scattering Transform. SIAM (Philadelphia 1981).
2. Arnold, V.I., Mathematical Methods of Classical Mechanics, (Springer–Verlag, 1978).
3. Magri, F., Lecture Notes in Physics 120, Boiti et al. eds. 233 (Springer–Verlag, 1980).
4. Fuchssteiner, B. and Fokas, A.S., Physica D4, 47 (1981).
5. Chen, H.H., Lee, Y.C. and Lin, J.E., Phys. Lett. 91A, 381 (1982).
6. Chen, H.H., Lee, Y.C. and Lin, J.E., Physica D9, 439 (1983).
7. Fuchssteiner, B., Progr. Theor. Phys. 70, 1508 (1983).
8. Stramp, W. and Oevel, W., Progr. Theor. Phys. 74, L922 (1985).

9. Li, Yi–shen and Zhu, Gou–cheng., to be published in J. Phys. A. Math. Gen.

10. Li, Yi–shen and Zhu, Gou–cheng, to be published.
11. Chen, H.H., Lee, Y.C. and Zhu. G.C., Plasma preprint UMLPF* 85–016, 1984.
12. Cheng, Yi and Li, Yi–shen, to be published.
13. Cheng, Yi and Li, Yi–Shen, to be published.
14. Chen, H.H. and Liu, S.C., Phys. Rev. Lett. 37, 693 (1976).
15. Calogero, F. and Degasperis, A., Lett. Nuovo Cimento 22, 138 (1978).
16. Li, Yi–shen, in Proc. of the 1980 Beijing Sym. On Diff. Geo and Diff. Eqs. Chern, S.S. and Wu, W.T. eds. Vol. 3 129 (Science Press, Beijing, 1980).
17. Li, Yi–Shen, Scienta Scinica A25, 911 (1981).
18. Lakshmanan, M. and Bullough, R.K., Phys. Lett. 80A, 287 (1980).
19. Cheng, Yi and Li, Yi–shen, to be published in J. Phys. A. Math. Gen.
20. Li, Yi–shen and Cheng, Yi, to be published.
21. Cavalcante, J. and Mckean, H.P., Physica D4, 253 (1982).
22. Li, Yi–shen and Cheng, Yi, to be published.
23. Leo, M., Leo, R.A. Solinani, G. and Solombrino, L., Lett. Math. Phys. 8, 267 (1984).
24. Flaschka, H., Progr. Theor. Phys. 51, 703 (1974).
25. Case, K.M., J. Math. Phys. 26, 1158 (1985).
26. Fokas, A.S. and Ablowitz, M.J. Stud., Appl. Math. 69, 211 (1983).

From Single Solitons to Auto-Bäcklund Transformations and Hereditary Symmetries

Benno Fuchssteiner

University of Paderborn
Germany

Dedicated to Professor Dr.F.Beck on the occasion of his 60th birthday

ABSTRACT

A method is given to determine auto-Bäcklund transformations and hereditary recursion operators out of the hidden intrinsic algebraic structure of **single** soliton solutions. New auto-Bäcklund transformations are presented. Bianchi permutability and soliton phase shift are explained in terms of commutative groups.

I. Introduction

For completely integrable systems we claim that ABT (auto-Bäcklund transformations) can be computed from the single solitons. This does not mean that equations admitting solitary waves do have auto-Bäcklund transformations, only that solitary waves of completely integrable systems do have hidden symmetries which yield ABT's.

As an introductory example we perform the necessary computations for the KdV

$$u_t = u_{xxx} + 6uu_x \,. \tag{1.1}$$

For this equation the single soliton with asymptotic speed c is given by

$$s(x + ct) = \frac{c}{2}\cosh^{-2}\{\frac{\sqrt{c}}{2}(x - x_0 + ct)\}. \tag{1.2}$$

The KdV is translation invariant. We consider translations by $+\beta$ and $-\beta$, respectively for (1.2)

$$\frac{\sqrt{c}}{2}x \rightarrow \frac{\sqrt{c}}{2}x+\beta \text{ and } \frac{\sqrt{c}}{2}x \rightarrow \frac{\sqrt{c}}{2}x-\beta \tag{1.3}$$

and denote the corresponding solutions by s_β and $s_{-\beta}$. We decompose into odd and even parts

$$V_+ = \frac{1}{2}(s_\beta + s_{-\beta}) , \ V_- = \frac{1}{2}(s_\beta - s_{-\beta}), \tag{1.4}$$

and, in order to abbreviate notation, we introduce

$$\begin{aligned} k &= \frac{\sqrt{c}}{2} \\ \xi &= (x - x_0 + ct) \\ N(k\xi) &= \{\cosh^2(k\xi)\cosh^2(\beta) - \sinh^2(k\xi)\sinh^2(\beta)\}^{-1} \\ &= \{\cosh^2(k\xi) + \sinh^2(\beta)\}^{-1}. \end{aligned} \tag{1.5}$$

Now, we consider explicitly s_β and we apply the addition-theorem for the cosh-function. Splitting up s_β into odd and even parts yields

$$\begin{aligned} s_\beta &= 2k^2 \cosh^{-2}(k\xi + \beta) \\ &= 2k^2\{\cosh(k\xi)\cosh(\beta) + \sinh(k\xi)\sinh(\beta)\}^{-2} \\ &= 2k^2N(k\xi)^2\{\cosh(k\xi)\cosh(\beta) - \sinh(k\xi)\sinh(\beta)\}^2 \\ &= V_+ + V_- \end{aligned} \tag{1.6}$$

where now the odd part V_- and the even part V_+ are given by

$$V_+ = 2k^2N(k\xi)^2\{\cosh^2(k\xi)\cosh^2(\beta) + \sinh^2(k\xi)\sinh^2(\beta)\} \tag{1.7}$$

$$V_- = -4k^2N(k\xi)^2\{\cosh(k\xi)\sinh(k\xi)\cosh(\beta)\sinh(\beta)\}. \tag{1.8}$$

Considered as functions in the variable ξ , the even part V_+ is a polynomial of second order in the common denominator $N(k\xi)$. Furthermore V_- is , apart from multiplication with a constant, the derivative of $N(k\xi)$. Thus we obtain a **relation** between V_+ and V_- . Let us write down this explicitly. Simple computations give

$$V_+ + \gamma N + \delta N^2 = 0 \tag{1.9}$$

where

$$\gamma = -2k^2\cosh(2\beta)$$

$$\delta = k^2\sinh^2(2\beta)$$

and

$$DN = (\ 2k \cosh(\beta) \sinh(\beta)\ )^{-1} V_- , \quad D = \frac{\partial}{\partial x} . \tag{1.10}$$

This gives

$$V_+ - \frac{2k \cosh(2\beta)}{\sinh(2\beta)} D^{-1}V_- + \{D^{-1}V_-\}^2 = 0. \tag{1.11}$$

Now, let β depend on k in such a way that

$$\lambda = -2k \coth(2\beta) \tag{1.12}$$

is independent of k . Then we have the following algebraic relation

$$B^{(\lambda)}(s_\beta , s_{-\beta}) := (s_\beta + s_{-\beta}) + \lambda D^{-1}(s_\beta - s_{-\beta}) + \frac{1}{2}\{D^{-1}(s_\beta - s_{-\beta})\}^2 = 0$$

where the coefficients are **independent** of $k=(\sqrt{c}\ )/2$. By translation invariance we then have

$$B^{(\lambda)}(s , s_{-2\beta}) = (s + s_{-2\beta}) + \lambda D^{-1}(s - s_{-2\beta}) + \frac{1}{2}\{D^{-1}(s - s_{-2\beta})\}^2 = 0 \tag{1.13}$$

This relation between s and $s_{-2\beta}$ we consider for general arguments

$$B^{(\lambda)}(u , \bar{u}) = u + \bar{u} + \lambda D^{-1}(u - \bar{u}) + \frac{1}{2}\{D^{-1}(u - \bar{u})\}^2 \tag{1.14}$$

u and $\bar{u}$ on the manifold S of C^∞-functions vanishing rapidly with all their derivatives at infinity. We consider u as fixed. The map

$$u \to \bar{u} = f_{B^{(\lambda)}}(u) \tag{1.15}$$

given implicitly by

$$B^{(\lambda)}(u\,,\,\bar{u}\,) = 0 \tag{1.16}$$

is injective around u if the implicit function theorem condition is fulfilled. This condition requires that for $B^{(\lambda)}$ the kernel of the variational derivative $B^{(\lambda)}{}_u$ with respect to u has to be trivial, i.e. for every v in the tangent space at u it must hold that whenever

$$B^{(\lambda)}{}_u[v] = 0 \;\text{ and }\; B^{(\lambda)}(u\,,\,\bar{u}\,) = 0 \tag{1.17}$$

then

$$v = 0\,. \tag{1.18}$$

Recall that the partial variational derivative $B^{(\lambda)}{}_u$ is defined to be

$$B^{(\lambda)}{}_u[v] = \frac{\partial}{\partial\varepsilon} B^{(\lambda)}(u + \varepsilon\, v\,,\,\bar{u}\,)_{|\varepsilon=0}. \tag{1.19}$$

Those λ violating this condition for u we call the **spectral** points of u . The other λ are said to be **non-spectral** points.

Obviously (1.15), or (1.16), map the soliton solution s onto $s_{-2\beta}$ where

$$2\beta = \operatorname{arcoth}\,(-\frac{\lambda}{\sqrt{c}}). \tag{1.20}$$

This relation is called the **phase shift relation**. In particular s and all its translations are mapped onto the zero function for $\lambda = \pm\sqrt{c}$. Hence $\pm 2k$ are spectral points for the one soliton having celerity $c = 4k^2$. All other λ's are nonspectral because for $B^{(\lambda)}(u\,,\,\bar{u}\,) = 0$ the operator $B^{(\lambda)}{}_u$ is a differential operator which is invertible on the vector space S.

The set of spectral points in general is finite because (1.17) is an ordinary differential equation for v having only for certain λ's solutions vanishing rapidly at infinity. We shall see this in detail later on.

Now we return to the special case where $u = s$ and $\bar{u} = s_{-2\beta}$. Recall that

$$cu_x = K(u) = u_{xxx} + 6uu_x \tag{1.21}$$

$$c\bar{u}_x = K(\bar{u}) = \bar{u}_{xxx} + 6\,\bar{u}\bar{u}_x \tag{1.22}$$

because both functions are one-solitons. Translation invariance of $B^{(\lambda)}(u\,,\,\bar{u}\,)=0$ given by the free parameter x_0 yields

$$B^{(\lambda)}{}_u\,[\,u_x\,] + B^{(\lambda)}{}_{\bar{u}}[\bar{u}_x] = 0 \tag{1.23}$$

and insertion of (1.21) and (1.22) gives

$$B^{(\lambda)}{}_u\,[K(u)] + B^{(\lambda)}{}_{\bar{u}}[K(\bar{u})] = 0. \tag{1.24}$$

Observe that for nonspectral points λ, where $\bar{u}$ locally is uniquely given by

$$B^{(\lambda)}(u\,,\bar{u}\,) = 0 \tag{1.25}$$

the equation (1.23) is a differential equation for u (or rather the integral of u). And if this equation is nontrivial then on the manifold **S** under consideration this equation has the same number of integration constants as equation (1.21).

But there is an essential difference between (1.21) and (1.24) + (1.25), coming from the important fact that the relation $B^{(\lambda)}$ is independent of c or k :

The solutions for (1.21) are a one-parameter family (parameter x_0) whereas the solutions for (1.24)+(1.25) are a two-parameter family (parameters x_0 and c). Thus the system (1.24) + (1.25) has to many integration parameters. This can only be if the system is trivial, i.e. if (1.24) is identically fulfilled whenever (1.25) holds. And, since the nonspectral points are dense, this must also hold for those λ which are spectral.

Thus we have found for **arbitrary** arguments: Whenever

$$B^{(\lambda)}(u\,,\bar{u}\,) = 0$$

then

$$B^{(\lambda)}{}_u\,[K(u)] + B^{(\lambda)}{}_{\bar{u}}[K(\bar{u})] = 0. \tag{1.26}$$

This relation is equivalent [4] to $B^{(\lambda)}(u,\bar{u}\,)$=0 being an auto-Bäcklund transformation for the KdV (1.1), i.e. whenever $u(t)$ is a solution of the KdV and $\bar{u}(t)$ is such that $B^{(\lambda)}\,(u(t),\bar{u}(t))$=0 then $\bar{u}(t)$ is again a solution of the KdV. Of course, (1.26) can also be proved by explicit computation since we obtain easily

$$\{D^3 + 3(u + \bar{u}\,)D\}\;B(u\,,\bar{u}\,) = 0. \tag{1.27}$$

This auto-Bäcklund transformation is related to the ABT

$$(u + \bar{u}) + \delta + \frac{1}{2}\{D^{-1}(u - \bar{u})\}^2 = 0 \tag{1.28}$$

which is usually found in the literature [1]. By the formal substitution

$$D^{-1} \;\rightarrow\; D^{-1} + constant$$

coming out of the existence of integration constants, (1.28) goes over into (1.26).

However, there is a notable difference between (1.28) and (1.25) since (1.25) is compatible with the boundary condition on the manifold **S** , i.e. with the requirement that u and $\bar{u}$ vanish at infinity, whereas (1.28) is not compatible with this requirement. This point is essential in the following where we turn our interest to the spectral points of u.

Since $B^{(\lambda)}(u\,,\bar{u}\,) = 0$ is an auto-Bäcklund transformation for (1.1) the property of being a spectral point is obviously invariant against this flow. So we have arrived at some kind of spectral problem for which (1.1) constitutes an isospectral flow (see [9]). However, this is a **nonlinear**

Spectral problem :
Given a solution u of (1.1), find those λ's such that there is some nonzero vectorfield ω and some $\bar{u}$ on the manifold under consideration such that

$$B^{(\lambda)}{}_u(u,\bar{u})[\omega] = 0 \tag{1.29}$$

and

$$B^{(\lambda)}(u,\bar{u}) = 0 \; . \tag{1.30}$$

We do not know of any criteria which give reasonable answers to the question under what circumstances such a nonlinear spectral problem is equivalent to a linear one. However, in case of the KdV this problem is easily linearized (see [9]):

Variational derivative of (1.14) with respect to u yields the operator:

$$B_u = I + (D^{-1}(u - \bar{u}))D^{-1} + \lambda D^{-1} \; . \tag{1.31}$$

And the spectral problem (1.29) reads as follows

$$0 = \omega + (D^{-1}(u - \bar{u}))D^{-1}\omega + \lambda D^{-1}\omega \; . \tag{1.32}$$

Abbreviation $D^{-1}\omega = v$ gives

$$D^{-1}(u - \bar{u}) = -(\frac{v_x}{v} + \lambda) \; . \tag{1.33}$$

Writing $u + \bar{u}$ as $2u - (u - \bar{u})$ and then replacing all terms $u - \bar{u}$ in (1.14) by (1.33) we obtain

$$2u + (\frac{v_x}{v} + \lambda)_x + \frac{1}{2}(\frac{v_x}{v} + \lambda)^2 - \lambda(\frac{v_x}{v} + \lambda) = 0$$

which is certainly a nonlinear eigenvalue equation. By multiplication with v^2 we obtain

$$2u\, v^2 + v_{xx}v - \frac{1}{2}v_x v_x = \frac{1}{2}\lambda^2 v^2 \; . \tag{1.34}$$

If this problem can be linearized there must be operators $A(v)$ and $\Psi(u)$ such that $A(v)v = Cv^2$ and $A(v)\Psi(u)v$ is equal to the left side of (1.34). Comparison of suitable terms yields:

$$D^{-1}vD\{v_{xx} + 2uv + 2\, D^{-1}\,(u\, v_x)\} = \lambda^2 D^{-1}\, vDv \; . \tag{1.35}$$

Hence $A(v) = D^{-1}vD$ and $\Psi(u) = D^2 + 2u + 2D^{-1}uD$.

Going back to $\omega = v_x$ we see that ω is a solution of (1.29) if and only if ω is an eigenvector of

$$\Phi(u) = D\Psi(u)D^{-1} = D^2 + 2u + 2DuD^{-1}. \tag{1.36}$$

And if λ is the spectral point given by (1.29) then λ^2 is the corresponding eigenvalue of $\Phi(u)$.

We like to know what the meaning of this operator $\Phi(u)$ is. Since the KdV (1.1) is an isospectral flow for this differential (or rather integro-differential) operator we look for the second component $\Lambda(u)$ of the corresponding Lax-pair

$$\frac{d}{dt}\Phi(u) = [\Lambda(u),\Phi(u)] \tag{1.37}$$

describing the time evolution of $\Phi(u)$ if u evolves according to the KdV. This operator $\Lambda(u)$ is easily found by using the fact that $B^{(\lambda)}(u\,,\,\bar{u})=0$ is an auto-Bäcklund transformation.

Take the time-derivative of (1.29) and replace the time derivatives of u and $\bar{u}$ by the right hand side of the KdV to obtain

$$B^{(\lambda)}{}_{uu}(u\,,\,\bar{u})[\omega\,,\,K(u)\,] + B^{(\lambda)}{}_{u\bar{u}}(u\,,\,\bar{u})[\omega\,,\,K(\bar{u})\,] + B^{(\lambda)}{}_{u}(u\,,\,\bar{u})[\omega_t] = 0. \tag{1.38}$$

Since (1.29) holds u can be changed in (1.30) infinitesimaly by ω without changing $\bar{u}$. Hence variational derivative of u by ω and $\bar{u}$ by zero in (1.24) yields that the first two terms in (1.38) are equal to

$$-B^{(\lambda)}{}_{u}[K\,'(u)[\omega]].$$

Hence, we can choose ω such that

$$\omega_t = K'(u)[\omega]. \tag{1.39}$$

Inserting this into the time derivative of the eigenvalue equation

$$\Phi(u)\omega = \lambda^2\omega, \tag{1.40}$$

where λ is time-independent we obtain

$$\frac{d}{dt}\Phi(u) = K'(u)\Phi(u) - \Phi(u)K'(u) \tag{1.41}$$

and*)

$$\Lambda(u) = K'(u) = D^3+6Du. \tag{1.42}$$

Hence $\Phi(u)$ must be a recursion operator ([12]) or a strong symmetry ([7]) for the KdV. In fact this is the well known hereditary Lenard operator [11] which generates all symmetry generators of the KdV recursively.

Also from our construction via single-solitons follows the meaning of the spectral points in terms of multi-solitons. Consider an N-soliton solution u of the KdV, i.e. a solution which for large t decomposes asymptotically into N single solitons such that the overlap of these single solitons becomes exponentially small. Then by simple asymptotic arguments the spectral points of u must be those points which are spectral for one of these asymptotically occurring single soliton solutions.

*)Actually on first sight this equation only holds when applied to an arbitrary eigenvector ω of $\Phi(u)$. But application of suitable local-global arguments then shows that this must be true in general (see the widely referenced paper [6]).

By (1.20) the spectral points of these single solitons are the $\lambda_n = \pm 2k_n = \sqrt{c_n}$ (c_n speed of the n-th soliton) which correspond to annihilations of these solitons by translating them by

$$\beta = \infty = \frac{1}{2}\operatorname{arcoth}(-\frac{\lambda}{2k}) \tag{1.43}$$

out of finite sight.

Thus two facts are shown, namely

i) that the eigenvalues of $\Phi(u)$ are the speeds of the asymptotic solitons,

ii) that whenever u is some N-soliton solution then for spectral λ_n the $\bar{u}$ appearing in $B^{(\lambda_n)}(u,\bar{u})=0$ is the $(N-1)$-soliton solution where the n-th soliton is missing.

REMARK:

For the explanation of **phase shift** consider a solution $u=u(x,t)$ of the KdV such that asymptotically for $t \to \pm\infty$ a soliton with speed $c_1 = 4k_1^2$ is emerging. Let $B^{(\lambda)}(u,\bar{u})=0$ be a Bäcklund transformation between this solution and another solution $\bar{u}$. Because of the local structure of $B^{(\lambda)}(\cdot,\cdot)$ and the asymptotic behavior of u the effect of this Bäcklund transformation on the emerging soliton is the same as if it were a single soliton. Hence the corresponding soliton emerging out of $\bar{u}$ is, compared to u, shifted by a translation of the amount

$$\frac{4\beta}{\sqrt{c_1}} = \frac{2}{\sqrt{c_1}}\operatorname{arcoth}(-\frac{\lambda}{\sqrt{c_1}}). \tag{1.44}$$

Since the square root is multivalued we have no information about the direction of this shift. So, comparing the asymptotic solitons emerging at $t=-\infty$ with those at $t=+\infty$ we see that they are either shifted in the same direction or in opposite directions. Therefore the total phase shift of this soliton, compared to the corresponding soliton of u, is either zero or

$$\pm\frac{4}{\sqrt{c_1}}\operatorname{arcoth}(-\frac{\lambda}{\sqrt{c_1}}) \tag{1.45}$$

In fact all cases are possible. If in addition another soliton with speed $c_2 = 4k_2^2$ is emerging one can show by simple arguments that if the parameter λ is in between the square roots of the two velocities c_1 and c_2 then the smaller soliton of $\bar{u}$ compared with that of u is forwarded and the faster one is retarded. This can be shown without looking at explicit solutions just by considering small perturbation of the Bäcklund transformation. Now consider $\lambda = \sqrt{c_2}$, i.e. the Bäcklund transformation which annihilates the second soliton with speed c_2. Assume the case $c_2 > c_1$. Then the annihilation is done by shifting this second soliton for $t \to +\infty$ into plus infinity. The phase shift of the soliton with speed c_1 is then

$$+\frac{4}{\sqrt{c_1}}\operatorname{arcoth}(\sqrt{c_2/c_1}) \qquad (1.46)$$

In case $c_2<c_1$ we have to change the sign of this phase shift. Now, take a pure multisoliton solution and annihilate successively all emerging solitons except the one with speed c_1 . Then the phase shifts have to be added. Because of Bianchi permutability (see section II) the order of annihilation does not play any role. And the phase shift of the soliton, emerging out of u with speed c_1 , compared to the corresponding single soliton, which is the result of all these annihilations, must be

$$\frac{4}{\sqrt{c_1}}\sum_i \varepsilon_i \operatorname{arcoth}(\sqrt{c_i/c_1}) \qquad (1.47)$$

where

$$\varepsilon_i = \begin{cases} -1 & \text{if } c_i > c_1 \\ +1 & \text{if } c_i < c_1 \end{cases} .$$

A result well known from the literature [10].

So the result of all our considerations is that looking at the hidden algebraic structure of the single soliton of the KdV we find the N-soliton structure (via the auto-Bäcklund transformation), the abelian symmetry group (via the hereditary recursion operator). Furthermore, via the operator Φ , one can find the bi-Hamiltonian structure ([4]) and the angle-variables of the system (via the master-symmetries [8]) constructed by the recursion operator, furthermore information about details of soliton interaction. To me this seems to be a most remarkable fact.

II. General principles

What was done in the last chapter, with respect to the translation group, for the KdV can be done in general for integrable systems with respect to arbitrary symmetry groups. Let us resume the general aspects.

Consider evolution equations given by differential equations and being of the form

$$u_t = K(u), \qquad (2.1)$$

where $u = u(x,t)$ and where u_t denotes partial derivative with respect to t (time). Eq. (2.1) is regarded as a dynamical system, i.e. $K(u)$ is to be a C^∞-vector field on the manifold of all admissible functions in the variable x . Of course, u may be vector valued, i.e. $u=(u_1, \cdots, u_n)$. Only for simplicity, the space variable x considered is is one dimensional.

Consider a finite-dimensional abelian symmetry group **G** for (2.1) having infinitesimal generators $G_1, \ldots, G_n$. Recall that a vectorfield G is a symmetry group generator for (2.1) if $[K,G]=0$ in the vectorfield Lie algebra. The commutator $[K,G]=L_K G$ is given by the Lie-derivative, which in case the manifold is a vectorspace is defined by

$$L_K G = \frac{\partial}{\partial \varepsilon}\Big|_{\varepsilon=0} \{G(u + \varepsilon K(u)) - K(u + \varepsilon G(u))\}. \tag{2.2}$$

In the example of section I the group **G** was the one-dimensional translation group and its generator was $G(u)=u_x$.

Now, we consider the manifold of **G**-soliton solutions of (2.1). That is the set of all special solutions $u(x,t)$ of (2.1) where the time-development is given by the action of some one-parameter subgroup of **G** , i.e. those $u(x,t)$ such that there is some $k=(k_1,k_2, \ldots ,k_n)$ with

$$k_1 G_1(u)+k_2 G_2(u) \cdots +k_n G_n(u)=K(u). \tag{2.3}$$

These solutions are easily determined: Take a solution $s(k,x)$ of (2.3) and the one-parameter subgroup $\{g(\tau),\tau\in \mathbb{R}\}$ of **G** having the infinitesimal generator

$$<k,G>=k_1 G_1+k_2 G_2+ \cdots +k_n G_n\,, \quad k=(k_1,\ldots,k_n)$$

Then

$$u(x,t)=g(t)(s(k,x)) \tag{2.4}$$

is the **G**-soliton with initial condition $u(x,t=0)=s(k,x)$. This solution we denote by $s(k)$ or $s(k,t)$, it plays the role of the function given in (1.2).

Observe that, because of invariance with respect to time translation, the solution $s(k,x)$ of (2.3) has at least one free integration parameter, i.e. whenever $\beta\in$ **G** then application of β to $s(k,x)$ yields

$$s_\beta(k,x) = \beta(s(k,x)) \tag{2.5}$$

which again is a solution of (2.3). For abbreviation we denote this solution by $s_\beta(k)$ and the one obtained by application of β^{-1} we denote by $s_{-\beta}(k)$.

Now, we proceed as before. We look for a relation $B^{(\lambda)}(\cdot,\cdot)$ such that

$$B^{(\lambda)}(s_\beta,s_{-\beta}) = 0 \tag{2.6}$$

holds independently of the special group parameter k we had chosen. Since we have to do with a several parameter group our $\lambda=(\lambda_1, \ldots ,\lambda_n)$ now also has several components.

The function $B^{(\lambda)}(u,\bar{u})$ is said to be **admissible** around u and $\bar{u}$ if i) and ii) are fulfilled:

i) $B^{(\lambda)}$ is invariant against **G**

ii) $B^{(\lambda)}$ defines a diffeomorphism around (2.6)

The relation $B^{(\lambda)}$ is said to be **G**-invariant if whenever $B^{(\lambda)}(u,\bar{u})=0$ then

$$B^{(\lambda)}(g(u)\ ,\ g(\bar{u}\ ))=0 \tag{2.7}$$

must hold for every $g\in\mathbf{G}$. An equivalent infinitesimal condition is that

$$B^{(\lambda)}{}_{u}\ [G(u)]+B^{(\lambda)}{}_{\bar{u}}[G(\bar{u})]=0 \tag{2.8}$$

holds for all infinitesimal generators G of one-parameter subgroups of **G** . And $B^{(\lambda)}$ is said to be a diffeomorphism around (2.6) if the implicit function theorem condition holds, i.e. if the operators

$$B^{(\lambda)}{}_{u}\ (s_\beta,s_{-\beta})[\cdot],\qquad B^{(\lambda)}{}_{\bar{u}}(s_\beta,s_{-\beta})[\cdot]$$

are invertible operators between the tangent spaces of **S** at s_β and $s_{-\beta}$, respectively.

PROPOSITION:
Let Λ be the set of those λ where for every $k=(k_1,\cdots,k_n)$ there is some $\beta=\beta(k)$ with

$$B^{(\lambda)}(s_\beta(k),\ s_{-\beta}(k))=0$$

such that $B^{(\lambda)}$ is admissible around these points. Then for every element λ in the closure of Λ

$$B^{(\lambda)}(u\ ,\ \bar{u}\ )=0 \tag{2.9}$$

is an auto-Bäcklund transformation for (2.1).

The arguments leading to this proposition are the same as in section 1. For the moment we put $u=s_\beta$ and $\bar{u}=s_{-\beta}$. Then, since B is **G**-invariant, we have

$$B^{(\lambda)}{}_{u}\ [G_i(u)]+B^{(\lambda)}{}_{\bar{u}}[G_i(\bar{u})]=0,\ \ i=1,\cdots,n.$$

Because of (2.3) this yields

$$B^{(\lambda)}{}_{u}\ [K(u)]+B^{(\lambda)}{}_{\bar{u}}[K(\bar{u})]=0. \tag{2.10}$$

Since around the special u and $\bar{u}$ the quantity $B^{(\lambda)}$ defines a diffeomorphism this is an ordinary differential equation admitting as solution all solutions of (2.3) - and this for any arbitrary set $k_1,\ldots,k_n$. Now, since (2.10) has the same degree as (2.3) it has to many integration parameters (namely in addition to those of (2.3) the parameters $k_1,\ldots,k_n$). Hence (2.10) must be an identity which holds for **all** $u,\bar{u}$ connected by

$$B^{(\lambda)}(u\ ,\ \bar{u}\ )=0. \tag{2.11}$$

Therefore it must be an auto-Bäcklund transformation. The isospectral problem given by the Bäcklund transformations obtained this way is the same as before:

Spectral problem :
Given a solution u of (2.1), find those λ's such that there is some nonzero

vectorfield ω *and some* $\bar{u}$ *on the manifold under consideration such that*

$$B^{(\lambda)}{}_u(u,\bar{u})[\omega] = 0 \tag{2.12}$$

and

$$B^{(\lambda)}{}_u(u,\bar{u}) = 0\ . \tag{2.13}$$

Those λ*'s are said to be spectral points of the Bäcklund transformation and the* ω *appearing in (2.12) are called eigenvectors. This spectral problem is invariant under the flow (2.1).*

Again, the time development of its eigenvector ω is given by

$$\omega_t = K'(u)[\omega]. \tag{2.14}$$

Hence, if the spectral problem can be linearized, then the corresponding linear operator $\Phi(u)$ is part of a Lax-pair

$$\frac{d}{dt}\Phi(u) = K'(u)\Phi(u) - \Phi(u)K'(u) \tag{2.15}$$

which shows in particular that its eigenvectors are solutions of the linearized equation (or symmetry equation) (see [7]). In case the general solution decomposes asymptotically into single soliton solutions the interpretation of spectral points and eigenvectors is the same as before.

The spectral λ's correspond to the speeds of the asymptotic solitons. This correspondence can be found by looking at the single soliton solution where we pick out those λ's which are limits of group actions having parameters going to infinity in such a way that the soliton is removed out of sight into infinity. Of course, these λ's are depending now on the parameter k, from (2.3), characterizing the soliton under consideration. The eigenvector itself is the soliton which is lost by going from u to $\bar{u}$. This explains why $B^{(\lambda)}(u\ ,\bar{u}\,) = 0$ for spectral λ connects N-solitons with $(N-1)$-solitons.

Also **Bianchi-permutability** has a very simple explanation, it is a consequence of the abelian structure of **G** . To see this consider nonspectral λ and denote for the moment those $\bar{u}$ coming from $B^{(\lambda)}(u,\bar{u})$=0 by

$$\bar{u}=U(\lambda,u). \tag{2.16}$$

Recall that if u is a single soliton then $\bar{u}$ simply is u moved by some action of an element in **G** . Since **G** is commutative and since $B^{(\lambda)}(\cdot,\cdot)$ is **G**-invariant we must have

$$U(\lambda_2,U(\lambda_1,u))=U(\lambda_1,U(\lambda_2,u)) \tag{2.17}$$

or

$$u=\phi(u), \tag{2.18}$$

where

$$\phi(u)=U(-\lambda_2,U(-\lambda_1,U(\lambda_2,U(\lambda_1,u)))). \tag{2.19}$$

Certainly

$$\bar{u}-\phi(u)=0 \tag{2.20}$$

is an auto-Bäcklund transformation for (2.1) which reduces to the identity mapping on the G-solitons. Hence, this must be the identity mapping everywhere and (2.17) must hold for general u. However, only for nonspectral λ's on first view. But this induces also for all limits of nonspectral λ by density.

Let me emphasize that the condition of admissibility is an essential one. Condition i) was necessary in order to arrive at (2.10) and the following example will serve to see that condition ii) is equally important.

EXAMPLE:

Consider the square root of the single soliton solution of the KdV

$$\sigma(x+ct)=\sqrt{\frac{c}{2}}\cosh^{-1}\{\frac{\sqrt{c}}{2}(x-x_0+ct)\}. \tag{2.21}$$

We perform a translation on the x-line by $(2\Theta/\sqrt{c})$ and obtain for σ_Θ (resulting from this translation)

$$\sigma_\Theta(x)=\frac{\sqrt{c/2}\cosh(\Theta)\cosh(y)-\sqrt{c/2}\sinh(\Theta)\sinh(y)}{(\cosh^2(\Theta)\cosh^2(y)-\sinh^2(\Theta)\sinh^2(y))} \tag{2.22}$$

where

$$y=\frac{1}{2}\sqrt{c}\,(x-x_0+ct).$$

Splitting up into odd- and even parts

$$\sigma_\Theta=V_++V_-,\quad V_+=\frac{1}{2}(\sigma_\Theta+\sigma_{-\Theta}),\quad V_-=\frac{1}{2}(\sigma_\Theta-\sigma_{-\Theta}) \tag{2.23}$$

we find

$$2\{D^{-1}(V_+V_-)\}^2-\sinh^2(\Theta)V_+^2+\cosh^2(\Theta)V_-^2=0 \tag{2.24}$$

or

$$\{D^{-1}(\sigma_\Theta^2-\sigma_{-\Theta}^2)\}^2+2\{\sigma_\Theta^2+\sigma_{-\Theta}^2-2\alpha\sigma_\Theta\sigma_{-\Theta}\}=0. \tag{2.25}$$

Another relation between σ_Θ and $\sigma_{-\Theta}$ is found by choosing Θ such that

$$\cosh(\Theta)\sinh(\Theta)=\alpha\sqrt{c}.$$

Then

$$2D^{-1}(\sigma_\Theta^2-\sigma_{-\Theta}^2)-\alpha\sigma_\Theta\sigma_{-\Theta}=0. \tag{2.26}$$

Recalling that σ_Θ^2 is a single soliton of the KdV we have found the following relation between a single soliton u and suitable translates

$$\{\frac{1}{2}[D^{-1}(u-\bar{u})]^2+u+\bar{u}\}^2-4\alpha^2 u\bar{u}=0 \tag{2.27}$$

$$4\{D^{-1}(u-\bar{u})\}^2-\alpha^2 u\bar{u}=0. \tag{2.28}$$

Both relations are auto-Bäcklund transformations for the KdV **only** for very special choices of the parameters α . The reason they are not Bäcklund transformations is obviously that condition ii) in the admissibility requirements is not fulfilled. This because the inverse of the variational derivative of B leads out of the manifold under consideration. One might try to repair this by enlarging the manifold under consideration in dropping the boundary conditions at infinity. But then the manifold of single solitons also enlarges (theta-function solutions), and for those additional single solitons the relations (2.27) and (2.28) do not hold.

Let me conclude this section by some additional comments.

REMARKS:

1) It should be noted that if there is any **G**-invariant auto-Bäcklund transformation

$$B^{(\lambda)}(u\ ,\bar{u})=0 \tag{2.29}$$

for (2.1) with nonspectral λ then it must be found by our procedure. To see this take u to be some **G**-soliton, i.e. a solution of (2.3). Since $B^{(\lambda)}$ is invariant with rspect to **G** and the one parameter group given by $K(u)$ we find

$$B^{(\lambda)}{}_u\,[K(u)-\sum k_iG_i(u)]+B^{(\lambda)}{}_{\bar{u}}[K(\bar{u})-\sum k_iG_i(\bar{u})]=0. \tag{2.30}$$

But $K(u)-\sum k_iG_i(u)=0$ since u is a **G**-soliton. Since λ is assumed to be nonspectral this relation must also hold for $\bar{u}$ instead of u. So $\bar{u}$ again is a **G**-soliton (having the same $k_1,\ldots,k_n$ as u). This implies that $\bar{u}$ results out of u by some action of an element in **G** . Hence, $B^{(\lambda)}(u,\bar{u})=0$ can be found as a relation connecting **G**-solitons.

2) One might ask what happens in the KdV-case if **G** , which was the group of x-translations, is enlarged by adding those one-parameter groups having the well known symmetries as generators. Then, in this case, the **G**-solitons are just the multisolitons (up to some order depending on the dimension of **G**) and the Bäcklund transformations found by our procedure are simply the iterations of the well known Bäcklund transformation.

3) The content of this section was that, in a certain sense, the auto Bäcklund relations are completely determined by the one-soliton-solutions. On first view this seems to be rather surprising, but in fact it is not if one thinks of exactly solvable systems in terms of linearizable flows. Because, according to what we know up to now, the variable transformation (for example the inverse- scattering transform) which transforms the given flow into a linear one $u_t=Au$, A linear, transforms the one-soliton solutions into eigenvectors of A and the multisoliton solutions into linear combinations of such eigenvectors. Taking further into account that the auto-Bäcklund transformations connect N-soliton-solutions and $(N+1)$-soliton-solutions,

one easily guesses that in terms of the new variable the auto-Bäcklund relation must be of the form $(A-\lambda)u_2=u_1$, where λ is the eigenvalue of that part of u_2 which corresponds to the disappearing soliton. But certainly, for such a linear relation we can check whether or not it is an auto-Bäcklund relation on the linear hull of the eigenvectors by checking if it has this property on all single eigenvectors. Now, going back to the original manifold, where the flow was nonlinear, this then corresponds to checking out the required properties on the one-soliton-solutions.

4) One computational advantage of our derivation of ABT's is that we know right away what happens to solitons occurring asymptotically in fields of interacting solitons. Since the group element $\beta=\beta(k)$ in the proposition was chosen in such a way that (2.9) holds, we now know from G-invariance of $B^{(\lambda)}$ that a single soliton $s(k)$ emerging asymptotically out of u is transformed by $B^{(\lambda)}(u,\bar{u})=0$ into a single soliton $s(k)_{-2\beta(k)}$ emerging out of $\bar{u}$. The information about the phase shift now can be build up in the same way as in case of the KdV, provided the Bäcklund transformation respects the local structure of emerging solitons.

EXAMPLES AND APPLICATIONS

III. The Benjamin-Ono equation

We are interested in the BO ([2] ,[13]) which is

$$u_t = K(u), \quad K(u) = (2uu_x + Hu_{xx}) \tag{3.1}$$

where H is the Hilbert transform

$$(Hf)(x) = \frac{1}{\pi}\int_{+\infty}^{-\infty} \frac{f(\xi)}{\xi-x}d\xi \quad \textit{(principal value integration)}$$

and where u is assumed to be an element of S, now S being the space of functions having the property that all derivatives are absolutely integrable on the real line. For convenience we have defined $H1=0$.

Observe that K commutes with $G(u)=u_x$, the infinitesimal generator of the translation group. Equation (3.1) has (see [1,p.204])

$$s(k,x) = \frac{ik}{k(x-x_0)+i} - \frac{ik}{k(x-x_0)-i}, \quad x_o \in \boldsymbol{R} \tag{3.2}$$

as one-soliton solutions, i.e. as solutions of $ku_x=K(u)$. This is easily seen from

$$iH\left(\frac{1}{x-\alpha}\right) = \begin{cases} \dfrac{1}{x-\alpha} & \text{if } \textit{imaginary part}(\alpha) < 0 \\ -\dfrac{1}{x-\alpha} & \text{if } \textit{imaginary part}(\alpha) > 0 \end{cases} \tag{3.3}$$

We restrict our attention to positive k. Replacing x_0 in (3.2) by $\beta+x_0$ we obtain

$$s_\beta(x) = \eta_\beta(x) + \eta^*_\beta(x) \tag{3.4}$$

where * denotes complex conjugation and where

$$\eta_\beta = \frac{ik}{k(x-\beta-x_0)+i} \,. \tag{3.5}$$

Decomposition of (3.4) into odd and even parts (with respect to β)

$$s_\beta = V_+ + V_-, \quad V_+ = \frac{1}{2}(s_\beta + s_{-\beta}), \quad V_- = \frac{1}{2}(s_\beta - s_{-\beta}) \tag{3.6}$$

gives

$$V_+ = \frac{1}{2}\{\eta_\beta + \eta^*_\beta + \eta_{-\beta} + \eta^*_{-\beta}\} \tag{3.7}$$

$$V_- = \frac{1}{2}\{\eta_\beta + \eta^*_\beta - \eta_{-\beta} - \eta^*_{-\beta}\}. \tag{3.8}$$

Observe that integration and exponentiation of a simple pole yields the same pole. Application of this together with a decomposition into partial fractions leads to:

$$\exp(2iD^{-1}V_-) = \frac{\eta_\beta \cdot \eta^*_{-\beta}}{\eta^*_\beta \eta_{-\beta}} = 1 + 2\frac{\beta}{(k\beta-i)}\{\eta_\beta + \eta^*_{-\beta}\}. \tag{3.9}$$

Now, observe that application of iH to V_- changes the sign of the second and the fourth term and leaves the others unchanged. Hence the second and third terms cancel in $V_+ + iHV_-$. This yields

$$V_+ + iHV_- = \eta_\beta + \eta^*_{-\beta}.$$

From this together with (3.9) we obtain

$$\exp(2iD^{-1}V_-) = 1 + 2\lambda(iHV_- + V_+), \tag{3.10}$$

where

$$\lambda = \beta/(k\beta-i). \tag{3.11}$$

Inserting

$$V_+ = \frac{1}{2}(u + \bar{u}), \quad V_- = \frac{1}{2}(u - \bar{u}), \quad u = s_\beta, \; \bar{u} = s_{-\beta}$$

we obtain

$$\exp(iD^{-1}(u-\bar{u})) = 1 + \lambda\{iH(u - \bar{u}) + (u + \bar{u})\}. \tag{3.12}$$

Since this relation is admissible it must be an auto-Bäcklund transformation for (3.1). The spectral point corresponding to a soliton with speed k one easily obtains by putting $\beta=\infty$ in (3.11). Hence, each soliton with speed k generates a spectral point $\lambda=1/k$.

In the literature [15] sometimes the more general Bäcklund transformation

$$\mu \exp(iD^{-1}(u-\bar{u})) = 1 + \lambda\{iH(u - \bar{u}) + (u + \bar{u})\} \tag{3.13}$$

is given. This relation is generated out of (3.11) by the formal substitution

$$D^{-1} \rightarrow D^{-1} + constant.$$

However, if and only if $\mu=1$ this auto-Bäcklund transformation stays on the required manifold of functions disappearing at infinity.

There are other auto-Bäcklund transformations for the Benjamin-Ono equation. Consider

$$iHV_{+} = \frac{1}{2}\{\eta_{\beta} - \eta^{*}_{\beta} + \eta_{-\beta} - \eta^{*}_{-\beta}\} \tag{3.14}$$

$$iHV_{-} = \frac{1}{2}\{\eta_{\beta} - \eta^{*}_{\beta} - \eta_{-\beta} + \eta^{*}_{-\beta}\}. \tag{3.15}$$

Solving equations (3.7),(3.8),(3.14) and (3.15) for the η's we obtain

$$\eta_{\pm\beta} = \frac{1}{2}(1+iH)(V_{+} \pm V_{-}) \tag{3.16}$$

and

$$\eta^{*}_{\pm\beta} = \frac{1}{2}(1-iH)(V_{+} \pm V_{-}). \tag{3.17}$$

Integration and exponentiation yields the following two relations

$$\begin{aligned} \exp(iD^{-1}(1+iH)V_{-}) &= \eta_{\beta}/\eta_{-\beta} \\ &= 1 + 2i\beta\eta_{\beta} = 1 + i\beta(1+iH)(V_{+} + V_{-}). \end{aligned} \tag{3.18}$$

$$\begin{aligned} \exp(iD^{-1}(1-iH)V_{-})) &= \eta^{*}_{-\beta}/\eta^{*}_{\beta} \\ &= 1 - 2i\beta\eta^{*}_{-\beta} = 1 - i\beta(1-iH)(V_{+} - V_{-}). \end{aligned} \tag{3.19}$$

Hence resubstitution by (3.11) yields a new auto-Bäcklund transformation consisting of the following pair of relations

$$\exp(\frac{i}{2}D^{-1}(1+iH)(u-\bar{u})) = 1 + i\beta(1+iH)u \tag{3.20}$$

$$\exp(\frac{i}{2}D^{-1}(1-iH)(u-\bar{u})) = 1 - i\beta(1-iH)\bar{u} \ . \tag{3.21}$$

This transformation certainly is not equivalent to the one given by (3.11) since it has no spectral points. The transformation splits into two parts corresponding to the eigenspaces of the operator H . This decomposition is the usual decomposition into functions being analytic in the upper and lower half of the complex plane, respectively.

IV Burger's equation

It is obvious that the BO contains Burgers equation as a special case. For example if a solution u of BO is analytic in the upper half of the complex plane then $Hu=iu$ and u therefore must be a solution of

$$u_t = (2uu_x + iu_{xx}). \tag{4.1}$$

The same holds true for the lower half plane, we only have to replace i by $-i$. This means that the flow of the BO restricted to certain invariant manifolds reduces to versions of Burgers equation.

However this does not mean that auto-Bäcklund transformations for BO automatically yield ABT's for Burgers equation because such an ABT is not necessarily compatible with the restriction to invariant manifolds. For this reason (3.12) cannot be used to obtain an ABT for Burgers equation. In contrast to that (3.20) + (3.21) give right away an ABT for Burgers equation since these are compatible with the restriction to functions analytic either in the upper or lower half of the complex plane. Let us carry out the details for the case of the upper half of the complex plane. In this case only (3.21) is meaningful. Replacing H by i we obtain

$$\exp(iD^{-1}(u-\bar{u})) = 1 + 2i\beta\bar{u} \tag{4.2}$$

as an ABT for (4.1). Performing the substitution $t \rightarrow it/\delta$, $x \rightarrow ix/\delta$ we obtain for arbitrary values of α that

$$\exp(\delta^{-1}D^{-1}(\bar{u}-u)) = \alpha\bar{u} + 1 \tag{4.3}$$

must be an ABT for

$$u_t = 2uu_x + \delta u_{xx} \tag{4.4}$$

which is the general form of Burgers equation. Those readers who do not feel at ease with the methods we proposed in this paper should take the opportunity to check that statement by a simple direct computation.

This example actually shows that our method constitutes a helpful tool. It enabled us to compute out of the symmetry structure of a special family of functions (3.2) an ABT for an evolution equation for which these functions were not even solutions.

In this case the recursion operator is easily obtained by the nonlinear spectral problem given by this auto-Bäcklund transformation (see [9]) although the boundary conditions at infinity have to be changed in the formulation of the spectral problem.

One might ask if in case of BO the nonlinear spectral problem also leads to a recursion operator. Indeed this is the case, however the procedure is very complicated and the recursion operator obtained is of a form involving u and other functions only determined by u via an implicit function.

V. A Nonlinear System

We investigate a nonlinear system of differential equations for $w_i = w_i(x,t)$, $i=1,2$, which is intimately connected to the isospectral flow of the KdV.

We introduce the differential operators

$$L(u) = D^2 + u \tag{5.1}$$

$$T(\phi) = 2\phi D + 6\phi_x \tag{5.2}$$

and we use the notation

$$w = \begin{bmatrix} w_1 \\ w_2 \end{bmatrix} \tag{5.3}$$

$$|w|^2 = w_1{}^2 + w_2{}^2 \tag{5.4}$$

$$M = \begin{bmatrix} 0 & -1 \\ 1 & 0 \end{bmatrix} \tag{5.5}$$

$$\tau(w) = \begin{bmatrix} T(w_1) & 0 \\ 0 & T(w_2) \end{bmatrix} \quad . \tag{5.6}$$

The system we are interested in is the following

$$\tau(w)Mw_t = \tau(w)L(\,|\,w\,|^2)w. \tag{5.7}$$

or in case that $\tau(w)$ is invertible

$$w_t = ML(\,|\,w\,|^2)w.$$

This equation has two obvious symmetry groups, namely the translation group with infinitesimal generator w_x and a gauge group (rotation in $\mathbb{R}_2$ or $\mathbb{C}_2$ leaving $|w|^2$ invariant). The infinitesimal generator of the gauge group is Mw .

Up to now, we have only used one-soliton manifolds defined by the infinitesimal generator of x-translation. In our present example we take the generator of the gauge group instead, i.e. we define for the moment the one-soliton manifold by

$$\kappa\tau(w)w = \tau(w)L(\,|\,w\,|^2)w \tag{5.8}$$

where the speed parameter is now called κ .

In order to compute the one-soliton manifold we observe that the single-soliton of the KdV is the well-known Bargmann potential [10] which is the same as saying that for the single-soliton of the KdV the square of the corresponding eigenvector of the Schrödinger operator is equal to u (see [5]). The reason for this fact is the following identity which holds for all $\phi = \phi(x)$ and $u = u(x)$

$$\Theta(u)\phi^2 = T(\phi)L(\,u\,)\phi \tag{5.9}$$

where

$$\Theta(u) = D^3 + 2Du + 2uD.$$

Because of the special form of $T(\phi)$ this yields that the relation

$$\Theta(u)\phi^2 = 4\alpha(\phi^2)_x \tag{5.10}$$

holds if and only if

$$T(\phi)(L(u)-\alpha)\phi = 0. \tag{5.11}$$

The function

$$s(c) = s(c,x) = \frac{c}{2}\cosh^{-2}\{\frac{\sqrt{c}}{2}(x - x_0 + ct)\} \tag{5.12}$$

given in (1.2) fulfills by definition

$$cs_x = \Theta(s)s \tag{5.13}$$

because the KdV is of the form $u_t=\Theta(u)u_x$. So, in case that $4\alpha=c$ the function $\phi^2=s(c)$ must be a solution of (5.10). Hence, $\phi = \sqrt{s(c)}$ solves (5.11) and we have

$$T(\phi)L(\phi^2)\phi = \frac{1}{4}cT(\phi)\phi\ , \quad \phi = \sqrt{s(c)}. \tag{5.14}$$

Comparing this with (5.8) we have found the one-soliton-manifold for (5.7). In order to see this we observe that the time evolution on the one-soliton manifold, now given by the gauge-group, must be of the form

$$w(x,t) = \exp(t\kappa M)\psi(x)\ , \quad \psi = \begin{bmatrix}\psi_1\\ \psi_2\end{bmatrix} \tag{5.15}$$

where the components of ψ must satisfy (5.8). If we require $\psi_2 = 0$ and consider $\kappa = c/4$ then because of (5.14) $\psi_1 = \phi$ is a solution of this equation. Looking at the kernel of $T(\phi)$ and knowing that the eigenvalues of the Schrödinger operator have multiplicity 1 we easily observe that the general solution of (5.8) must come out of this solution by an application of the gauge group. Hence the general solution of (5.8) must be

$$w = \begin{bmatrix}\cos(\rho)\sqrt{s(c,x)}\\ \sin(\rho)\sqrt{s(c,x)}\end{bmatrix}, \quad \kappa = c/4. \tag{5.16}$$

Since (1.14) gives an ABT for the functions $s(c,x)$ we are now able to rephrase this in order to obtain an ABT for the solution w of (5.7). We apply the translation (1.3) to the solution (5.12) and decompose into odd and even parts. Using the notation of section I and looking at formula (1.6) we find

$$V_+ = \sqrt{2}\ kN(k\xi)\{\cosh(k\xi)\cdot\cosh(\beta)\} \tag{5.17}$$

$$V_- = -\sqrt{2}\ kN(k\xi)\{\sinh(k\xi)\cdot\sinh(\beta)\}. \tag{5.18}$$

A straightforward computation yields

$$(V_+)_x = V_-\ k\coth(\beta)\ \{2\cosh^2(k\xi)\ N(k\xi)-1\} \tag{5.19}$$

$$= V_- \, k \coth(\beta) \, \frac{\cosh^2(k\xi) - \sinh^2(\beta)}{\cosh^2(k\xi) + \sinh^2(\beta)}$$

$$= V_- \sqrt{k^2 \coth^2(\beta) - \frac{4k^2 \cosh^2(k\xi) \cosh^2(\beta)}{(\cosh^2(k\xi) + \sinh^2(\beta))^2}}$$

$$= V_- \sqrt{\lambda^2 - 2V_+^2}, \quad \lambda = k \coth(\beta).$$

So in case of $\rho=0$ we have found for the special solution (5.16) the following relation between w and its translation $\overline{w}$

$$(w_1 + \overline{w}_1)_x = (w_1 - \overline{w}_1)\sqrt{\lambda^2 - \frac{1}{2}(w_1 + \overline{w}_1)^2}. \tag{5.20}$$

Since this relation is not gauge-invariant it cannot be the ABT we are looking for. But, certainly the ABT must be the gauge-invariant relation coming out of this

$$(w_i + \overline{w}_i)_x = (w_i - \overline{w}_i)\sqrt{\lambda^2 - \frac{1}{2}| \, w + \overline{w} \, |^2}, \quad i = 1,2. \tag{5.21}$$

However this relation is still unsatisfactory because we have not yet used the complete single soliton manifold given by the two one-parameter groups taken into account (translation and gauge). Considering this extended manifold we should get an auto-Bäcklund transformation with two free parameters instead of one.

For obtaining the second parameter we let the effect of the application of the gauge group depend on x since this can be compensated by adding to the equation under consideration a linear combination of infinitesimal generators of the translation group. To be precise: Replacing (5.16) by

$$w = \begin{bmatrix} \cos(\gamma x + \rho)\sqrt{s(c,x)} \\ \sin(\gamma x + \rho)\sqrt{s(c,x)} \end{bmatrix} \tag{5.22}$$

we see that this must be a solution to

$$\kappa w + 2\gamma M w_x + \gamma^2 w = L(| \, w \, |^2) w, \tag{5.23}$$

hence a solution on the soliton manifold given by gauge and translation. Insertion of (5.22) into (5.21) shows that the contributions coming from the x-derivatives of $\cos(\gamma x)$ and $\sin(\gamma x)$ can be compensated by replacing w_{1x} and w_{2x} by $w_{1x} + \gamma w_2$ and $w_{2x} - \gamma w_1$, respectively. The same has to be done for the components of $\overline{w}$. This yields

$$(w + \overline{w})_x = \gamma M(w + \overline{w}) + (w - \overline{w})\sqrt{\lambda^2 - \frac{1}{2}| \, w + \overline{w} \, |^2} \tag{5.24}$$

as an auto-Bäcklund transformation for (5.7).

VI. The nonlinear Schrödinger equation (NLS)

Since the NLS is the restriction of (5.7) to a special invariant submanifold we easily find an ABT for the NLS.

Consider solutions of (5.7) where both components vanish rapidly at $\pm\infty$. Then on this manifold the kernel of $\tau(w)$ consists only of the zero-element. So in this case (5.7) has the form

$$w_{1t} = -w_{2xx} - (w_1^2 + w_2^2)w_2 \qquad (6.1)$$
$$w_{2t} = +w_{1xx} + (w_1^2 + w_2^2)w_1 .$$

Consider real valued solutions of this equations and look at the corresponding evolution equation for $\psi = w_1 + iw_2$ instead of (6.1).

This equation is easily found to be

$$-i\psi_t = (D^2 + \psi\psi^*)\psi \qquad (6.2)$$

since M corresponds to multiplication with i . Hence, all those auto-Bäcklund transformations (5.23) which connect real valued solutions vanishing rapidly at $\pm\infty$ are ABT's for the NLS.

Rewriting (5.23) in complex form we obtain

$$(\psi + \overline{\psi})_x = i\gamma(\psi + \overline{\psi}) + (\psi - \overline{\psi})\sqrt{\lambda^2 - \frac{1}{2}|\psi + \overline{\psi}|^2} \qquad (6.3)$$

as ABT for (6.1). This is exactly the ABT known from the literature ([3] , [14]). In other words: The ABT (5.23) is the analytic extension of the well-known ABT for the NLS.

VII. The inverse problem

We can also run in the opposite direction through our recipes. For example, if we have a one-parameter family of functions admitting a relation (independent of the parameter) between these functions and corresponding translated functions, then, in the admissible case, this relation is an ABT for the evolution equation(s) having the translations of our one-parameter family of functions as one-soliton manifold.

Let us illustrate this for the most simple example. Consider a translation-invariant evolution equation having

$$s_k(x,t) = \cos\left(\frac{1}{k}(x-kt)\right) \qquad (7.1)$$

as one-soliton solutions. Such an equation is easily found

$$u_t^2 + u^2 = 1. \qquad (7.2)$$

Then looking at the special form of (7.1) one immediately sees that there should be a one-parameter family of ABT for (7.2), since translations of (7.1) can be expressed in cosine-functions.

Let us carry out the details. Consider

$$s(k,x) = \cos(x/k) \tag{7.3}$$

$$s_\beta(k,x) = \cos((x+\beta)/k) = \cos(\beta/k)\cos(x/k) - \sin(\beta/k)\sin(x/k). \tag{7.4}$$

This yields the relation

$$(s_\beta - s\cos(\beta/k))^2 = \sin^2(\beta/k)(1 - s^2). \tag{7.5}$$

Since this relation is independent of k it must be an ABT for the equation (7.2). Thus, replacing $\cos(\beta/k)$ by λ we have found an ABT for

$$(u - \lambda\bar{u})^2 = (1 - \lambda^2)(1 - \bar{u}^2) \tag{7.6}$$

or

$$u^2 + \bar{u}^2 - 2\lambda u\bar{u} = 1 - \lambda^2. \tag{7.7}$$

Actually, there is one dubious point: Equation (7.2) is not exactly of the form (2.1) which was considered in the second section. Since the square root is a multivalued function it cannot be transformed into (2.1). That means the notions we have given so far cannot be applied to this situation. But an intuitive feeling for what goes on tells us that certainly our notions and methods carry over to evolution equations which are given in (eventually multivalued) implicit form.

To phrase the result of this section vaguely: Two-parameter families of functions having strong symmetry properties can be considered as one-soliton solutions of evolution equations admitting auto-Bäcklund relations.

BIBLIOGRAPHY

[1] Ablowitz, M. J. and H. Segur, Solitons and the Inverse Scattering Transform, Siam studies in applied Mathematics, SIAM, Philadelphia, 1981.

[2] Benjamin, T. B., "Internal Waves of finite amplitude and permanent form," J.Fluid Mech., vol. 29, pp. 559-592, 1967.

[3] Chen, H.H., "Relation between Bäcklund Transformations and Inverse Scattering Problems," in: Lecture Notes in Mathematics (R.M. Miura ed.) Springer Verlag, Berlin-Heidelberg-New York, vol. 515, pp. 241-252, 1974.

[4] Fuchssteiner, B. and A. S. Fokas, "Symplectic Structures, Their, Bäcklund Transformations and Hereditary Symmetries," Physica, vol. 4 D, pp. 47-66, 1981.

[5] Fuchssteiner, B., "Pure soliton solutions of some nonlinear Partial differential equations," Comm. math. Phys., vol. 55, pp. 187-194, 1977.

[6] Fuchssteiner, B., "Application of spectral gradient methods to nonlinear evolution equations," preprint, 1979.

[7] Fuchssteiner, B., "Application of Hereditary Symmetries to Nonlinear Evolution equations," Nonlinear Analysis TMA, vol. 3, pp. 849-862, 1979 .

[8] Fuchssteiner, B., "Mastersymmetries, Higher-order time-dependent Symmetries and conserved Densities of Nonlinear Evolution Equations," Progr. Theor.Phys., vol. 70, pp. 1508-1522, 1983.

[9] Fuchssteiner, B. and R.N. Aiyer , "Multisolitons, or the discrete Eigenfunctions of the Recursion Operator of non-linear Evolution Equations: II. Background," J. Phys., vol. A 19, 1986.

[10] Lamb, G. L., Elements of Soliton theory, Pure and Applied Mathematics series, Wiley, New York-Chichester-Brisbane-Toronto, 1980.

[11] Lax, P.D., "Integrals of nonlinear Equations of Evolution and Solitary Waves," Commun.Pure Apl. Math., vol. 21, pp. 351-375, 1976.

[12] Olver, P.J., "Evolution Equations posessing infinitely many Symmetries," J.Math.Phys., vol. 18, pp. 1212-1215, 1977.

[13] Ono, J.H., "Algebraic Solitary waves in stratified Fluids," J.Phys.Soc. Japan, vol. 39, pp. 1082-1091, 1975.

[14] Roger, C. and W. F. Shadwick, Bäcklund Transformations and their Applications, Mathematics in Science and Engineering Vol. 161, Academic Press, New York - London - Paris - Sydney - Tokyo - Toronto, 1982.

[15] Satsuma, J., M.J. Ablowitz, and Y. Kodama, "On an internal wave equation describing a stratified fluid with finite depth," Physics Letters, vol. 73 A, pp. 283-286, 1979.

EXACT SOLUTIONS OF BURGERS' EQUATION WITH REACTION TERMS

Junkichi Satsuma
Department of Applied Physics, Faculty of Engineering,
University of Tokyo, Bunkyo-ku, Tokyo 113
JAPAN

ABSTRACT

Exact solutions are found for Burgers' equation with reaction terms, by reducing them to bilinear forms through dependent variable transformations. Included are travelling wave solutions and the one describing the coalescence of two travelling waves.

1. INTRODUCTION

It is well known that Burgers' equation,

$$u_t = u_{xx} + \alpha u u_x, \tag{1}$$

describes the far field of wave propagation in nonlinear dissipative systems. This equation admits a solution in the form of a travelling wave,

$$u = \frac{2k}{\alpha} + \frac{2k}{\alpha} \tanh k(x+2kt), \tag{2}$$

where k is arbitrary. The existence of such a solution is due to the balancing between two effects, dissipation and convection. It is a remarkable fact that initial value problems of (1) can be solved exactly by reducing it to the linear diffusion equation,

$$f_t = f_{xx}, \tag{3}$$

through the Hopf-Cole transformation,

$$u = \frac{2}{\alpha} (\log f)_x . \tag{4}$$

The solution shows that, if many travelling waves exist initially, then they coalesce into one travelling wave asymptotically.

The nonlinear diffusion equation of the Fisher type,

$$u_t = u_{xx} + F(u), \tag{5}$$

where F(u) is a polynomial of u, is proposed as a model equation describing a diffusion process with reaction effect. Equation (5) also admits a travelling wave solution. In this case, the existence is due to the balancing between the effects of diffusion and reaction.

Attempts to find exact solutions of (5) have been made for the cases that the leading terms of F(u) are quadratic or cubic in u. Ablowitz and Zeppetella[1)] considered the equation,

$$u_t = u_{xx} + u(1-u). \tag{6}$$

They applied the Painlevé analysis which has been successfully used in the soliton theory and obtained the explicit expression of a travelling wave,

$$u = \frac{1}{[1 + \exp\{(\pm x - 5t)/\sqrt{6}\}]^2} . \tag{7}$$

Kawahara and Tanaka[2)] investigated the equation,

$$u_t = u_{xx} + u(1-u)(u-\gamma). \tag{8}$$

This equation has three travelling wave solutions which connect two constant states, u = 0,1, u = 0,γ and u = γ,1, respectively. The explicit forms are as follows;

$$u = \frac{1}{2} + \frac{1}{2} \tanh \frac{1}{2\sqrt{2}} [\pm x - (\sqrt{2}\gamma - \frac{1}{\sqrt{2}})t], \tag{9a}$$

$$u = \frac{\gamma}{2} + \frac{\gamma}{2} \tanh \frac{\gamma}{2\sqrt{2}} [\pm x - (\sqrt{2} - \frac{\gamma}{\sqrt{2}})t], \tag{9b}$$

$$u = \frac{1+\gamma}{2} + \frac{1-\gamma}{2} \tanh \frac{1-\gamma}{2\sqrt{2}} [\pm x + \frac{1+\gamma}{\sqrt{2}} t]. \tag{9c}$$

They employed Hirota's method which has also been powerful in the soliton theory and obtained an exact solution which describes the coalescence of two travelling waves,

$$u = \frac{\exp(\eta_1) + \gamma \exp(\eta_2)}{1 + \exp(\eta_1) + \exp(\eta_2)}, \tag{10}$$

where

$$\eta_1 = \frac{1}{\sqrt{2}}[x - (\sqrt{2}\gamma - \frac{1}{\sqrt{2}})t] + \eta_1^{(0)}, \tag{11a}$$

$$\eta_2 = \frac{\gamma}{\sqrt{2}}[x - (\sqrt{2} - \frac{\gamma}{\sqrt{2}})t] + \eta_2^{(0)}, \tag{11b}$$

and where $\eta_1^{(0)}$ and $\eta_2^{(0)}$ are arbitrary phase constants.

Compared with the equations describing nonlinear dispersive systems, there are not so many nonlinear diffusion equations for which exact solutions are known. Hence, it is an interesting problem to search for exact solutions of various types of nonlinear diffusion equations. The solutions may play a certain role to understand the structure of general nonlinear diffusive systems.

Motivated by the above reason, we try to find exact solutions of the equation

$$u_t = u_{xx} + \alpha u u_x + \beta F(u), \tag{12}$$

which may be termed Burgers' equation with reaction term. It is not clear at present whether (12) has any physical meaning. We apply Hirota's method to treat (12).

In §2, we study the case, $F(u) = u(1-u)$. By reducing (12) into a bilinear form, we obtain a travelling wave solution. In §3, we study the case, $F(u) = u(1-u)(u-\gamma)$. In this case, we find a solution describing the coalescence of two travelling waves. Finally in §4, we give a concluding remark.

2. SOLUTIONS FOR $F(u) = u(1-u)$

In this section, we consider the equation

$$u_t = u_{xx} + \alpha u u_x + \beta u(1-u). \qquad (13)$$

By the dependent variable transformation,

$$u = g/f \qquad (14)$$

(13) is transformed into

$$g_t f - g f_t = g_{xx} f - 2g_x f_x - g f_{xx} + \alpha g g_x - \beta g(g-f) + \frac{g f_x (2f_x - \alpha g)}{f} \qquad (15)$$

Let us demand that (15) should be in a bilinear form with respect to f and g. Except for trivial cases, we find that, if the condition

$$g = \frac{2}{\alpha} f_x \qquad (16)$$

or

$$u = \frac{2}{\alpha} (\log f)_x \qquad (17)$$

is satisfied, then (15) reduces to the bilinear form,

$$f_{xt} f - f_x f_t = f_{xxx} f - f_{xx} f_x - \frac{2\beta}{\alpha} f_x^2 + \beta f_x f. \qquad (18)$$

A travelling wave solution which tends to zero as $x \to +\infty$ or $-\infty$ is obtained by assuming

$$f(x,t) = 1 + e^{kx-\omega t+\eta^{(0)}} \qquad (19)$$

where $\eta^{(0)}$ is an arbitrary phase constant. Substituting (19) into (18), we find that (19) is a solution if $k = \alpha/2$ and $\omega = -\alpha^2/4 - \beta$. The explicit expression for u is

$$u = \frac{1}{2} + \frac{1}{2} \tanh \frac{1}{4}\alpha \{x + \frac{1}{2} (\alpha + \frac{4\beta}{\alpha})t\}, \qquad (20)$$

where the phase constant is omitted. This solution describes a travelling wave which monotonically changes from 0 to 1, and has the velocity $-(\alpha + 4\beta/\alpha)/2$. It propagates in the negative direction if $\beta > -\alpha^2/4$ and in the positive direction if $\beta < -\alpha^2/4$. Especially if $\beta = -\alpha^2/4$, the solution is in an equilibrium state.

In the limit of $\beta \to 0$, (20) reduces to a specific one of (2) with the wave number $k = \alpha/4$. On the other hand, in the limit of $\alpha \to 0$, (20) loses the character of a travelling wave and simply becomes an oscillating solution. The solution found in ref.1 is not recovered from (18). In order to obtain it, we put $g = 1$ in (15). Then (15) reduces to

$$f_t f - f_{xx} f - \beta f(1-f) + 2f_x^2 - \alpha f_x = 0. \tag{21}$$

If we assume the solution expressed by an exponential function, we find that

$$f(x,t) = 1 + e^{\eta} + \frac{1}{4} e^{2\eta}, \tag{22}$$

$$\eta = \pm\sqrt{\beta/6}\ x - \frac{5}{6} \beta t + \eta^{(0)} \tag{23}$$

satisfies (21) only when $\alpha = 0$. The solution (22) is equivalent to (7).

It is conjectured that for (15) with $\beta \neq 0$, (19) and (22) are the only solutions which have simple analytical structure.

3. SOLUTIONS FOR $F(u) = u(1-u)(u-\gamma)$

In this section, we consider the equation

$$u_t = u_{xx} + \alpha u u_x + \beta u(1-u)(u-\gamma). \tag{24}$$

Without loss of generality, we may assume $0 < \gamma < 1$. By the same argument as in the previous section, we find that (24) reduces to

$$f_{xt} f - f_x f_t = f_{xxx} f + (\alpha A - 3) f_{xx} f_x + \beta(\gamma + 1) A f_x^2 - \beta\gamma f_x f \tag{25}$$

through the dependent variable transformation,

$$u = A(\log f)_x. \tag{26}$$

The coefficient A in (26) should satisfy

$$\beta A^2 + \alpha A - 2 = 0. \tag{27}$$

The transformation (26) coincides with (4) if $\beta = 0$ and is equivalent to the one employed in ref.2 if $\alpha = 0$.

Assuming a solution of the form,

$$f = e^{\delta x}(1 + e^{kx-\omega t+\eta^{(0)}}) \tag{28}$$

and substituting (28) into (25), we find that (28) gives nontrivial solutions for the following six sets of the parameters δ, k and ω;

$$\delta = 0, \quad k = \frac{1}{A}, \qquad \omega = \beta\gamma - \frac{1}{A^2}, \tag{29a}$$

$$\delta = \frac{1}{A}, \ k = -\frac{1}{A}, \qquad \omega = \beta\gamma - \frac{1}{A^2}, \tag{29b}$$

$$\delta = 0, \quad k = \frac{\gamma}{A}, \qquad \omega = \beta\gamma - \frac{\gamma^2}{A^2}, \tag{30a}$$

$$\delta = \frac{\gamma}{A}, \ k = -\frac{\gamma}{A}, \qquad \omega = \beta\gamma - \frac{\gamma^2}{A^2}, \tag{30b}$$

$$\delta = \frac{1}{A}, \ k = -\frac{1-\gamma}{A}, \ \omega = \frac{1-\gamma^2}{A^2}, \tag{31a}$$

$$\delta = \frac{\gamma}{A}, \ k = \frac{1-\gamma}{A}, \qquad \omega = -\frac{1-\gamma^2}{A^2}. \tag{31b}$$

The parameter δ designates the asymptotic value of u at $x = +\infty$ or $-\infty$. Among(29a)-(31b), (·a) gives the same u as (·b), respectively. Consequently, we obtain the following three travelling wave solutions of (24);

$$u = \frac{1}{2} + \frac{1}{2}\tanh\frac{1}{2A}\{x - (\beta\gamma A - \frac{1}{A})t\}, \tag{32a}$$

$$u = \frac{\gamma}{2} + \frac{\gamma}{2} \tanh \frac{\gamma}{2A} \{x - (\beta A - \frac{\gamma}{A})t\} \ , \tag{32b}$$

$$u = \frac{1+\gamma}{2} + \frac{1-\gamma}{2} \tanh \frac{1-\gamma}{2A} \{x + \frac{1+\gamma}{A} t\} \ , \tag{32c}$$

where the phase constants are omitted. These solutions reduce to the specific travelling wave solutions of Burgers' equation if $\alpha \to 0$ and to the travelling wave solutions (9a)-(9c) if $\beta \to 0$.

Let us evaluate the deviations of the phase part of (32a),

$$\eta = \frac{1}{2A} \{x - (\beta\gamma A - \frac{1}{A})t\} \ ,$$

from the limiting cases $\beta = 0$ and $\alpha = 0$.

If we choose $\alpha = 1$ for simplicity and assume $0 < \beta << 1$, we obtain

$$\eta \approx \frac{1}{4(1-2\beta)} [x + \{ \frac{1}{2} + \beta(1 - 2\gamma)\}t]. \tag{33}$$

Compared to the solution (2) with $\alpha = 1$ and $k = 1/4$, we see that the width of wave increases by $(1-2\beta)$ times and the velocity changes by $\beta(1-2\gamma)$.

If we choose $\beta = 1$ and assume $0 < \alpha << 1$, we obtain

$$\eta \approx \frac{1}{2\sqrt{2}(1\mp \alpha/\ 2\sqrt{2})} [\pm x - \{\sqrt{2}\gamma - \frac{1}{\sqrt{2}} \mp \frac{1}{4} (\ 1 + 2\gamma)\alpha\}t]. \tag{34}$$

Compared to the solution (9a), we see that the width of wave changes by $(1\mp\alpha/2\sqrt{2})$ times and the velocity by $\mp(1+2\gamma)\alpha/4$ for the upper and lower signs, respectively.

A remarkable property of (25) is that it admits an exact solution showing the coalescence of two travelling waves. In fact, by direct substitution we find that

$$f = 1 + \exp(\eta_1) + \exp(\eta_2), \tag{35}$$

$$\eta_1 = \frac{1}{A}\{x - (\beta\gamma A - \frac{1}{A})t\} + \eta_1^{(0)}, \tag{36a}$$

$$\eta_2 = \frac{\gamma}{A}\{x - (\beta A - \frac{\gamma}{A})t\} + \eta_2^{(0)}, \tag{36b}$$

satisfies (25). This solution describes a nonstationary state that a travelling wave connecting 0 and γ, and another wave connecting γ and 1 merge together to make a travelling wave connecting 0 and 1. It is noted that this solution is an extension of (10) with (11).

4. CONCLUDING REMARK

In the soliton theory, the bilinear forms of nonlinear evolution equations written by Hirota's operators play important roles in obtaining exact soliton solutions, in studying the relationship among the nonlinear evolution equations and in finding the linear problems associated with the equations.[3] The structure of bilinear equations has been made clear to a considerable extent by means of the method of algebraic analysis.[4]

The bilinear forms of equations given in this paper are the extended ones in the sense that they are not written only by Hirota's operators. It is a characteristic of the extended bilinear equations that they have the solutions of the type (37). It may be an interesting problem to study the structure of such equations.

REFERENCES

1) Ablowitz, M.J. and Zeppetella, A., Bull. Math. Bio. 41, 835 (1979).
2) Kawahara, T. and Tanaka, M., Phys. Letters 97A, 311 (1983).
3) Hirota, R., Proceeding of RIMS Symposium on Non-Linear Integrable Systems — Classical Theory and Quantum Theory, ed. by Jimbo, M. and Miwa, T. (World Science Publishing Co., Singapore, 1983) 15.
4) Date, E., Kashiwara, M., Jimbo, M. and Miwa, T., ibd., 39.

INVESTIGATION OF THE PERIODIC LIOUVILLE EQUATION

E. R. Tracy and A. J. Neil
Department of Physics
College of William and Mary
Williamsburg, VA 23185

H. H. Chen
Department of Physics and Astronomy
University of Maryland
College Park, MD 20742

C. H. Chin
Department of Electrophysics
National Chiao-Tung University
Hsin-Chu, Taiwan

ABSTRACT

The Liouville equation with periodic boundary conditions is discussed. In previous papers it was shown how to solve this equation by treating it as a singular limit of the sinh-Poisson equation. A few simple cases are examined here in detail to illustrate the approach.

1. INTRODUCTION:

The Liouville equation:

$$\left(\frac{\partial^2}{\partial x^2} + \frac{\partial^2}{\partial y^2}\right)u(x,y) = \nabla^2 u = \frac{\lambda^2}{8}\, e^{-u} \tag{1}$$

arises as a model equation in many different areas of mathematical physics. It appears in the study of two dimensional equilibria in both inviscid fluids[1] and ideal magnetohydrodynamics[2]. Equation (1) has also been studied in connection with instantons and solitons in field theory [3,4], the quantum theory of bosonic strings[5], and as a solvable model in the study of critical phenomena[6].

The Liouville equation has the following general solution:

$$u(x,y) = -\ln \left[\frac{F'(x+iy)G'(x-iy)}{(1+\frac{\lambda^2}{16} F(x+iy)G(x-iy))^2} \right] \qquad (2)$$

where: $F'(t) = \frac{dF}{dt}$, $G'(t) = \frac{dG}{dt}$, $i^2 = -1$, and F and G are arbitrary functions.

Unfortunately this general form has not proven to be useful for two important reasons:

1) It is difficult to use Eq. (2) to satisfy specified boundary conditions.
2) It is not known how to guarantee, using Eq. (2), that solutions of Eq. (1) will be real for all x and y.

Both of these problems must be solved for Eq. (1) to be useful in physical applications.

A well known special solution to Eq. (1) is:

$$u(x,y) = 2\ln \left[\frac{\lambda}{4\sqrt{1-\epsilon^2}} (\cosh x + \epsilon \cos y) \right] \qquad (3)$$

For $\lambda > 0$ and $0 \le \epsilon^2 < 1$. This is called the "island chain"[2] or "cat's eyes"[1] solution. Figure 1 is a contour plot of this solution, from this figure it is evident how the solution got its nickname.

Until the work of refs. (3) and (4) appeared the island chain solution was (to our knowledge) the only real solution ever extracted from Eq. (2). This is curious since Eq. (1) can be shown to be the compatibility condition for the following Lax pair of linear operators:

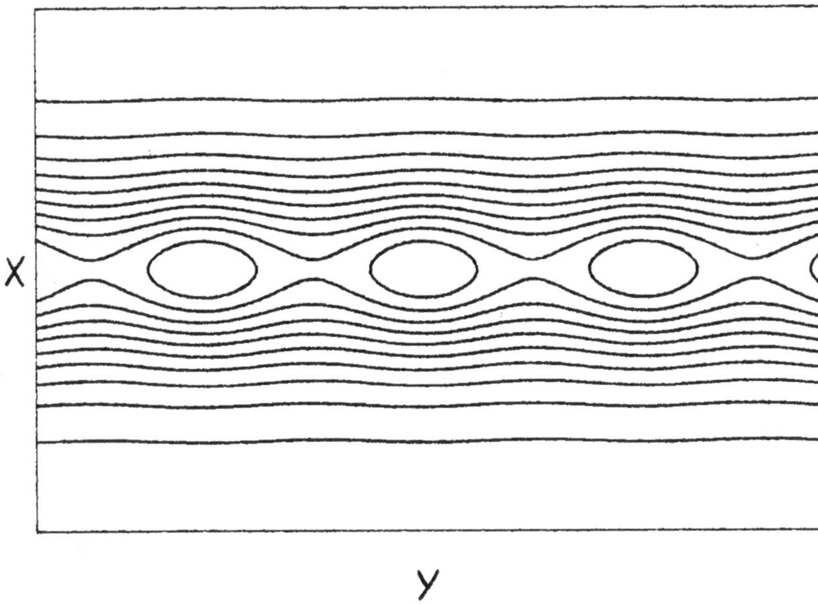

Figure 1. Profiles of constant ϕ

$$L_1\psi = [\sigma_y \frac{\partial}{\partial x} - \frac{1}{4}(u_y+iu_x)\sigma_x + \frac{e^{-u}}{32\sqrt{E}}(1+\sigma_z)-\sqrt{E}]\psi = 0$$

$$L_2\psi = [-i\sigma_y \frac{\partial}{\partial y} - \frac{1}{4}(u_y+iu_x)\sigma_x - \frac{e^{-u}}{32\sqrt{E}}(1+\sigma_z)-\sqrt{E}]\psi = 0 \; . \tag{4}$$

Here σ_x, σ_y and σ_z are the Pauli matrices and ψ is a two-component spinor.

Knowing a Lax representation for a nonlinear system not only implies solvability but also usually provides one with a constructive method for generating solutions. Thus everything might seem to be in good shape.

However, as is pointed out in refs. (3) and (7), this Lax pair is <u>not</u> useful for constructing solutions. The reason is that the scattering data is incomplete and must somehow be supplemented.

In refs. (3) and (4) the authors abandon the above Lax pair and opt for a nonlocal Lax pair which allows them to construct solutions. References (3) and (4) are concerned with infinite line boundary conditions.

In this paper, and in related papers[7,8], we are concerned with periodic boundary conditions. The nonlocal approach of ref. (3) and (4) can be avoided by treating the Liouville equation as a singular limit of the sinh-Poisson equation:

$$\nabla^2\phi + \frac{\tilde{\lambda}^2}{4} \sinh\phi = 0 \tag{5}$$

which was previously studied by Ting, Chen and Lee[9]. By making the substitutions $\tilde{\lambda}^2 = \lambda^2 e^{-\beta}$, $\phi = u-\beta$ and then taking $\beta \to +\infty$ Eq. (1) is recovered. Since there is no ambiguity in the characterization of the sinh-Poisson scattering data, by carefully following what happens to this data in the Liouville limit one can resolve the ambiguities present in the Liouville system. This was proven in ref. (8).

The incompleteness of the Lax pair (eqs. (4)) and the connections with sinh-Poisson are discussed in detail in refs. (7) and (8). The focus of this paper is to show how the approach in ref. (8) "works" in a few simple cases which we examine in detail. Needed results from refs. (7) and (8) will be stated without proof.

First we show how to generate real periodic and quasi-periodic solutions of the sinh-Poisson equation. The simplest case is examined in detail.

Next we show how to take the Liouville limit, and a few simple cases are examined.

2. REAL SOLUTIONS OF THE SINH-POISSON EQUATION

The following result is proven in ref. (8): Given

i) a set of complex constants organized in the following manner:

a) N inversion pairs $E_1, E_2 \ldots E_{2N}$ such that $E_{N+j} = \tilde{\lambda}^4/256E_j^*$ for $j = 1 \ldots N$.

b) M inversion pairs. $E_{2N+1}, E_{2N+2} \ldots E_{2N+2M}$ such that:

$$E_{2N+k} = \frac{\tilde{\lambda}^2}{16}\alpha_k \qquad E_{2N+M+k} = \frac{\tilde{\lambda}^2}{16}\frac{1}{\alpha_k^*} \qquad k = 1 \ldots M .$$

These constants are called the "main spectrum".

ii) Associated with each pair of main spectra there is an "auxiliary variable" denoted by μ_j, $j = 1,2 \ldots N+M$.

iii) These auxiliary variables satisfy the following equations of motion:

$$\mu_{j-} = \left(\frac{\partial}{\partial x} + i\frac{\partial}{\partial y}\right)\mu_j = \frac{2i\kappa_j \left(\prod_{k=0}^{2(N+M)} (\mu_j - E_k)\right)^{1/2}}{\prod_{m \neq j} (\mu_j - \mu_m)} \tag{6}$$

$$\mu_{j+} = \left(\frac{\partial}{\partial x} - i\frac{\partial}{\partial y}\right)\mu_j = \frac{i\tilde{\lambda}^2}{8\sqrt{Q}} \prod_{m \neq j}\mu_m \frac{\kappa_j \left(\prod_{k=0}^{2(N+M)} (\mu_j - E_k)\right)^{1/2}}{\prod_{n \neq j}(\mu_j - \mu_n)}$$

where: $Q = \prod_{k=1}^{2(N+M)} E_k$ and $E_o = 0$. The constants κ_j are sheet indices and take values of ± 1.

iv) The initial values of μ_j (i.e. μ_j^o) must be chosen so that:

$$W(z) = \prod_{k=1}^{2(N+M)} (z - E_k) - \prod_{j=1}^{N+M} (z - \mu_j^o)\left(z - \frac{\tilde{\lambda}^4}{256\mu_j^{o*}}\right) \tag{7}$$

is the perfect square of a finite polynomial in z:

$$W(z) = f^2(z) \qquad\qquad f(z) = \sum_{k=1}^{N+M} f_k z^k$$

for some f_k. This choice of $\mu_j{}^o$ will guarantee the reality of solutions of the sinh-Poisson equation.[8]

v) The sheet indices, κ_j, are fixed by the relation:

$$f(\mu_j) = \kappa_j \sqrt{\prod_{k=1}^{2(N+M)} (z-E_k)} \qquad \text{for each } j = 1,\ldots N+M.$$

vi) Finally, the solution of the Sinh-Poisson equation is given by:

$$\phi(x,y) = -\ln\left(\frac{\prod_{j=1}^{N+M} \mu_j(x,y)}{Q^{1/2}}\right) \tag{8}$$

From the above it is also possible to prove that:

$$f_{N+M} = \frac{i}{2}\,\phi_- \tag{9}$$

To clarify this approach, let's examine the simplest case in complete detail.

A. The N=1, M=0 Solution of the Sinh-Poisson Equation

In this situation we have $E_2 = \tilde{\lambda}^4/256E_1^*$. There is one auxiliary variable and its equations of motion are:

$$\begin{aligned} \mu_- &= 2i\kappa\sqrt{\mu(\mu-E_1)\left(\mu-\frac{\tilde{\lambda}^4}{256E_1^*}\right)} \\ \mu_+ &= 2i\left(\frac{E_1^*}{E_1}\right)^{1/2} \mu_- \end{aligned} \tag{10}$$

since $(E_1^*/E_1)^{1/2}$ is a pure phase, write it as $\exp(-i\theta)$ with $E_1 = E\exp(i\theta)(E>0)$. The polynomial W(z) (eq. (7)) becomes:

$$W(z) = (z-Ee^{i\theta})(z - \frac{\tilde{\lambda}^4}{256E}e^{i\theta}) - (z-\mu^o)(z - \frac{\tilde{\lambda}^4}{256\mu^{o*}})$$

This must equal $f^2(z) = f_1{}^2 z$. Thus:

$$-(E + \frac{\tilde{\lambda}^4}{256E})e^{i\theta} + (\mu^o + \frac{\tilde{\lambda}^4}{256\mu^{o*}}) = f_1{}^2 \tag{11}$$

$$\frac{\tilde{\lambda}^4}{256} e^{2i\theta} - \frac{\tilde{\lambda}^4}{256}\frac{\mu^o}{\mu^{o*}} = 0 \tag{12}$$

Here we have merely equated coefficients of similar powers in z. Equation (12) requires:

$$\mu^o = \pm \rho^o e^{i\theta} \qquad\qquad \rho^o > 0 \tag{13}$$

Take the positive choice. From Eq. (8) it is easy to see that the negative choice causes $\phi(x,y)$ to be complex. Changing variables in Eqs. (10) to $\mu(x,y) = \rho(x,y)e^{i\theta}$ we find:

$$\rho_- = 2ie^{\frac{i\theta}{2}} \kappa \sqrt{\rho(\rho-E)(\rho-\tilde{\lambda}^4/256E)}$$

$$\rho_+ = 2ie^{\frac{-i\theta}{2}} \kappa \sqrt{\rho(\rho-E)(\rho-\tilde{\lambda}^4/256E)}$$

This implies:

$$\rho(x,y) = \rho((\cos\theta/2)x + (\sin\theta/2)y)$$
$$= \rho(x')$$

where $x' = (\cos\theta/2)x + \sin(\theta/2)y$. Therefore

$$\rho_{\pm} = e^{\mp i\frac{\theta}{2}} \frac{d\rho}{dx'}$$

and, by eq. (8), a similar result holds for $\phi(x,y)$. Therefore: $\phi_- = e^{i\theta/2}\, d\phi/dx'$ where $d\phi/dx'$ must be real. Equation (11) becomes:

$$e^{i\theta}[-(E + \frac{\tilde{\lambda}^4}{256E}) + (\rho^o + \frac{\tilde{\lambda}^4}{256\rho^o})] = f_1{}^2 = 1/4(\phi_-)^2$$
$$= -1/4\, e^{i\theta}(\phi_{x'})^2$$

where we have used eq. (9). This leads to

$$E + \frac{\tilde{\lambda}^4}{256E} - \rho^o - \frac{\tilde{\lambda}^4}{256\rho^o} \geq 0$$

which is true if and only if:

$$\frac{\tilde{\lambda}^4}{256E} \leq \rho^o \leq E$$

This condition leads to:

$$\frac{d\rho}{dx'} = 2\kappa\sqrt{\rho(E-\rho)(\rho - \frac{\tilde{\lambda}^4}{256E})} \qquad (14)$$

In this case κ can be chosen as either +1 or -1. Since $\rho(x')$ oscillates between E and $\tilde{\lambda}^4/256E$ the choice of κ merely affects the initial direction of motion. When $\rho(x')$ hits either E or $\tilde{\lambda}^4/256E$ then κ must flip sign, to insure $\rho(x')$ turns around.

The solution of eq. (14) is:

$$\rho(x,y) = E\mathrm{dn}^2(\sqrt{E}x'|k)$$

where: $x' = \cos(\theta/2)x + \sin(\theta/2)y, \quad k^2 = 1 - \frac{\tilde{\lambda}^4}{256E^2}$

and $\mathrm{dn}(\sqrt{E}x'|k)$ is a Jacobi elliptic function[10,11] with modulus k. Therefore:

$$\phi(x,y) = \ln[\frac{\tilde{\lambda}^2}{16E}\ \mathrm{dn}^{-2}(\sqrt{E}x'|k)]$$

Two things are worth noting about this N=1, M=0 solution of sinh-Poisson. First, it is periodic in one direction in the (x,y) plane and constant in the other. Second, if we rotate the spectral data, i.e. change the phase θ in $E_1 = Ee^{i\theta}$, then the physical solution also rotates, but by an angle $\theta/2$ in the (x,y) plane. This is a general result, true for arbitrary N and M.

B. The Liouville Limit for N=1, M=0

The reader will recall from the discussion in the introduction that we can pass from the sinh-Poisson equation to the Liouville equation by substituting $\tilde{\lambda}^2 = \lambda^2 e^{-\beta}$ and $\phi = u-\beta$ and then taking $\beta \to +\infty$. Performing this limit on the N=1, M=0 solution above we find that as $\beta \to +\infty$:[11]

$$k \to 1$$

$$\mathrm{dn}^2(\sqrt{E}x'|k) \to \mathrm{sech}^2(\sqrt{E}x')$$

where sech is the hyperbolic secant. Therefore the solution of the Liouville equation which this tends to is:

$$u(x,y) = \ln(\frac{\lambda^2}{16E}\cosh^2(\sqrt{E}x'))$$

It is easy to directly verify that this is a solution to eq. (1).

3. THE LIOUVILLE LIMIT IN GENERAL

In ref. (8) the Liouville limit for general values of N and M was carried out. In summary, it was shown there that the auxiliary variables μ_j split into three distinct categories depending on how they scale with $\tilde{\lambda}$. In order for the limit to be well defined N must be odd (see ref. (8)).

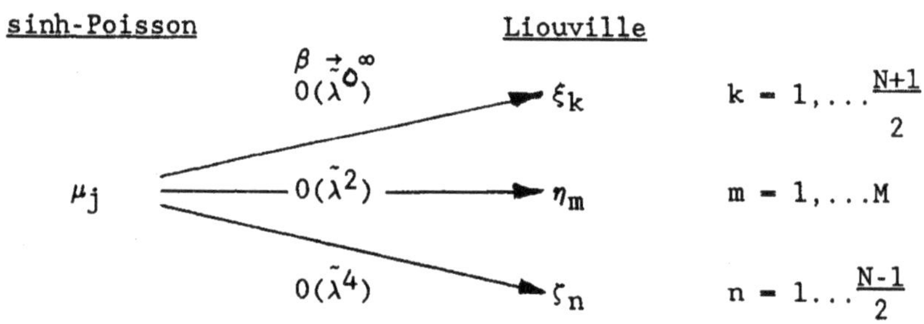

The total number of auxiliary variables remains fixed but the equations of motion (eqs. (6)) also separate into three distinct classes. These equations are:

$$\xi_{j-} = \frac{2i\xi_j}{\prod\limits_{m\neq j}(\xi_j - \xi_m)} R_1(\xi_j) \qquad j = 1 \ldots \frac{N+1}{2} \tag{15}$$

$$\xi_{j+} = H_1(x_+)\Big(\prod_{m\neq j}\xi_m\Big)\xi_{j-}$$

$$\eta_{m-} = 0 \qquad m = 1 \ldots M \tag{16}$$

$$\eta_{m+} = H_2(x_+)\,\frac{\prod\limits_{r=1}^{M}\eta_r}{\prod\limits_{s\neq m}(\eta_m - \eta_s)}\,R_2(\eta_m)$$

$$\zeta_{n-} = 0 \qquad\qquad n = 1\ldots\frac{N-1}{2} \qquad (17)$$

$$\zeta_{n+} = 2(-1)^{\frac{N-1}{2}+M} \frac{\prod_{k=1}^{N} E_k^* \prod_{r\neq n}\zeta_r}{\prod_{s\neq n}(\zeta_n-\zeta_s)} R_3(\zeta_n)$$

where:

$$R_1(\xi) = \Big(\prod_{k=1}^{N+1/2}(\xi-E_k)\Big)^{1/2}$$

$$R_2(\eta) = \Big(\prod_{m=1}^{M}(\eta-\alpha_m)(\eta-1/\alpha_m^*)\Big)^{1/2}$$

$$R_3(\zeta) = \Big(\prod_{k=1}^{N-1/2}(\zeta - 1/E_k^*)\Big)^{1/2}$$

The equations are written this way to emphasize that in passing to the Liouville limit the original Riemann surface associated with eqs. (6) has separated into <u>three distinct surfaces</u>.

We will not need the general forms of H_1 and H_2 here (see ref. (8) for details). It is important to point out, however, that H_1 and H_2 are well defined functions of the η's and ζ's and therefore the eqs. (15)-(17) are a <u>closed</u> set of equations.

In ref. (7) it is shown that by proceeding directly from the Lax pair for the Liouville system (eqs. (4)) one arrives at eqs. (15), but $H_1(x_+)$ appears as an arbitrary function. Equations (16) and (17) do not appear because the η's and ζ's are "hidden" from the Liouville scattering operator.

4. THE N=1, M=1 SOLUTION

When N=1 there are no ζ's. With M=1 eqs. (16) become[8]:

$$\begin{aligned} \eta_- &= 0 \\ \eta_+ &= 2\,\sqrt{E_1^*\,\alpha_1^*/\alpha_1}\;\eta\sqrt{(\eta-\alpha_1)(\eta-1/\alpha_1^*)} \end{aligned} \tag{18}$$

The reader can verify that a solution of the reality constraint for sinh-Poisson can always be taken as: $\mu_j{}^o = E_k$ (one auxiliary variable at each branch point). This is true for all values of $\tilde{\lambda}$ so it also insures reality for Liouville. In this present case we choose:

$$\begin{aligned} \xi^o &= \xi(0,0) = E_1 \\ \eta^o &= \eta(0,0) = 1/\alpha_1^* \end{aligned}$$

with this choice of initial conditions eqs. (18) have the following solution:

$$\eta(x_+) = \frac{2e^{i\theta'}}{(|\alpha_1|+1/|\alpha_1|)+(|\alpha_1|-1/|\alpha_1|)\cosh(2\sqrt{E_1^*}\,x_+)}$$

with: $x_+ = \frac{(x+iy)}{2}$ and $\sqrt{\alpha_1/\alpha_1^*} = e^{i\theta'}$.

Notice that, except for the overall $\exp(i\theta')$ factor, the phase of α_1 plays no role in the final solution.

In this situation $H_1(x_+)$ is given by[8]:

$$H_1(x_+) = \left(\frac{E_1^*}{E_1}\,\frac{\alpha_1^*}{\alpha_1}\right)^{1/2}\eta(x_+)$$

Earlier we mentioned that by "rotating" in spectral space we could rotate the solution $\phi(x,y)$ in the (x,y) plane. This means, without loss of generality, that E_1 can be chosen to be real and positive. In principle it could still be true that the relative phase between E_1 and α_1 is important. However that turns out not to be the case.

Plugging this result for $H_1(x_+)$ into eqs. (15), after some tedious algebra we get:

$$\xi(x,y) = E_1\left(\frac{2\epsilon\cosh(2\sqrt{E_1}x_+)+(1+\epsilon^2)}{(\cosh(\sqrt{E_1}x)+\epsilon\cos(\sqrt{E_1}y))^2}\right) \qquad E_1>0$$

where $\epsilon = \frac{|\alpha_1|-1}{|\alpha_1|+1}$. This finally leads to:

$$u(x,y) = \ln\left(\frac{\lambda^2}{16E_1}\ \frac{(\cosh\sqrt{E_1}x+\epsilon\cos(\sqrt{E_1}y))^2}{(1-\epsilon^2)}\right)$$

which is precisely the island chain solution (eq. (3)). In fact Fig. 1 was generated by solving eqs. (15)-(16) for N=1, M=1. The initial conditions chosen, however, were slightly different because the equations of motion, eqs. (15)-(17), are numerically singular if the auxiliary variables are exactly at a branch point.

5. SUMMARY

Real periodic solutions of the Liouville equation can be constructed by treating this system as a singular limit of the sinh-Poisson equation. A constraint must be forced on the variables which appear in the theory to guarantee reality. The required constraint for general cases was presented, then a few simple cases were examined in detail to illustrate the approach.

REFERENCES:

1. Holm, D. D., Marsden, J. E., Ratiu, T. and Weinstein, A., Physics Reports 123, Nos. 1 & 2 (1985) 1-116.
2. Finn, J. M. and Kaw, P. K., Phys. Fluids, 20, 72-78 (1977).
3. D'Hoker, E. and Jackiw, R., Phys. Rev. D 26, 12, 3517-3542 (1982); D'Hoker, E., Freedman, D. Z. and Jackiw, R., Phys. Rev. D 28, 10, 2583-2598 (1983).
4. Andreev, V. A., Teor. Mat. Fiz. 29, 213 (1976) [Theor. Math. Phys. 29, 1027 (1976)].
5. Polyakov, A. M., Phys. Lett. 103B, 3, 207-213 (1981).
6. Gervais, J. L. and Neven, A., Nucl. Phys. B257, 59 (1985).
7. Chin, C. H., Tracy, E. R. and Chen, H. H., submitted to Phys. Lett.
8. Tracy, E. R., Chin, C. H. and Chen, H. H., to appear in Physica D.
9. Ting, A. C., Chen, H. H. and Lee, Y. C., Phys. Rev. Lett. 53, 1348 (1984).
10. Byrd, P. F. and Friedman, M. D., Handbook of Elliptic Integrals for Engineers and Scientists, Springer-Verlag (New York, 1971).
11. Abramowitz, M. and Stegun, I. A., Handbook of Mathematical Functions, Dover Publications, Inc. (New York, 1972).

Fredholm determinant methods for continuous and discrete soliton equations

Christoph Pöppe
Universität Heidelberg, Sonderforschungsbereich 123
Im Neuenheimer Feld 294, D-6900 Heidelberg, West Germany
(Sept 86 – June 87: School of Mathematics
University of Minnesota, Minneapolis, MN 55455, USA)

ABSTRACT

The Fredholm determinant method is a new and rigorous way to investigate soliton equations and to construct their solutions. It consists essentially of establishing a comparatively simple relation between the soliton equation and its linearized form. The inverse scattering method, Hirota's method of constructing N-soliton solutions, and Bäcklund transformations are given a new and unified interpretation; moreover, some new results can be obtained. This paper gives a survey of the Fredholm determinant method as applied to the hierarchy of Korteweg-deVries (KdV) equations and the Toda lattice equation.

1. Introduction

There are many different ways to describe what a soliton equation is since soliton equations have many features in common [1,3]: explicit formulas for classes of special solutions called N-soliton solutions, Bäcklund transformations, an infinite number of conservation laws etc. The most important common property, however, is the existence of the so-called spectral transform: it maps the solution of the soliton equation (which is generally a nonlinear evolution equation) at every instant of time to a solution of a corresponding linear equation which is so simple that it can easily be solved explicitly. The spectral transform is uniquely invertible; so it is possible to recover the solution of the soliton equation from its image under the spectral transform. In particular, the initial value problem for a soliton equation can be solved *in principle* by 1) applying the spectral transform, 2) solving the linear problem, and 3) applying the inverse spectral transform. However, both transforms are not very explicit though they involve only linear problems. The spectral transform consists essentially of computing the asymptotic values of canonical solutions of an ordinary differential equation on the real line where the solution of the nonlinear evolution equation enters as a coefficient. The inverse transform involves solving a family of Fredholm integral equations. Moreover, up to now, there is no good way of characterizing the class of soliton equations (though the Painlevé criterion [1,p.233] looks very promising).

As a step towards overcoming the drawbacks of this (very powerful) method, I would like to present the "Fredholm determinant transform". Though it can be interpreted as a reformulation of the inverse spectral transform, it is much more explicit; it has a larger domain of application; and for the proofs a new formalism is given which may help understanding why the spectral transform works.

The Fredholm determinant transform was first described for the sine-Gordon and KdV equations [7,8]; recently, Wolfgang Bauhardt and myself [2] gave a Fredholm determinant formalism for the Toda lattice [10] and related equations.

2. FREDHOLM DETERMINANT FORMALISM

for KdV

Let $f(x,t)$ be a solution of

$$f_t + 8f_{xxx} = 0$$

($f \to 0$ for $x \to -\infty$ sufficiently fast).
For every x and t, define the integral operator

$$F_{(x,t)}\phi(s) := \int_{-\infty}^{0} f(s+\sigma+2x,t)\phi(\sigma)\, d\sigma$$

Then, the Fredholm determinant

$$p(x,t) := \det(1+F)$$

solves Hirota's bilinear KdV eq. [5]

$$(D_{xt} + D_x^4)p \cdot p = 0$$

i.e.

$$pp_{xt} - p_x p_t + pp_{xxxx} - 4pp_{xxx} + 3p_{xx}^2 = 0. \quad (*)$$

This transforms by

for Toda lattice

Let $f_n(t)$ be a solution of

$$\dot{f}_n = f_{n+1} - f_{n-1}$$

($f_n \to 0$ for $n \to \infty$ sufficiently fast).
For every n and t, define the infinite matrix operator

$$(F_n(t)\phi)_k := \sum_{j=0}^{\infty} f_{n+k+j}\phi_j$$

Then, the Fredholm determinant

$$p_n(t) := \det(1+F_n)$$

solves Hirota's discrete bilinear KdV eq. [4]

$$((D_t - 2\exp hD_x)\sinh \frac{h}{2} D_x)p \cdot p = 0$$

i.e.

$$\dot{p}_{n+1}p_n - p_{n+1}\dot{p}_n = p_{n+2}p_{n-1} - p_{n+1}p_n. \quad (**)$$

This transforms by

$$u = -\partial_x^2 \log p$$

to a solution of

$$u_t = 6uu_x - u_{xxx}$$

(KdV equation)

$$y_n = -\log \frac{p_{n+2}}{p_n}$$

to a solution of

$$\dot{y}_n = 2 - e^{-y_{n+1}+y_n} - e^{-y_n+y_{n-1}}$$

(one-sided Toda lattice equation) which implies

$$\ddot{y}_n = e^{-y_n+y_{n-2}} - e^{-y_{n+2}+y_n}$$

(Toda lattice equation).

Sketch of proof for KdV. Let K be a Fredholm integral operator defined by

$$K\phi(s) := \int_{-\infty}^{0} k(s,\sigma)\phi(\sigma)\, d\sigma$$

such that $\det(1+K)$ exists. If K depends differentiably on some parameter t, then

$$\frac{\partial}{\partial t} \log \det(1+K) = \operatorname{tr} K_t(1+K)^{-1}$$

(this is obvious for finite-dimensional matrices). Thus, proving (*) amounts to proving an identity of the form

$$\operatorname{tr} A + \operatorname{tr} B_1 \operatorname{tr} C_1 + \operatorname{tr} B_2 \operatorname{tr} C_2 + \cdots = 0 \quad .$$

Hence, we need a formalism to compute products of traces of special operators. Here it is: Define

$$[K] := k(0,0) \quad .$$

Then, the following rules hold: (F, G, H being "additive" operators, i.e. those of the special form defined above, K, L like K above)

$$[F] = \operatorname{tr} F_x$$

$$[FKG] = \frac{1}{2} \operatorname{tr}(F_x KG + FKG_x)$$

$$[KF][GL] = \frac{1}{2}[K(F_x G + FG_x)L]$$

("product rule")

$$[(1+F)^{-1}G(1+H)^{-1}] = \frac{1}{2} \operatorname{tr}\big(G_x(1+H)^{-1} - F_x(1+F)^{-1}G(1+H)^{-1} + (1+F)^{-1}G_x - (1+F)^{-1}G(1+H)^{-1}H_x\big)$$

in particular

$$[(1+F)^{-1}F] = \operatorname{tr} F_x(1+F)^{-1} \quad .$$

The last rules allow to express all the occurring traces as square brackets of some operator; the product rule then serves to transform products of square brackets into single square brackets. Now, verification of (*) is just a piece of algebra with operators.

Sketch of proof for Toda. It is possible to give a discrete version of the above proof. Let K be an infinite matrix operator defined by

$$(K\phi)_m := \sum_{j=0}^{\infty} k_{mj}\phi_j \quad .$$

Then,

$$[K] := k_{00}$$

has properties similar to the above formulas. However, differentiating in x has to be replaced by the application of shift operators $V_+\phi = (0, \phi_0, \phi_1, \ldots)$ and $V_-\phi = (\phi_1, \phi_2, \ldots)$ which makes the proof quite lengthy. On the other hand one can take advantage of the fact that there exists a rank-one projection

$$P = 1 - V_-V_+, \qquad P\phi = (\phi_0, 0, 0, \ldots)$$

such that $[K] = \operatorname{tr} PK$. This helps to prove the identities

$$\frac{p_{n+2}}{p_n} = \det(1 - pF_n(1+F_n)^{-1})$$

and

$$\frac{\dot{p}_n}{p_n} = -\operatorname{tr}(P(1+F_n)^{-1}F_{n-1}) \quad .$$

Moreover, for rank-1 operators X we have $\det(1+X) = 1 + \operatorname{tr} X$; every operator of the form PK has rank 1. This, again with some algebra with operators, suffices to prove the identity (**) which, in this case, involves traces and determinants.

Remark. A side result of the comparison between the KdV and the Toda lattice equations is the fact that the KdV equation is **not**, as is common folklore, a limiting case of the Toda lattice, but of the one-sided Toda lattice.

For KdV and sine-Gordon, almost all known facts can be deduced from the linear base equation (where they are almost trivial) via the Fredholm determinant transform.

In particular:

3. N-SOLITON SOLUTIONS

Those correspond to Fredholm operators F with finite rank, e.g. F generated by

$$f(x,t)=\sum_{j=1}^{N}\exp(\frac{a_j}{2}x-a_j^3t+\eta_j)$$

Due to additivity and standard properties of the exp function, F maps for every x into the finite-dimensional space spanned by the functions $\{\exp(a_j/2)s,\ j=1,\ldots,N\}$. In this case, the Fredholm determinant reduces to an ordinary determinant which can be explicitly computed. This reproduces the well-known N-soliton solution formulas.

In fact, there exists a power series expansion for the Fredholm determinant of an integral operator [9]: If K is defined by $K\phi(s):=\int_a^b k(s,\sigma)\phi(\sigma)d\sigma$ and is "trace class" such that a determinant is defined at all, then

$$p=\det(1+\lambda K)=\sum_{m=0}^{\infty}\lambda^m p_m\,,$$

where the series converges in the whole complex plane, and

$$p_m=\frac{1}{m!}\Big(\int_a^b\Big)^m\det\left(k(\sigma_j,\sigma_l)\right)_{j,l\le m}d\sigma_1\ldots d\sigma_m\,,$$

($\left(k(\sigma_j,\sigma_l)\right)_{j,l\le m}$ denotes the matrix whose entries are $k(\sigma_j,\sigma_l)$). For an operator of rank N, $p_m=0$ for $m>N$, so the above representation gives rise to a closed expression for p:

$$p_m=\sum_{\substack{P\subset\{1,\ldots,N\}\\ \#P=m}}\ \prod_{j<k\in P}\left(\frac{a_j-a_k}{a_j+a_k}\right)^2\prod_{j\in P}\exp(a_jx-a_j^3t+\eta_j)\quad \text{for} m\le N.$$

This reproduces Hirota's (see e.g. [5]) well-known N-soliton solution formulas (and helps to explain the "miracle" why his power series terminate): His method amounts to substituting a formal power power series expansion into e.g. Hirota's bilinear KdV equation, ending up with a series of equations, one for each coefficient, that are obviously fulfilled by the p_m defined above.

The rank of F remains finite if one differentiates the "fundamental solution"

$$f(x,t)=\exp(\frac{a}{2}x-a^3t+\eta)$$

one or several times w.r.t. a; those solutions (and linear combinations of them for different values of a) give rise to explicit formulas for "degenerate soliton" (multiple poles in the scattering data) and, in the limit, to rational solutions.

Analogous statements can be shown to apply to the Toda lattice. Here

$$f_n(t) = \sum_{j=1}^{N} \exp(a_j n + 2(\sinh a_j)t + \eta_j)$$

generates finite-rank operators F_n. It is even possible to show that the N-soliton solutions obtained in this way tend to N-soliton solutions of KdV in the continuous limit if the a_j are scaled appropriately.

4. BÄCKLUND TRANSFORMATIONS

As the base equation itself, the "Bäcklund transformation" relating different solutions of the base equation to eachother is a linearization of the Bäcklund transformation on the levels of the functions resp. p and u.

Let f, g be two solutions of the linearized KdV equation $f_t + 8f_{xxx} = 0$ related by

$$g_x + f_x = \frac{a_0}{2}(g - f)$$

Then their Fredholm determinant transforms are Bäcklund transforms of eachother w.r.t. the parameter a_0. The proof follows the same recipe as above.

The simplicity of the linear Bäcklund transformation equation allows explicit solution for g when f is known. This can be used to prove many statements about the effect of as Bäcklund transformation in a very easy and straightforward way: Each Bäcklund transformation adds a soliton with speed parameter a_0 to a given solution, with the exception of one singular case; every other soliton component suffers a phase shift that can be explicitly computed. Repetitive application of a Bäcklund transformation with always the same parameter results in degenerate soliton or rational solutions. Finally, the famous superposition principle that allows to compute in a purely algebraic way a double Bäcklund transform whenever two simple transforms are known, can be deduced from an analogous linearized formula.

5. KDV HIERARCHY

Not only the KdV equation itself can be traced back to its linearized form; the same applies to all members of the KdV hierarchy as well as their "modified" counterparts.

In particular, each solution of Hirota's bilinear KdV equation can be transformed to a solution of KdV as well as to a solution of the modified KdV equation. Those two solutions are related by the well-known Miura transform [6] .

The higher order KdV equations can be obtained as Fredholm determinant transforms of the simple constant coefficient equations

$$f_t = (\frac{\partial}{\partial x})^{2n+1} f \quad .$$

We expect that the Fredholm determinant technique will be powerful enough to prove (and discover) the same kind of results (possibly more) for the discrete case and, of course, for any other soliton equation.

REFERENCES

1. M.J.Ablowitz,H.Segur, "Solitons and the Inverse Scattering Transform", SIAM, Philadelphia, 1981.
2. W.Bauhardt,C.Pöppe, *The Fredholm Determinant Method for Discrete Integrable Evolution Equations*, to be published.
3. F.Calogero,A.Degasperis, "Spectral Transform and Solitons", North-Holland, Amsterdam, 1982.
4. R.Hirota, *Nonlinear Partial Difference Equations. I. A Difference Analogue of the Korteweg-de Vries Equation*, J.Phys.Soc.Jap **43** (1977), 1424–1433.
5. R.Hirota, *Direct Methods in Soliton Theory*, in "Solitons", R.Bullough, P.Caudrey (eds.), Springer, Berlin, 1980, pp. 157-176.
6. R.M.Miura, *The Korteweg-de Vries equation: A survey of results*, SIAM Rev. **18** (1976), 412–459.
7. C.Pöppe, *Construction of Solutions of the Sine-Gordon Equation by Means of Fredholm Determinants*, Physica D **9** (1983), 103–139.
8. C.Pöppe, *The Fredholm Determinant Method for the KdV Equations*, Physica D **13** (1984), 137–160.
9. F.Smithies, "Integral Equations", Cambridge University Press, Cambridge, 1965.
10. M.Toda, "Theory of Nonlinear Lattices", Springer, Berlin, 1981.

PAINLEVE SERIES, RECURSION OPERATORS AND BAECKLUND TRANSFORMATIONS

Walter Strampp and Fritz Renner
Fachbereich 17
Mathematik
Gesamthochschule Kassel
3500 Kassel
W.-GERMANY

ABSTRACT

A test for the integrability of a nonlinear equation is the Painleve' analysis introduced by Weiss, Tabor, and Carnevale.
In this paper it is shown that symmetries and recursion operators can be obtained from the Painleve' expansion. Furthermore, we discuss expansions for hierarchies of equations.

1. INTRODUCTION

An ode is said to be integrable if there exists a sufficient number of integrals. The ode is said to possess the Painleve' property when all movable singularities of its solutions are poles. Now, a test for the integrability is to look whether the ode possesses the Painleve' property[1)-4)].

A pde is integrable if there exists an infinity of symmetries or integrals in involution. The Painleve' property has been introduced first by Ablowitz et al[4)] by taking recourse to ode's. Weiss et al[5)] gave a direct extension of the Painleve' property to pde's. The idea is to introduce a singularity manifold and to look for a single-valued expansion of solutions about it.

The purpose of this paper is to give a method for obtaining symmetries from the Painleve' expansion and for deriving Baecklund transformations for the whole hierarchy of an integrable equation.

2. PAINLEVE-EXPANSIONS

We consider evolution equations

$$u_t = K(u) \quad (1)$$

where K is a polynomial in u and its spacial derivatives. Let us try to expand solutions into series[5-7)]

$$u = \phi^{-a} \sum_{j=0}^{\infty} u_j \phi^j . \quad (2)$$

Now, eq. (1) is said to possess the Painleve' property[5-7)] if

A1: There exist integers a>0, b>0 such that inserting of (2) into (1) yields

$$\sum_{j=0}^{\infty} F_j(u_j, u_{j-1}, \ldots, u_0, \phi)\phi^{-a-b+j} = 0 \quad (3)$$

A2: The functions u_j can be determined recursively from the equations

$$F_j = 0 \quad (4)$$

The functions u_j are determined uniquely with the exception of b resonance functions which can be choosen arbitrarily. The series can be truncated if

A3: By setting the resonance functions equal to zero we can achieve that u_j=0, j>a, in addition to eqs. (4).

3. PAINLEVE-EXPANSIONS AND SYMMETRIES

From the series expansion we can see easily that the following holds

Proposition: Assume that A1, A2, and A3 hold. Then u_{a-1} will be an Infinitesimal Transformation about u_a .

Example: Consider the KdV

$$u_t + uu_x + u_{xxx} = 0$$

possessing the Painleve' expansion[5),6)]

$$u = \phi^{-2} \sum_{j=0}^{\infty} u_j \phi^j .$$

The truncated expansion [5),6)] leads to $u_0=-12\phi_x^2$, $u_1=12\phi_{xx}$ and to the conditions

$$\phi_x\phi_t+\phi_x^2\phi u_2+4\phi_x\phi_{xxx}-3\phi_{xx}^2=0 \quad (5.1)$$

$$\phi_{xt}+\phi_{xx}u_2+\phi_{4x}=0 \quad (5.2)$$

$$(\phi_{xt}+\phi_{xx}u_2+\phi_{4x})_x=0 \quad (5.3)$$

$$u_{2t}+u_2u_{2x}+u_{2xxx}=0 \quad (5.4)$$

Obviously, (5.4) means that u_2 solves the KdV while (5.3) states that u_1 is a solution to the KdV linearized about u_2. Thus, by any solution ϕ of (5.1) and (5.2) we obtain an Infinitesimal Transformation about u_2.

4. CONNECTION WITH RECURSION OPERATORS

We are used to calculate symmetries of eq. (1) by recursion operators which are characterized by

$$R(u)_t = (K'(u),R(u)) \quad (6)$$

Condition (6) can be seen as integrability condition of the following equations

$$R(u)w=\mu w\ ,\quad K'(u)w=w_t \quad (7)$$

having the shape of Lax-equations

$$L(u)v=\lambda v\ ,\quad B(u)v=v_t\ . \quad (8)$$

Now, we aim to obtain eigenfunctions of eqs. (6) in terms of the singularity manifold ϕ of Painleve' expansions. This is done by first applying the following algorithm [5-7]: a) Find the Painleve' expansion. b) Truncate it at the constant level term. c) Find an equation for the singularity manifold by elimination of u_a. d) Solve this equation by expressing the singularity manifold through eigenfunctions of the Lax-equations (8).

Assume that we have performed algorithm a)-d). Then, the above proposition gives an expression of solutions of the equation $K'(u)v=v_t$ in terms of the singularity manifold. In any case where algorithm a)-d) is working we can show furthermore that the equation $R(u)v=\lambda v$ is satisfied too.

In the KdV case the following holds. If v_1, v_2 satisfy the corresponding Lax-equations then $\phi=v_1/v_2$ will satisfy eqs. (5.1), (5.2). By the proposition we obtain $w_t=K'(u)w$, where $w=(v_1/v_2)_{xx}$. Finally it can be shown that $R(u)w=4\lambda w$.

5. THE AKNS-SYSTEM

As a further example we shall consider the following nonlinear Schroedinger equation

$$u_{1t} = iu_{1xx} - 2iu_1^2 u_2 \;, \quad u_{2t} = -iu_{2xx} + 2iu_1 u_2^2 \tag{9}$$

which possesses the Painleve' property. The truncated expansion becomes

$$u_1 = u_{10}\phi^{-1} + u_{11} \;, \quad u_2 = u_{20}\phi^{-1} + u_{21} \tag{10}$$

where

$$u_{10} = v_1^2 \;, \quad u_{20} = v_2^2 \;, \quad \phi = 1 - D^{-1}(v_1 v_2) \tag{11}$$

and v_1 and v_2 are eigenfunctions of the AKNS eigenvalue-problem

$$\begin{aligned} &v_{1x} - u_{11}v_2 = \lambda v_1 \;, \quad -v_{2x} + u_{21}v_1 = \lambda v_2 \\ &v_{1t} = Av_1 + Bv_2 \;, \quad v_{2t} = Cv_1 - Bv_2 \;. \end{aligned} \tag{12}$$

Of course, the truncated expansion (10), (11) can be considered as Baecklund transformation for eq. (9).

Next, we study the Painleve' expansion for eqs. (12). This leads to the fact that the following vector field

$$(v_1^2, v_2^2, v_1 D^{-1}(v_1 v_2), v_2 D^{-1}(v_1 v_2))$$

is a symmetry of eqs. (12). The flow generated by this symmetry can be determined and its field functions part yields again the Baecklund transformation (10), (11).

6. BAECKLUND-TRANSFORMATIONS FOR HIERARCHIES

We now state the following question: how can the Baecklund-transformation (truncated Painleve' expansion) be modified to hold for the whole hierarchy of AKNS equations. This problem is solved by using the following method[8)] for generating Lax-pairs by recursion operators.

Let R be a hereditary recursion operator and L, B_0 a Lax-pair for $u_t=K_0(u)$. Consider the extended system

$$\begin{aligned} u_t &= K_0(u) \\ v_t &= B_0(u)v \end{aligned} \tag{13}$$

and assume that

$$R_E(u,v) = \begin{pmatrix} R(u) & 0 \\ A(u,v) & L(u) \end{pmatrix}$$

is a hereditary recursion operator of (13), where A is linear in v. Then we can construct the following hierarchies

$$K_n(u) = R(u)^n K_0(u)$$

$$B_{n+1}(u) = A(u,v)K_n(u)+L(u)B_n(u)v$$

and L, B_n is a Lax-pair for $u_t=K_n(u)$.

In the AKNS-case (9) R_E is obtained as

$$R_E = \begin{pmatrix} D-2u_1D^{-1} & -2u_1D^{-1}u_1 & 0 & 0 \\ 2u_2D^{-1} & -D+2u_2D^{-1}u_1 & 0 & 0 \\ -v_2-v_1D^{-1}u_2 & -v_1D^{-1}u_1 & 2D & -2u_1 \\ v_2D^{-1}u_2 & v_1+v_2D^{-1}u_1 & 2u_2 & -2D \end{pmatrix}.$$

Now, if the vector field S is a symmetry of the Lax-equations $L(u)v=v$, $B_0(u)v=v_t$ then it follows from the construction of the hierarchies and a further mild condition that S is a symmetry of $L(u)v=\lambda v$, $B_n(u)v=v_t$.

Thus, in the AKNS-case we obtain that the vector field $S=(v_1^2, v_2^2, v_1 D^{-1}(v_1 v_2), v_2 D^{-1}(v_1 v_2))$ is a symmetry of the hierarchy of Lax-equations. A consequence of this is that the truncated Painleve' series (Baecklund transformation) for the hierarchy of AKNS-equations is given by

$$u_1 = u_{11} + (v_1^2/(1-D^{-1}(v_1 v_2))),\quad u_2 = u_{21} + (v_2^2/(1-D^{-1}(v_1 v_2)))$$

where v solves $L(u_1)v=\lambda v$, $B_n(u_1)v=v_t$.

REFERENCES

1) Kowalewskaya, S., Acta Math. 14,81(1890).
2) Chang, Y.F., Tabor, M. and Weiss, J., J.Math.Phys. 23, 531(1982).
3) Bountis, T., Segur, H. and Vivaldi, F., Phys. Rev. A25, 1257(1982).
4) Ablowitz, M.J., Ramani, A. and Segur, H., J.Math. Phys.21, 715(1980).
5) Weiss, J., Tabor, M. and Carnevale G., J.Math.Phys. 24, 522(1983).
6) Weiss, J., J.Math.Phys. 25, 13(1984).
7) Weiss, J., J.Math.Phys. 24, 1405(1983).
8) Oevel, W., Mastersymmetries: weak action/angle structure for hamiltonian and nonhamiltonian systems. Preprint Paderborn (1986).

Symmetries and Involution Systems:

Some Experiments in Computer Algebra

Fritz Schwarz
GMD, Institut F1
Postfach 1240
5205 St. Augustin
West Germany

Abstract An algorithm is described which allows to determine for an arbitrary system of partial differential equations all its integrability conditions, i.e. to complete it to an involutive system. Various examples of involutive systems obtained by applying this algorithm are given (Burgers- and generalizations, nonlinear Schrödinger, KdV, mKdV). Further it is argued that this algorithm leads to a unifying picture of several apparently unrelated concepts like e.g. Cole-Hopf transformations, Miura-transformations and Bäcklund transformations. The relation between the symmetries of an involutive system and its subsystems is discussed in detail. A prerequisite for performing the extensive calculations which are almost always involved in this analysis are several computer algebra packages which have been developed for this purpose.

1 Introduction

Nonlinear partial differential equations (p.d.e.'s) have been an active field of research for more than 200 years now. Nevertheless there are numerous unsolved problems left and unexpected answers remain to be discovered as it has been elucidated by the spectacular discovery of soliton solutions about twenty years ago by Zabusky and Kruskal [1]. The present state of research in this area may be learned e.g. from the textbooks [2], [3] and [4].

There are two fundamental concepts related to differential equations both of which have its origin in the work of Sophus Lie. The first is that of a symmetry of a differential equation. By symmetry we always mean a so-called point- or Lie-symmetry. Lie arrived at it in trying to systematize the integration theory of differential equations. Roughly speaking a symmetry is a transformation which leaves the differential equation invariant. Equivalently it may be characterized by saying that it transforms solutions into each other, i.e. it permutates solutions corresponding to different arbitrary elements like e.g. initial values. Symmetries provide an important insight into the geometrical structure of the solution set of any differential equation. This is especially interesting if the differential equations are nonlinear and its solution cannot be obtained explicitly. Furthermore, symmetries are an important tool for solving a differential equation or for obtaining special solutions [5].

The concept of an involutive system of partial differential equations is related to the question whether for a given system there may exist additional equations which are algebraically independent of it. It is misleading to believe that a system of differential equations is easier to solve if it comprises fewer equations. On the contrary, many important solution procedures are based on the construction of additional equations. An involution system is a system which comprises the maximal number of equations which are algebraically independent of each other and which has the same solution set as the system one originally started with.

To illustrate the concept of an involution system consider the special case of a system of p.d.e.'s of first order for a single unknown function z depending on n variables $x_1, \ldots x_n$. Let the equations of the system be written as

$$F_\nu(x_1, \ldots x_n, z, p_1, \ldots p_n) = 0, \quad \nu = 1 \ldots n \tag{1}$$

with the usual notation $p_i = \frac{\partial z}{\partial x_i}$ for partial derivatives. The Jacobi bracket is defined by

$$[F_\nu, F_\mu] = \sum_{k=1}^{N} \{(\frac{\partial F_\nu}{\partial x_k} + p_k \frac{\partial F_\nu}{\partial z}) \frac{\partial F_\mu}{\partial p_k} - (\frac{\partial F_\mu}{\partial x_k} + p_k \frac{\partial F_\mu}{\partial z}) \frac{\partial F_\nu}{\partial p_k}\} \tag{2}$$

which reduces to the Poisson bracket if the $F's$ do not depend explicitly on the function z. The given system of p.d.e.'s is said to be an involution system if

$$[F_\mu, F_\nu] \equiv 0 \;\; for \;\; \mu, \nu = 1 \ldots N. \tag{3}$$

Any given system (1) which is not involutive may be replaced by an equivalent involution system if all brackets (2) which do not vanish are added to the system. As an example consider the following system which is due to Graindorge [12]:

$$\begin{gathered} F_1 = x_1^2 p_1 - 2x_5 p_2 + (x_1^2 x_4 - 2x_5) p_3 - 2x_1 x_4 p_4 \\ F_2 = 2x_5 p_4 + x_1^2 p_5 \end{gathered} \tag{4}$$

In the first step the bracket

$$[F_1, F_2] = 2x_1(F_2 + x_1 F_3) \;\; with \;\; F_3 = p_2 + (1 - x_5) p_3 \tag{5}$$

is formed leading to the new equation $F_3 = 0$. In the second step the equation $F_4 = 0$ follows from

$$[F_2, F_3] = -x_1^2 F_4 \;\; with \;\; F_4 = p_3. \tag{6}$$

The four equations $F_1 = 0, \ldots F_4 = 0$ are an involution system the general solution of which is easily obtained as $\phi(x_1^2 x_4 - x_5^2)$. Two things are to be learned from this example. On the one hand, the general solution is much more easily obtained from the involutive system than from the system (4) which has originally been given. Secondly, even without determining the solution explicitly, from the involutive system its degree of arbitraryness is seen to comprise a single arbitrary function which has been denoted by ϕ.

In the general case if the number of dependent variables is arbitrary, or what amounts to the same thing if the order is higher than one, the construction of an involution system by adding all integrability conditions is more complicated. However there is also a complete theory available which has been developed by the French mathematicians Riquier [6] and Janet [7], [8],[9] early in this century. It is completely equivalent to the theory of Cartan [10] and Kähler [11] who treat the same problem in terms of differential forms.

At the beginning of the next section an algorithm will be described which is based on the theory of Riquier and Janet. It will be applied to obtain several involutive systems starting from well known examples of nonlinear evolution equations. The symmetries of an involutive system and its subsystems are investigated in Section 3. They are used to show the consistency of the results obtained in Section 2. In the final Section 4 some general remarks concerning the further developement of the present subject are made.

2 Construction of Involution Systems

Any system of partial differential equations is characterized by the number m of unknown functions $u^\alpha, \alpha = 1 \ldots m$ which occur and the number n of arguments $x_i, i = 1 \ldots n$ on which these function may depend. They are called ***dependent*** and ***independent*** variabels respectively. Furthermore, the number of differential equations N and the order d_i of the highest derivative with respect to any independent variable are of fundamental importance. In the most general case these quantities are not correlated. Partial derivatives are denoted by

$$u^\alpha_{i_1 \ldots i_n} = \frac{\partial^{i_1 + \ldots i_n} u^\alpha}{\partial x_1^{i_1} \ldots \partial x_n^{i_n}} \qquad (1 \le \alpha \le m) \tag{7}$$

where $i_1 = \ldots i_n = 0$ is identified with u^α. A system of p.d.e.'s is given either implicitly

$$\omega_\nu(x_i, u^\alpha, u^\alpha_i, u^\alpha_{ij}, \ldots) = 0, \qquad (\nu = 1 \ldots N) \tag{8}$$

or in resolved form

$$u^\alpha_{i_1 \ldots i_n} = \Psi^\alpha_{i_1 \ldots i_n}(x_1, \ldots x_n, u^1 \ldots u^m, \ldots u^\beta_{k_1 \ldots k_n \ldots}), \tag{9}$$

i.e. each equation is resolved with respect to one derivative which occurs in it. It has to obey certain constraints which will be explained shortly. Let us make the additional assumption that the equations (9) are polynomial in all variables.

One of the basic concepts of the theory of Riquier and Janet is to establish a complete ordering between the dependent variables and their derivatives. For our purpose it is enough to make the following definition. Any derivative which may be obtained by appropriate differentiation from any other derivative of the same function follows the latter in any ordering. In any differential equation, ***leading derivatives*** are those which cannot be obtained by differentiation of some other derivative occuring in the same equation.

Consider now the given system of p.d.e.'s in a resolved form satisfying the following constraints. Any equation is resolved for one of its leading derivatives. The first members of any two equations are distinct. No first member or a derivative thereof appears in the second member of any equation. A system with these properties is called ***orthonomic***. The first members of all equations of an orthonomic system and its derivatives are called ***principal derivatives***. All other derivatives are called ***parametric***.

The definition of an orthonomic system as given in the preceding paragraph shows that not all resolutions with respect to leading derivatives are allowed. It may happen for example that the principal derivatives occuring in a second member of some equation cannot be replaced by parametric ones because after that substitution the derivative at the left hand side is not a leading derivative any more. We do not go into the details of this problem here. Instead we simply presume that there is a criterion available which allows it to exclude all those resolutions which lead to inconsistent substitution rules and call the remaining ones ***allowed resolutions***.

After an orthonomic system has been constructed from the given system of p.d.e.'s, the possible integrability conditions are identified. To this end all pairs of left hand sides have to be considered. If appropriate differentiation of either member leads to identical derivatives, the same differentiation has to be applied to the corresponding right hand sides. They have to be simplified after performing all substitutions which follow from the system. If the result in both cases is identical to each other, the integrability condition is satisfied. Otherwise the difference of the respective right hand sides has to be added to the system as a new differential equation. If no further integrability conditions may be obtained in this way, the system is in involution. From this proceeding the following algorithm INVSYS for constructing an involution system has been designed.

Algorithm INVSYS (Construction of involution system). Given a set of N p.d.e.'s

$$\omega_\nu(x_i, u^\alpha, u_i^\alpha, u_{ij}^\alpha, \ldots) = 0, \qquad (\nu = 1 \ldots N)$$

in n independent variables x_i and m dependent variables u^α. The integrability conditions which must be added to the system to make it involutive are returned.

S1. Find the leading derivatives in each p.d.e., determine the allowed resolutions which follow from them and store these resolutions on a list.

S2. Choose a resolution from the list constructed in S1 and solve each differential equation for the corresponding derivative. Substitute all principal derivatives occuring at a right hand side by parametric derivatives if this may be done consistently. Otherwise discard the respective resolution and go to S4.

S3. Identify the integrability conditions by appropriate differentiations of first members of pairs of differential equations. For each condition determine the principal derivatives which have to be substituted and perform these substitutions. If an integrability condition does not vanish identically, store it on a list.

S4. If the list of resolutions is not empty go to S2.

S5. If the list of integrability conditions is not empty, remove multiple occurences of identical conditions, add them to the original system of p.d.e.'s and go to S1.

The algorithm INVSYS has been realized in terms of a REDUCE computer algebra package of the same name. It will be described in detail elsewhere. Here we want to present some of the results which have been obtained by applying it. Consider first the following system of four equations

$$u_t + uu_x + u_{xx} = 0 \tag{10a}$$

$$v_t + v_{xx} = 0 \tag{10b}$$

$$uv - 2v_x = 0 \tag{10c}$$

$$4v_t + v(u^2 + 2u_x) = 0 \tag{10d}$$

where u and v are functions of t and x. The first three equations of this system are the Burgers equation, the diffusion equation and the Cole-Hopf transformation respectively. The latter two equations form a Bäcklund tranformation for the former two. The first result obtained with the help of the algorithm INVSYS is that the system (10) is in involution. An interesting question to ask is whether there are subsystems from which the algorithm INVSYS does generate the full system (10). It turns out that this is in fact the case. From equations (10a,b,c) for example equation (10d) is generated. That means, the second member (10d) of the Bäcklund transformation is obtained as an integrability condition for (10a,b,c). It does therefore not contain new information on the Burgers-diffusion system, it simply means that any solution of the system (10a,b,c) satisfies also eq.(10d). This should be compared with the remarks following eq.(6) in Section 1. The complete answer for any subsystem comprising at least two equations is shown in Table 1. From this Table it follows that the subsystems comprising the three equations (10a,c,d) and (10b,c,d) do generate the full system too. The same is true for the subsystems (10b,c) and (10c,d) comprising only two equations. Therefore these subsystems are completely

System	Integrability conditions
(10)	None, involution system
(10a),(10b),(10c)	(4)
(10a),(10b),(10d)	$(uv-2v_x)(uu_x+u_{xx})=0,\quad u^2(uv-2v_x)(\frac{v_t}{v_x})_x=0,$ $(uv-2v_x)u_t=0,\quad v^2(uv-2v_x)(\frac{v_{xx}}{v})_x=0,$ $[u_t-2(\frac{v_{xx}}{v})_x]v^2=0,\quad [u_t-2(\frac{v_x}{v})_t]v^2=0,$
(10a),(10c),(10d)	(2), $2(v_x-v\frac{\partial}{\partial x})(v_t+v_{xx})=0$
(10b),(10c),(10d)	(1), $(u+\frac{\partial}{\partial x})(u_t+uu_x+u_{xx})=0$
(10a),(10c)	$2(v_x-v\frac{\partial}{\partial x})(v_t+v_{xx})=0$
(10a),(10d)	$(\frac{uv-2v_x}{v})_t=0,\quad v(v\frac{\partial}{\partial t}-v_t)(v_t+v_{xx})-v^2(uv-2v_x)(\frac{v_x}{v})_t=0$
(10b),(10c)	(1), (4)
(10b),(10d)	$v(u_t+uu_x+u_{xx})_x+uvu_t+2v_x(\frac{1}{2}u^2+u_x)_x=0$
(10c),(10d)	(1), (2)

Table 1: The integrability conditions for the subsystems of the Burgers-diffusion system (10). In the first column the equations of the subsystem are given. Boldface notation means that the corresponding subsystem generates the full system.

equivalent to the full system. Any other subsystem leads out of the system (10) because new and more complicated integrability conditions are required to make it involutive. This is especially true for the system consisting of the three equations (10a,b,d).

Similar results are obtained for the three-dimensional generalization of the Burgers-diffusion system (10). Let $\mathbf{u}$ denote a three dimensional vector function now depending on three space coordinates and the time, v a scalar function of the same arguments. Cole [13] has shown that the two sets of equations

$$\mathbf{u}_t+\mathbf{u}\cdot grad\,\mathbf{u}+\Delta\mathbf{u}=0, \tag{11a}$$

$$v_t+\Delta v=0 \tag{11b}$$

are transformed into each other by

$$\mathbf{u}v-2\,grad\,v=0. \tag{11c}$$

The algorithm INVSYS determines the following four integrability conditions which make the full system (11a,...e) involutive:

$$4v_t+v(|\mathbf{u}|^2+2\,div\,\mathbf{u})=0, \tag{11d}$$

$$rot\,\mathbf{u}=0. \tag{11e}$$

Eqs.(11a...d) are a straightforward generalization of the one-dimensional system (10a...d); (11e) is an obvious consequence of (11c) and has no counterpart in one space dimension. The four equations (11c,d) may be considered as the generalization to three space dimensions of the Bäcklund transformation (10c,d). The structure of subsystems generating the full system is similar to that of eqs.(10).

The Korteweg deVries equation, the modified Korteweg deVries equation and the Miura transformation (12a,b,c) relating them are completed to the following system of four equations by the algorithm INVSYS:

$$u_t + 6uu_x + u_{xxx} = 0 \tag{12a}$$

$$v_t - 6v^2 v_x + v_{xxx} = 0 \tag{12b}$$

$$v_x + u + v^2 = 0 \tag{12c}$$

$$v_t - 2u(u + v^2) - u_{xx} + 2u_x v = 0 \tag{12d}$$

Eqs.(12c,d) are the Bäcklund transformation which has been found by Estabrook and Wahlquist [15]. The system (12) is involutive. Similar as for the Burgers-diffusion system, eq.(12d) is generated as an integrability conditon for eq.s(12a,b,c). Estabrook and Wahlquist have also determined an auto Bäcklund transformation for the potential version of the Korteweg deVries equation [14]. The corresponding involution system is

$$u_t - 6u_x^2 + u_{xxx} = 0 \tag{13a}$$

$$v_t - 6v_x^2 + v_{xxx} = 0 \tag{13b}$$

$$u_x + v_x - (u - v)^2 + k = 0 \tag{13c}$$

$$u_t + v_t - 4[(u_x + k)k + (u_x - k)(u - v)^2 + u_x^2 + u_{xx}(v - u)] = 0 \tag{13d}$$

$$u_t + v_t - 4[(v_x + k)k + (v_x - k)(u - v)^2 + v_x^2 + v_{xx}(u - v)] = 0 \tag{13e}$$

The full system may be generated by various of its subsystems, e.g. the last two equations turn out to be integrability conditions for the former three.

The non-linear Schrödinger equation

$$iw_t + w_{xx} + 2w|w|^2 = 0 \tag{14}$$

which is in fact a system of two equations for the real and the imaginary part of the complex function $w = u + iv$, has to be supplemented by the fourth order equation

$$w_{tt} + w_{xxxx} + 8w_{xx}|w|^2 + 4w_x|w|_x^2 + 4w|w_x|^2 + 4w|w|^4 = 0 \tag{15}$$

to form an involution system. This means that the latter equation is integrable just like the Schrödinger equation (14) itself.

3 Symmetries of Involution Systems

The second concept which has been mentioned in the Introduction, i.e. the symmetries of a system of differential equations, will be applied now to show the consistency of the results described in the last Section. If an involution system is generated from a subsystem the solution set of either system should be the same. This is an immediate consequence of the definitions of a symmetry and the involutiveness.

As an example let us consider again the Burgers-diffusion system (10). The six generators $U_1, \ldots U_6$ which are defined by

$$U_1 = \frac{\partial}{\partial t}, \quad U_2 = \frac{\partial}{\partial x}, \quad U_3 = v\frac{\partial}{\partial v},$$

$$U_4 = 2t\frac{\partial}{\partial t} + x\frac{\partial}{\partial x} - u\frac{\partial}{\partial u}, \quad U_5 = 2t\frac{\partial}{\partial x} + 2\frac{\partial}{\partial u} - xv\frac{\partial}{\partial v}, \tag{16}$$

$$U_6 = 4t^2\frac{\partial}{\partial t} + 4tx\frac{\partial}{\partial x} + 4(x - tu)\frac{\partial}{\partial u} - (2t + x^2)v\frac{\partial}{\partial v}$$

System	Symmetries
(10)	$U_1, U_2, U_3, U_4, U_5, U_6, V_1$
(10a),(10b),(10c)	$U_1, U_2, U_3, U_4, U_5, U_6, V_1$
(10a),(10b),(10d)	U_1, U_2, U_3, U_4
(10a),(10c),(10d)	$U_1, U_2, U_3, U_4, U_5, U_6, V_1$
(10b),(10c),(10d)	$U_1, U_2, U_3, U_4, U_5, U_6, V_1$
(10a),(10b)	$U_1, U_2, U_3, U_4, U_5, U_6, V_2$
(10a),(10c)	$U_1, U_2, U_4, U_5, U_6, V_3$
(10a),(10d)	U_1, U_2, V_4
(10b),(10c)	$U_1, U_2, U_3, U_4, U_5, U_6, V_1$
(10b),(10d)	U_1, U_2, U_3, U_4
(10c),(10d)	$U_1, U_2, U_3, U_4, U_5, U_6, V_1$
(10a)	$U_1, U_2, U_4, U_5\vert_{v=0}, U_6\vert_{v=0}$
(10b)	$U_1, U_2, U_3, U_4\vert_{u=0}, U_5\vert_{u=0}, U_6\vert_{u=0}, V_2$

Table 2: The symmetry group of the Burgers-diffusion system and its subsystems. The generators in the second column are those given in eqs.(16) and (17).

and the four generators $V_1, \dots V_4$

$$V_1 = (2\phi_x - \phi u)\frac{\partial}{\partial u} + \phi v \frac{\partial}{\partial v} \quad with \quad \phi_t + \phi_{xx} = 0,$$
$$V_2 = \phi\frac{\partial}{\partial v}, \quad V_3 = f(t)v\frac{\partial}{\partial v}, \quad V_4 = g(x)v\frac{\partial}{\partial v}, \quad f(t), g(x) \ \ arbitrary \tag{17}$$

which depend on various functions of t and x play a role thereby. The structure of the finite group generated by (16) is determined by the commutator table

$$[U_2, U_4] = U_2, \quad [U_2, U_5] = -U_3, \quad [U_2, U_6] = 2U_5, \tag{18}$$

$$[U_4, U_5] = U_5, \quad [U_4, U_6] = 2U_6. \tag{19}$$

Comparing Tables 1 and 2 the following regularities are observed. All those subsets of eqs.(10) which generate the full set of equations have an identical symmetry group comprising the six generators $U_1 \dots U_6$ supplemented by the generator V_1. which is also the symmetry group of the full involutive system (10). This was to be expected because by definition all these systems are completely equivalent to each other. Any other subsystem has a symmetry group which is different from that because its general solution satisfies additional equations which are not contained in the system (10). As an example consider the system comprising the two eqs.(10a,d). Their integrability conditions as given in Table 1 are satisfied for all solutions of the full system (10) and in addition many more functions due to the derivatives which occur, i.e. the solution set of the full system is a subset of the general solution of eqs.(10a,d). By definition this is a violation of involutiveness and as a consequence the symmetry group has drastically changed as it may be seen from Table 2. This is in marked distinction for example to eqs.(10b,c). Their integrability conditions are identical to the remaining two equations (10a,d) which is reflected by the fact that their symmetry group is identical to that of the full system.

A similar analysis may be performed for the other involution systems which have been described in the previous Section. In all cases a completely consistent picture is obtained between the results obtained from the algorithm INVSYS and the symmetry analysis.

4 Concluding Remarks

Involutiveness of a system of p.d.e.'s has turned out to be a basic concept in the realm of nonlinear evolution equations. In many cases it leads in a straightforward way to a Bäcklund transformation which is obtained by completing the given system by its integrability conditions. Sometimes new equations are found which are known to be integrable because they are part of an involutive system of integrable equations. An important application of the algorithm INVSYS is the package for automatically determining symmetries where it is used to find simpler equations so that the system can continue its solution algorithm.

To apply the algorithm INVSYS to larger problems which include more variables it has to be made more efficient. An example which is beyond the scope of its present ability is the system consisting of the Kadomtschev-Petviashvili (KP) equations and the mofified KP equation in the potential form

$$\begin{aligned} &u_{tx} + u_{4x} + 6u_x u_{xx} + 3u_{yy} = 0, \\ &v_{tx} + v_{4x} - \tfrac{3}{2} v_x^2 v_{xx} + 3v_{yy} - 3v_{xx} v_y = 0 \end{aligned} \tag{20}$$

which has been given by Konopelchenko[19]. The two equations are transformed into each other by

$$\begin{aligned} u_x &= \tfrac{1}{2} v_{xx} + \tfrac{1}{4} v_x^2 + \tfrac{1}{2} v_y \\ u_y &= \tfrac{1}{6} v_t + \tfrac{1}{6} v_{xxx} - \tfrac{1}{2} v_{xy} - \tfrac{1}{12} v_x^3 - \tfrac{1}{2} v_x v_y \end{aligned} \tag{21}$$

The last equation is an integrability condition of the former three. The algorithm INVSYS is not able in its present version to compelete the system (20) and (21) to an involution system due to storage space and CPU time overflow.

The importance of an involution system goes far beyond the more practical question of finding explicit solutions. It is a basic concept when the algebraic properties of differential equations are investigated. A fundamental problem occuring in connection with mathematical structures of any kind is that of the existence of a normal form. Only if the existence of a normal form is theoretically established and an algorithm for obtaining it in a finite number of steps has been designed, it is possible to classify problems into equivalent subsets. To obtain any results for a system of nonlinear differential equations in this direction it will certainly be necessary to extend it first to an involutive system.

In nonlinear analysis in general these problems have only very marginally been solved to date. For example in polynomial ideal theory which may be considered as the theory of differential equations of order zero, this question has been answered only recently in terms of the Gröbner basis of a polynomial ideal and the so-called Buchberger algorithm for obtaining it [20]. In this field too the tremendeous amount of calculations can only be handled by efficient computer algebra systems.

References

[1] N. J. Zabusky, M. D. Kruskal, "Interactions of solitons in a collisionless plasma and the recurrence of initial states", Phys. Rev. Lett. **15**, 240-243(1965).

[2] R. K. Bullough and P. J. Caudrey Eds.,"Solitons", Springer, Berlin, 1980.

[3] M. J. Ablowitz and H. Segur, "Solitons and the Inverse Scattering Transform", SIAM, Philadelphia, 1981.

[4] F. Calogero and A. Degasperis, "Spectral Transform and Solitons", North Holland, Amsterdam, 1982.

[5] F. Schwarz, "Symmetries of Differential Equations: From Sophus Lie to Computer Algebra", internal GMD report.

[6] C. Riquier, "Les systèmes d'équations aux dérivées partielles", Gauthier-Villars, Paris, 1910.

[7] M. Janet, "Sur les systèmes d'équations aux dérivées partielles", Journal de Mathematique 8III, 65-151(1920).

[8] M. Janet, "Les systèmes d'équations aux dérivées partielles", Gauthier-Villars, Paris, 1927.

[9] M. Janet, "Leçons sur les systèmes d'équations aux dérivées partielles", Gauthier-Villars, Paris, 1929.

[10] E. Cartan, "Les systèmes différentiels extérieur et leurs applications géométriques", Hermann, Paris, 1971.

[11] E. Kähler, "Einführung in die Theorie der Systeme von Differentialgleichungen", Chelsea Publishing Company, New York, 1949.

[12] Graindorge, Mémoires Liège, (1873).

[13] J. D. Cole, "On a Quasilinear Parabolic Equation Occuring in Aerodynamics", Quartely of Applied Mathematics **IX**, 225-236 (1951).

[14] H. D. Wahlquist, F. D. Estabrook, "Bäcklund Transformations for Solutions of the Korteweg-deVries Equation", Phys. Rev. Lett. **31**, 1386-1390 (1973).

[15] H. D. Wahlquist, F. D. Estabrook, "Prolongation Structures of Nonlinear Evolution Equations", Jour. Math. Phys. **16**, 1-7 (1975).

[16] D. K. Knuth, P. B. Bendix, "Contribution to Computational Problems in Abstract Algebra", J. L. Leech, Ed., Pergamon, 1970.

[17] F. Schwarz, "The Riquier-Janet Theory and its Application to Nonlinear Evolution Equations", Physica **11D**, 243-251 (1984).

[18] G. Huet, "Equations and Rewrite Rules, contribution to Formal Language Theory", Perspectives and Open Problems, R. V. Book Ed., Academic Press 1980 and JACM **27**, 797(1980).

[19] B. G. Konopelchenko, "On the Gauge-Invariant Description of the Evolution Equations Integrable by the Gelfånd-Dikij Spectral Problems", Phys. Lett. **92A**, 323-327(1982).

[20] B. Buchberger, "Gröbner Bases: An Algorithmic Method in Polynomial Ideal Theory", contribution to "Multidimensional Systems Theory", N. K. Bose (ed.), D. Reidel Publishing Company, Dordrecht, 1985.

Symplectic Geometry and Soliton Theory

Jean Pierre Francoise

Department of Mathematics, Building 89
University of Arizona
Tucson, AZ. 85721 U.S.A. (*)

ABSTRACT

The Analysis of an integrable System using Symplectic Geometry leads to the notion of the Action-angle coordinates. We consider here that notion in detail first for integrable systems with finitely many degrees of freedom and then for soliton equations.

(*) Permanent Address: Departement de Maths 425, Universite de Paris - Sud 91405 Orsay (France).

1. THE ACTION-ANGLE THEOREM AND THE ARNOL'D FORMULA

Let $f = (f_1, \ldots, f_m): (V^{2m}, \omega) \to \mathbf{R}^m$ be an integrable Hamiltonian System given by analytic functions f_j such that $\{f_i, f_j\} = 0$ for the Poisson bracket associated to the symplectic form ω and $df_1\wedge\ldots\wedge df_m \not\equiv 0$. Let us assume that the smooth fibers of f are compact and connected and denote by D the set of critical values of f. Let $c_0 \in \mathbf{R}^m D$; there is a neighborhood $U = f^{-1}(T)$, $c_0 \in T\subset\mathbf{R}^m D$ which retracts by deformation on the fiber $f^{-1}(c_0)$. On U there is a 1. form η such that $\omega|_U = d\eta$ [10, prop. 2.2]. Let $\gamma_j(c)$, $c \in T$, $j = 1, \ldots, m$, be a set of generators of $H_1(f^{-1}(c), Z)$, $c \in T$; we define locally on U the actions p_j by the Arnol'd formula [3]:

$$p_j = f^*(\int_{\gamma_j(c)} \eta)$$

The action-angle theorem states that there are associated angles q_i, $i = 1, \ldots, m$ such that:

$$\omega|_U = \sum_{i=1}^{m} dp_i \wedge dq_i \qquad\qquad \int_{\gamma_j(c)} dq_i = \delta_{ij}$$

The Arnol'd formula is very useful to give an expression of the actions and to relate the monodromy of the actions to the Picard-Lefschetz monodromy.

2. ALGEBRAICALLY COMPLETELY INTEGRABLE SYSTEMS

M. Adler and P. van Moerbeke defined [1] A.C.I. systems as those Hamiltonian systems for which the invariant real torii are part of Abelian varieties and the linearization of the flows is compatible with the addition law on the complex torii.

Most often, there is a family of Riemann surfaces, $\mathbf{C}_c$ generically smooth of genus g, such that the Hamiltonian flows X_j of the f_j are linear on the Jacobian or the Prym variety of $\mathbf{C}_c$. Given an A.C.I. system, there are several possible associated curves of eventually different genus (for instance the Kowalevskaya top ([1], [11]) or the Lagrange top ([13]).

On U, we can define functions $w = (w_1, \ldots, w_m)$ so that (f, w) gives a system of coordinates, using the retraction. We can write the symplectic form

$$w = \sum_{i=1}^{m} df_i \wedge \eta_i$$

with $\eta_i = \sum_{k=1}^{m} a_{ik}\, df_k + b_{ik}\, dw_k.$

We denote by ψ and call period matrix, the matrix whose entries are defined by

$$\psi_{ij} = \left(\int_{\gamma_j(c)} \eta_i\right).$$

Stokes' formula relates the actions to the period matrix,

$$f^*(\psi_{ij}) = \frac{\partial p_i}{\partial f_j}.$$

THEOREM 2.1 [7]

If $g > m$, the period matrix ψ is given by integrals of Abelian differentials of the first kind of $\mathbf{C}_c$.

Examples

The Kowalevskaya top has been considered in detail in [7]. The Kowalevskaya curve is of genus $g = m = 2$. The integrals are $F_1 = H$ (energy) and $F_2 = K$ (Kowalevskaya integral). We find

$$p_1 = \int_{-\infty}^{-\sqrt{K}} F\, dw, \quad p_2 = \int_{-\sqrt{K}}^{e} F\, dw$$

where $F = -\dfrac{1}{\sqrt{8(w + H)}} \cos^{-1}\left|1 + \dfrac{2}{d^2}\dfrac{(w^2 - K)(w + H)}{(w + H - 2\ell^2}\right|$

d is the free parameter and ℓ is a value of the momentum, and e is a real root of the polynomial

$\phi_c(w) = (w + H)(w^2 + d^2 - K) - 2d^2\ell^2$.

The Neumann system is of the same type and has been considered as well in [7]. It is equivalent to periodic KdV with a finite zone potential.

I have investigated two other examples which are not of this type, the spherical pendulum [7] and the Calogero-Moser model.

3. THE ACTION "ANGLES" WHEN THERE ARE NO INVARIANT TORII

Very often one refers to action-angles as a system of coordinates (p, q) such that the first integrals are functions of the p's and that the symplectic form is $dp \wedge dq$. Of course such a system of coordinates is not unique: In the case where there are invariant torii, the arbitrariness relies on a choice of generators of the homology group of the fibers of the moment mapping. But in the case where there are no invariant torii, then one has to live with this non-uniqueness. I suggest that the terminology of conjugate variables is more adapted to that situation because there are of course no angles. Good illustrations of the difference between these two situations are given by the periodic Toda lattice versus the open Toda lattice, and the periodic finite-zone potential KdV versus the Schwartz class potential KdV.

4. THE KDV EQUATION WITH A POTENTIAL IN THE SCHWARTZ CLASS

For this case, the conjugate variables are given by the scattering data. We derive here a formula that looks very similar to the ones of §2.

Let $L = \frac{d^2}{dx^2} + u$ be the Schrödinger operator. The first integrals of KdV are given by

$$f_n = Tr(L^{n+1/2})$$

The actions (in Faddeev's sense) are generated on the continuous part by the function

$$\frac{k}{4\pi}\ln|a(k)| = \frac{k}{4\pi}\ln\{Tr[L^{1/2}\frac{1-kL}{1+kL}]\}$$

as seen from formula 3.65 of [12]. L. Faddeev [5] has developed the symplectic aspects of the KdV, NLS, sine-Gordon, and three-wave equations.

5. THE PERIODIC KDV, NLS, SINE-GORDON

S. P. Novikov and his co-workers, B. A. Dubrovin and A. P. Veselov extended the Flaschka-McLaughlin computation of the perodic KdV action-angles to KdV, NLS, and sine-Gordon finite-zone potentials [14] [4].

A. P. Veselov and S. P. Novikov announced a computation of the actions of the Kowalevskaya top which uses the isomorphism between the top and particular solutions of KdV. Actually, the symplectic forms are not the same. For further research, what follows could be useful:

6. ACTIONS AND CONJUGATE STRUCTURES

When a given integrable system preserves several distinct symplectic structures, one can try to compare the actions for these different structures.

The proof of the following theorem uses J. Moser's path method and relative cohomology techniques, and it will be given elsewhere.

Theorem 6.1

Suppose that a system $f = (f_1, \ldots, f_m)$ is in involution for two distinct symplectic structures ω or ω'. The actions of (f, ω) and (f, ω') are the same if and only if there is an isotopy of ω to ω' which preserves f.

REFERENCES

[1] M. Adler, P. van Moerbeke
"Completely integrable systems, Kac-Moody Lie algebras and curves", Adv. in Math 38 (3) 1980, 267-317

"Linearization of Hamiltonian Systems, Jacobi varieties and Representation theory", Adv. in Math 38 (3) 1980, 318-379.

[3] V. I. Arnol'd, A. Avez
"Problémes ergodiques de la mécanique classique" Gauthiers-Villars Paris 1967.

[4] B. A. Dubrovin, S. P. Novikov
"Algebro-geometric Poisson brackets for real finite zone solutions of the sine-Gordon equation and the non-linear Schrödinger equation". Dokl. Akad. Nauk SSSR 267 (1982) 1295-1300.

[5] L. Faddeev
"A Hamiltonian Interpretation of the Inverse Scattering Method" in Solitons, Ed. R. K. Bullough, P. J. Caudrey, Topics in Current Physics 17, 1980, 339-354. Springer.

[6] H. Flaschka, D. W. McLaughlin
"Canonically conjugate variables for the Korteweg-de Vries equation and the Toda lattice with periodic boundary conditions." Progr. Theoret. Phys. 55 (1976) 438-456.

[7] J. P. Francoise
"Calculs explicites d'Action-angles",
Sém. de Mathématiques Supérieures de l'Univ. de Montreal 102, 1986.

"Monodromy and the Kowalevskaya Top", to appear in Asterisque, Equations différentielles singuliéres

"The Arnol'd Formula for Algebraically completely integrable systems", to appear.

[10] V. Guillemin, S. Sternberg
"The Gelfand-Četlin system and the quantification of the complex Flag Manifold". J. of Functional Anal. 52 (1), 1983, 106-128.

[11] L. Haine
"Geodesic Flow on SO(4) and Abelian surfaces". Math. Ann. 283, 1983, 635-672.

[12] A. C. Newell
"Solitons in Mathematics and Physics CBMS-NSF Regional Conference Series in Applied Mathematics, SIAM 1985

[13] P. van Moerbeke, T. Ratiu
"The Lagrange rigid body problem," Ann. de l' Inst. Fourier 23 (9) 1982, 211-234.

[14] A. P. Veselov, S. P. Novikov
"Poisson brackets and complex tori", Proc. of the Steklov Inst. of Math., 165 1985, 53-65.

NONLINEAR MODULATION OF A TRANSVERSALLY TRAPPED MODE

Francesco CALOGERO

Dipartimento di Fisica, Università di Roma "La Sapienza"
Piazzale Aldo Moro, I-00185 ROMA
ITALY

Pierre C. SABATIER

Laboratoire de Physique Mathématique, U.S.T.L.
Place E. Bataillon, F-34060 MONTPELLIER
FRANCE

ABSTRACT

It is shown that, for a large class of nonlinear PDEs in 2+1 dimensions, whose linear parts account for unidirectional dispersive wave propagation of transversally trapped modes, weak nonlinear effects yield consistently a modulation of the wave motion that is generally described by the nonlinear Schrödinger equation. The example of water waves traveling along a beach ("edge waves") is tersely discussed.

1. INTRODUCTION

For a large class of unidimensional propagation equations, characterized by a dispersive linear part admitting the "plane wave" solution $\cos[kx - \omega(k)t]$ and by a "small" nonlinear part, it has been shown rather generally that the main nonlinear effect is to induce an amplitude modulation of the carrier plane wave whose evolution is governed by a nonlinear Schrödinger equation, expressed in terms of "slow" space and time variables.[1,2,3,4] In this paper we report tersely on the extension of these results to equations in 2+1 dimensions whose linear part admits solutions, $\rho(k,y)\cos[kx - \omega(k)t]$, describing unidirectional wave propagation of transversally trapped modes. Here $\rho(k,y)$ represents a trapped mode, characterized by appropriate boundary conditions at the "y-boundaries" of the "physical domain" Ω, while $\omega(k)$ depends now of course, in addition to the wave number k for the ("longitudinal") x-direction, also on the particular ("transversal") eigenmode.

Specifically we consider here evolution PDEs of the following type :

$$\mathcal{D}u \equiv u_{tt} + \beta(y)\, d_x u - \frac{\partial}{\partial y}\alpha(y)\frac{\partial u}{\partial y}$$

$$+ \sum_{n=2} \varepsilon^{n-1} F^{(n)}\left(u, \frac{\partial}{\partial x}, \frac{\partial}{\partial y}, \frac{\partial}{\partial t}\right) = 0 \quad , \tag{1a}$$

with

$$d_x \equiv \sum_{m=1}^{M} (-)^m \gamma_m \frac{\partial^{2m}}{\partial x^{2m}} \quad , \tag{1b}$$

completed by suitable conditions at the two values of y (which do not depend on x or t, and need not be finite) that limit the physical domain Ω in the transversal direction. We assume the constants γ_m to be real, the function $\beta(y)$ to be real and continuous and the function $\alpha(y)$ to be differentiable and positive inside Ω. We moreover assume the nonlinear terms $F^{(n)}$ to be homogeneous polynomials of degree n in u and its partial derivatives :

$$F^{(2)} = \sum_{j,f,\ell,m} a(0,j,f,\ell,m) \frac{\partial^{j+f} u}{\partial x^j \partial y^f} \frac{\partial^{\ell+m} u}{\partial x^\ell \partial y^m}$$

$$+ \sum_{j,f,\ell,m} a(1,j,f,\ell,m) \frac{\partial^{j+f+1} u}{\partial x^j \partial y^f \partial t} \frac{\partial^{\ell+m} u}{\partial x^\ell \partial y^m} \quad , \tag{2}$$

$$F^{(3)} = \sum_{j,f,\ell,m,p,q} b(0,j,f,\ell,m,p,q) \frac{\partial^{j+f} u}{\partial x^j \partial y^f} \frac{\partial^{\ell+m} u}{\partial x^\ell \partial y^m} \frac{\partial^{p+q} u}{\partial x^p \partial y^q}$$

$$+ \sum_{j,f,\ell,m,p,q} b(1,j,f,\ell,m,p,q) \frac{\partial^{j+f+1} u}{\partial x^j \partial y^f \partial t} \frac{\partial^{\ell+m} u}{\partial x^\ell \partial y^m} \frac{\partial^{p+q} u}{\partial x^p \partial y^q} . \tag{3}$$

We assume the constants a and b to be all real. Note that we have included only terms containing at most first order derivatives with respect to t ; indeed higher t-derivatives can be eliminated using (1a). The nonlinear terms that we have now displayed are the only ones that do contribute in a treatment restricted to the search for an "ε^3-solution" of (1), namely a function $u(x,y,t)$ such that

$$\mathcal{D}u = O(\varepsilon^3) \tag{4}$$

inside the physical domain, as $\varepsilon \to 0$, and such that u reduces to a trapped mode, $\rho(k,y)\cos[kx - \omega(k)t]$, in the $\varepsilon = 0$ limit. As it will be clear in the following, such a treatment is adequate to obtain the results that we report in this paper. A more complete treatment shall be presented elsewhere.

2. MAIN PROBLEM AND ANSATZ

Let D indicate the linear part of the differential operator $\mathcal{D}$ (or, equivalently, the $\varepsilon = 0$ limit of $\mathcal{D}$; see (1)). Then $u = f(.,y)\cos[kx - \omega(k)t]$ is a solution of $Du = 0$ iff

$$D_k f \equiv [\beta(y)\, d(k) - \frac{d}{dy}\alpha(y)\frac{d}{dy}]\, f(.,y) = \omega^2(k)\, f(.,y) \tag{5}$$

where

$$d(k) = \sum_{m=1}^{M} \gamma_m k^{2m} \tag{6}$$

(see (1b)). Let k be a given real wave number. We assume that the boundary conditions are such that the spectrum of D_k is a denumerable set of simple real eigenvalues $\omega_m^2(k)$ $(m = 0,1,2,\ldots)$, and the corresponding orthonormal eigenfunctions $\rho_m(k,y)$ provide a complete basis in $L^2(\Omega)$:

$$[\beta(y)\, d(k) - \frac{d}{dy}\alpha(y)\frac{d}{dy}]\, \rho_m(k,y) = \omega_m^2(k)\, \rho_m(k,y) \ , \tag{7}$$

$$< \rho_m , \rho_\ell > = \delta_{\ell m} \ . \tag{8}$$

Here, and below as well, we use the notation

$$< f,g > \equiv \int_\Omega dy\, f^*(y)\, g(y) \ . \tag{9}$$

Given our previous assumptions on $\beta(y)$ and $\alpha(y)$, these conditions are generally satisfied provided Ω is the y-interval $[y_o,y_1]$ and the boundary conditions at y_o and y_1 are homogeneous (y_o and y_1 need not be finite ; but if they diverge, then α and β must behave appropriately to preserve the denumerable character of the spectrum ; see for instance the specific example treated below).

Let us now focus on a specific mode, say $\rho_{m_o}(k,y)$, corresponding to the positive eigenvalue $\omega^2_{m_o}(k)$. In the following, for notational simplicity, we set $m_o = 0$, namely we write ω_o, ρ_o in place of ω_{m_o} and ρ_{m_o}, and ω_m, ρ_m with $m = 1,2,3,\ldots$ for $m \neq m_o$.

For our ε^3-solution of (1) we now introduce the ansatz

$$u(x,y,t) = \phi \exp(iz) + \varepsilon[\tfrac{1}{2}\phi_o + \phi_2 \exp(2iz)] + \varepsilon^2 \phi_3 \exp(3iz) + c.c., \tag{10}$$

with

$$\phi(y,\xi,\tau) = p(y)\varphi(\xi,\tau) + i\varepsilon\tilde{p}(y)\tilde{\varphi}(\xi,\tau) + \varepsilon^2\Phi(y,\xi,\tau) , \tag{11}$$

$$\phi_n(y,\xi,\tau) = p_n(y)\,\varphi_n(\xi,\tau) + \varepsilon\,\Phi_n(y,\xi,\tau),\ (n\neq 1) , \tag{12}$$

and

$$z = kx - \omega t , \tag{13}$$

$$\xi = \varepsilon(x - vt) , \tag{14a}$$

$$\tau = \varepsilon^2 t . \tag{14b}$$

Here and below the abbreviated notation + c.c., see (10), indicates addition of the complex conjugate. Note that, since u is by assumption real, also real must be ϕ_o, p_o, φ_o and Φ_o (the factor $\tfrac{1}{2}$ in front of ϕ_o in (10) is introduced in view of this fact and of the addition of the complex conjugate contribution).

Our final position shall be

$$\omega = \omega_o(k) , \quad p(y) = \rho_o(k,y) ; \tag{15}$$

and our main task is to show that $\varphi(\xi,\tau)$, namely the function that modulates the leading contribution in (10), evolves according to the nonlinear Schrödinger equation,

$$i\,\varphi_\tau + \lambda\,\varphi_{\xi\xi} + \mu\,|\varphi|^2\,\varphi = 0\,, \tag{16}$$

with λ and μ constants which are evaluated below together with the other quantities that appear in the ansatz (10). Note that, in the $\varepsilon = 0$ limit, $u(x,y,t)$, as given by (10), (11), (12) and (15), does reduce to the solution, $\rho_o(k,y)\cos[kx - \omega_o(k)t]$, of the linear part of (1), describing unidirectional wave propagation in the x direction of a mode trapped in the y direction (indeed $\varphi(\xi,\tau)$, being a function of the "slow" variables ξ and τ, see (14), becomes, in the $\varepsilon = 0$ limit, merely a constant, that can be set to unity for notational convenience).

3. CALCULATIONS

In this section we outline the calculations that must be performed, after the ansatz (10) is inserted in the nonlinear PDE (1).

3.1 Nonlinear Terms

We assume (subject to a posteriori verification) that the functions appearing in the ansatz (10) are sufficiently differentiable. Then the following formulae follow readily from (10 ÷ 14) :

$$\frac{\partial^{j+f} u}{\partial x^j\,\partial y^f} = \alpha^{(1)}_{jf}\,\varphi\,\exp(iz) + \varepsilon[\beta^{(1)}_{jf}\,\varphi_\xi\,\exp(iz) + \gamma^{(1)}_{jf}\,\tilde{\varphi}\,\exp(iz) + \tfrac{1}{2}\,\alpha^{(o)}_{jf}\,\varphi_o + \alpha^{(2)}_{jf}\,\varphi_2\,\exp(2iz)] + \text{c.c.} + O(\varepsilon^2)\,, \tag{17}$$

$$\frac{\partial^{j+f+1} u}{\partial x^j\,\partial y^f\,\partial t} = \dot{\alpha}^{(1)}_{jf}\,\varphi\,\exp(iz) + \varepsilon[\dot{\beta}^{(1)}_{jf}\,\varphi_\xi\,\exp(iz) + \dot{\gamma}^{(1)}_{jf}\,\tilde{\varphi}\,\exp(iz) + \dot{\alpha}^{(2)}_{jf}\,\varphi_2\,\exp(2iz)] + \text{c.c.} + O(\varepsilon^2)\,, \tag{18}$$

with

$$\alpha^{(1)}_{jf} = (ik)^j\,p^{(f)}\,,\quad \beta^{(1)}_{jf} = j\,(ik)^{j-1}\,p^{(f)}\,,\quad \gamma^{(1)}_{jf} = i\,(ik)^j\,\tilde{p}^{(f)}\,,$$

$$\alpha^{(o)}_{jf} = \delta_{jo}\,p_o^{(f)}\,,\quad \alpha^{(2)}_{jf} = (2ik)^j\,p_2^{(f)}\,,$$

$$\dot{\alpha}^{(1)}_{jf} = k\omega\,(ik)^{j-1}\,p^{(f)}\ ,\quad \dot{\beta}^{(1)}_{jf} = -i\,(ik)^{j-1}\,(kv+j\omega)\,p^{(f)}\ ,$$

$$\dot{\gamma}^{(1)}_{jf} = \omega\,(ik)^{j}\,\tilde{p}^{(f)}\ ,\quad \dot{\alpha}^{(2)}_{jf} = -2i\omega\,(2ik)^{j}\,p_2^{(f)}\ . \tag{19}$$

Here the superscript (f) appended to p, $\tilde{p}$, p_o and p_2 indicates of course differentiation f times with respect to y.

From these formulae and (2 ÷ 3) there clearly obtains

$$\begin{aligned} F^{(2)} = {} & f_{1,-1}\,|\varphi|^2 + f_{1,1}\,\varphi^2\exp(2iz) + \\ & + \varepsilon[g_{-1,1}\,\varphi^*\,\varphi_\xi + (f_{0,1}\,\varphi_0\varphi + f_{2,-1}\,\varphi_2\,\varphi^*)\exp(iz) + \\ & + (g_{1,1}\,\varphi\,\varphi_\xi + h_{1,1}\,\varphi\,\tilde{\varphi})\exp(2iz) + \\ & + f_{2,1}\,\varphi_2\,\varphi\exp(3iz)] + \text{c.c.} + O(\varepsilon^2)\ , \end{aligned} \tag{20}$$

where the superscript * denotes complex conjugation,

$$F^{(3)} = f_{1,1,-1}\,\varphi\,|\varphi|^2\exp(iz) + f_{1,1,1}\,\varphi^3\exp(3iz) + \text{c.c.} + O(\varepsilon). \tag{21}$$

The coefficients f,g,h that appear in these equations are quadratic (for $F^{(2)}$) or cubic (for $F^{(3)}$) combinations of the quantities α, β, γ defined above ; they are easily computed inserting (17) and (18) in (2) and (3). Note that these coefficients are generally complex, with the exception of $f_{1,-1}$, that is real. They are of course functions of y, but do not depend on ξ or τ.

3.2 Linear Terms

As indicated above, we denote with D the linear part of the differential operator $\mathcal{D}$, see (1). It is then easily seen that (10 ÷ 14) imply

$$\begin{aligned} \exp(-iz)\,D[\phi\exp(iz)] = {} & \varphi[\beta\,d(k) - \frac{d}{dy}\,\alpha\,\frac{d}{dy} - \omega^2]\,p + \\ & + i\,\varepsilon\{\tilde{\varphi}\,[\beta d(k) - \frac{d}{dy}\,\alpha\,\frac{d}{dy} - \omega^2]\tilde{p} + \varphi_\xi[2v\omega - \beta\,d'(k)]p\} + \\ & + \varepsilon^2\,\{[\beta\,d(k) - \frac{d}{dy}\,\alpha\,\frac{d}{dy} - \omega^2]\Phi - \tilde{\varphi}_\xi[2v\omega - \beta\,d'(k)]\tilde{p} + \\ & + \varphi_{\xi\xi}[v^2 - \tfrac{1}{2}\,\beta\,d''(k)]p - 2i\omega\,\varphi_\tau\,p\} + O(\varepsilon^3)\ , \end{aligned} \tag{22}$$

and, for $n = 0,2,3,$

$$\exp(-inz)\, D[\psi_n \exp(inz)] = \varphi_n \left[\beta\, d(nk) - \frac{d}{dy}\,\alpha\,\frac{d}{dy} - n^2\,\omega^2\right] p_n +$$

$$+ \varepsilon\{[\beta\, d(nk) - \frac{d}{dy}\,\alpha\,\frac{d}{dy} - n^2\,\omega^2]\,\Phi_n +$$

$$+ i[2nv\omega - \beta\, d'(nk)]p_n\, \partial\varphi_n/\partial\xi\} + O(\varepsilon^2) \quad . \tag{23}$$

Here of course $\varphi \equiv \varphi(\xi,\tau)$, $\tilde{\varphi} \equiv \tilde{\varphi}(\xi,\tau)$, $\Phi \equiv \Phi(y,\xi,\tau)$, $p \equiv p(y)$, $\tilde{p} \equiv \tilde{p}(y)$ (see (11)), $\varphi_n \equiv \varphi_n(\xi,\tau)$, $\Phi_n \equiv \Phi_n(y,\xi,\tau)$, $p_n \equiv p_n(y)$ (see (12)), $\alpha \equiv \alpha(y)$, $\beta \equiv \beta(y)$ (see(1a)), and the primes appended to d(k) (see (6)) denote differentiation with respect to k.

4. RESULTS

Using (20 ÷ 23) it is now easy to ascertain the relations that are implied by (4), with (1) and (10). First of all one has (15). Then one gets, after a few intermediate steps,

$$v \equiv v(k) = [2\,\omega_o(k)]^{-1}\, d'(k) < \rho_o,\ \beta\,\rho_o > \quad , \tag{24}$$

$$\varphi_o = |\varphi|^2,\ \varphi_2 = \varphi^2 \ ,\ \varphi_3 = \varphi^3 \ ,\ \tilde{\varphi} = \varphi_\xi \ , \tag{25}$$

$$\Phi_o(y,\xi,\tau) = \varphi^*(\xi,\tau)\ \varphi_\xi(\xi,\tau)\ \tilde{p}_o(y) + c.c. \ , \tag{26}$$

$$\Phi_2(y,\xi,\tau) = \varphi(\xi,\tau)\ \varphi_\xi(\xi,\tau)\ \tilde{p}_2(y) \ , \tag{27}$$

$$\Phi(y,\xi,\tau) = \sum_{m=1}^{\infty} \rho_m(k,y)\ \Phi^{[m]}\ (\xi,\tau) \ , \tag{28}$$

$$\Phi^{[m]} = [\omega_m^2 - \omega_o^2]^{-1}\ [A_m\ \varphi_{\xi\xi} + B_m\ |\varphi|^2\ \varphi] \ , \tag{29}$$

$$A_m = < \rho_m,\ [2v\omega_o - \beta\, d'(k)]\tilde{p} - [v^2 - \tfrac{1}{2}\,\beta\, d''(k)]\rho_o > \ , \tag{30}$$

$$B_m = - < \rho_m,\ f_{0,1} + f_{2,-1} + f_{1,1,-1} > \ , \tag{31}$$

$$[\frac{d}{dy}\,\alpha\,\frac{d}{dy}]\ p_o = f_{1,-1} \ , \tag{32}$$

$$[\beta\, d(k) - \frac{d}{dy}\,\alpha\,\frac{d}{dy} - \omega_o^2]\tilde{p} = -[2v\omega_o - \beta\, d'(k)]\ \rho_o \ , \tag{33}$$

$$[\beta\, d(2k) - \frac{d}{dy}\,\alpha\,\frac{d}{dy} - 4\,\omega_o^2]p_2 = -\ f_{1,1} \ , \tag{34}$$

$$[\beta\ d(3k) - \frac{d}{dy}\,\alpha\,\frac{d}{dy} - g\ \omega_o^2]p_3 = -f_{2,1} - f_{1,1,1} \quad , \tag{35}$$

$$[\frac{d}{dy}\,\alpha\,\frac{d}{dy}]\tilde{p}_o = g_{-1,1} \quad , \tag{36}$$

$$[\beta\ d(2k) - \frac{d}{dy}\,\alpha\,\frac{d}{dy} - 4\ \omega_o^2]\tilde{p}_2 = -g_{11} - h_{11} - 2[4v\omega - \beta\ d'(2k)]p_2 , \tag{37}$$

and finally

$$2i\ \omega_o\ \varphi_\tau + A_o\ \varphi_{\xi\xi} + B_o\ |\varphi|^2\varphi = 0 \quad . \tag{38}$$

This last equation coincides with the nonlinear Schrödinger equation (16), with the identification

$$\lambda = A_o/(2\ \omega_o) \ , \ \mu = B_o/(2\ \omega_o) \ , \tag{39}$$

and of course A_o and B_o defined by (30) and (31), namely (after some rearrengements)

$$A_o = -v^2 + v\omega_o\ d''(k)/d'(k) + d'(k)[\langle\rho_o,\beta\rho_o\rangle\ \langle\rho_o,\tilde{p}\rangle - \langle\rho_o,\beta\tilde{p}\rangle] , \tag{40}$$

$$B_o = - \langle\ \rho_o,\ f_{0,1} + f_{2,-1} + f_{1,1,-1}\ \rangle \ . \tag{41}$$

Note that (24 ÷ 31) are explicit formulae, while (32 ÷ 37) are inhomogeneous Sturm-Liouville equations, that must be supplemented by the boundary conditions appropriate to the specific problem at hand, in order to determine uniquely the relevant functions of y, namely p_o, $\tilde{p}$, p_2, p_3, $\tilde{p}_o$ and $\tilde{p}_2$; actually only the first 3 of these are needed to compute the coefficients, see (40) and (41), of the nonlinear Schrödinger equation (38), since it is easily seen that the functions $f_{0,1}$, $f_{2,-1}$ and $f_{1,1,-1}$ involve only $p = \rho_o$, p_o and p_2 (see (2), (3) and (18÷20)). Of course it must be assumed that, with these boundary conditions, the operators $\beta\ d(nk) - \frac{d}{dy}\,\alpha\,\frac{d}{dy}$ do not have $n^2\ \omega_o^2$ as eigenvalue, at least for $n = 0, 2, 3$, so that (32) and (36) (corresponding to $n = 0$), (34) and (37) (corresponding to $n = 2$), and (35) (corresponding to $n = 3$), be solvable, implying that the ansatz (10) is valid. As for (33), it is solvable even though ω_o^2 is an eigenvalue of the differential operator $\beta\ d(k) - \frac{d}{dy}\,\alpha\,\frac{d}{dy}$, because, as implied by (24), the corresponding eigenfunction $\rho_o(k,y)$ is orthogonal to the r.h.s. of (33) ; and note moreover that the fact that $\tilde{p}(y)$, as computed from this equation, remains undetermined up to the addition of a term $C\,\rho_o(k,y)$, with C an arbitrary constant, is of little consequence, since clearly such an additional term yields eventually a vanishing contribution,at least as regards the coefficients of the nonlinear Schrödinger equation (38) (see (40)). Indeed it is easily seen, from (33), that (40) can be more explicitly rewritten as follows :

$$A_o = -[v(k)]^2 + v(k)\,\omega_o(k)\,d''(k)/d'(k) - \\ -[d'(k)]^2 \sum_{m=1}^{\infty} |< \rho_m, \beta\rho_o >|^2/[\omega_m^2(k) - \omega_o^2(k)]. \qquad (42)$$

Note that the coefficients ω_o and A_o in the nonlinear Schrödinger equation (38) are real, while B_o need not be real. However, only if B_o is real is this nonlinear Schrödinger equation integrable ; and moreover, only if the product $A_o\,B_o$ is positive, it can support solitons.

5. WAVES ALONG A BEACH

The problem of the propagation of water waves along a (longitudinally straight and transversally sloping) beach reduces, in the long wave approximation[5], to (1) with $\beta(y) = \alpha(y)$, $\gamma_m = \delta_{m1}$ (see (1b)) and (see (2) and (3))

$$F^{(2)} = (u_{xx} + u_{yy})u_t + (u_x^2 + u_y^2)_t \quad , \qquad (43)$$

$$F^{(3)} = \tfrac{1}{2}(u_{xx} + u_{yy})(u_x^2 + u_y^2) + u_x^2 u_{xx} + 2u_x u_y u_{xy} + u_y^2 u_{yy}. \qquad (44)$$

Thanks to the positivity of $\alpha(y)$ (that represents here the depth of the sea bottom) all the eigenvalues ω_m^2 are positive. Thanks to the special form of $F^{(2)}$ the coefficient $f_{1,-1}$, see (20), vanishes, hence p_o vanishes as well, see (32). Hence $f_{0,1}$ also vanishes. The other quantities relevant to the evaluation of the coefficients in the nonlinear Schrödinger equation (38) read :

$$f_{1,1} = i\,\omega[3k^2p^2 - pp'' - 2(p')^2] \,, \tag{45}$$

$$f_{2,-1} = i\,\omega[p\,p_2'' - 2p_2p'' - 6k^2pp_2 - 2p'p_2'] \,, \tag{46}$$

$$f_{1,1,-1} = \tfrac{1}{2}[9(p')^2\,p'' - 3k^4p^3 + k^2p^2p'' + k^2p\,(p')^2] \,, \tag{47}$$

where the primes indicate of course y-differentiation. Since p is real (see (15)), $f_{1,1}$ is imaginary ; hence p_2 is imaginary (see(34)); hence $f_{2,-1}$, as well as $f_{1,1,-1}$, are real. Hence the coefficient B_o of the nonlinear term in the Schrödinger equation (38) is real, being merely the component along ρ_o of the sum of $f_{2,-1}$ and $f_{1,1,-1}$ (see (41)).

5.1 Specific Example

Let $\alpha(y) = \alpha y$, with α a positive constant. This corresponds to a sea bottom with the constant slope α/g (g = acceleration of gravity). The physical domain Ω is the half-axis ($y_o = 0$, $y_1 = \infty$). The boundary conditions are $yu_y = 0$ at $y = 0$ and $yu \to 0$ as $y \to \infty$. It is then seen that the spectrum satisfies the conditions required above, indeed it reads :

$$\omega_m^2(k) = \alpha k\,(2m+1), \quad m = 0,1,2,... \tag{48}$$

$$\rho_m(k,y) = (2k)^{\frac{1}{2}}\exp(-ky)\,L_m(2ky), \; m = 0,1,2,... \tag{49}$$

with $L_m(z)$ the Laguerre polynomial of degree m. Note the simple dependence of the eigenvalues on α and k (that merely determine the overall scale), and the fact that the eigenmodes (49) turn out to be independent of α.

Let us restrict attention here only to the evolution of the first mode, setting $m_o = 0$, so that

$$\omega_o = (\alpha k)^{\frac{1}{2}}, \; \rho_o(k,y) = (2k)^{\frac{1}{2}}\exp(-ky) \,. \tag{50}$$

There occurs then a drastic simplification, since $f_{1,1}$, as given by (45), turns out to vanish ; hence p_2 vanishes (see (34), and note that $4\,\omega_o^2$ is not an eigenvalue of $\beta\, d(2k) - \frac{d}{dy}\,\alpha\,\frac{d}{dy}$, since $4\alpha k \neq 2\alpha k\,(2m+1)$, $m = 0,1,2,...$) ; hence $f_{2,-1}$ vanishes as well (see (46)). The computation of the remaining quantities is moreover plain and we find

$$v = \tfrac{1}{2}\,(\alpha/k)^{\frac{1}{2}}\ ,\ A_0 = -\tfrac{1}{4}\alpha/k\ ,\ B_0 = -4\,k^5\ . \tag{51}$$

Note that the nonlinear Schrödinger equation (38) with these parameters possesses the soliton solution

$$\varphi(\xi,\tau) = C\,\exp[i\,a(\xi-b\tau)]\ /\ \cosh\,[q(\xi-V\tau)]\quad , \tag{52}$$

where

$$C = (\alpha/8)^{\frac{1}{2}}\,q\,k^{-3}\,\exp(i\gamma),\ a = -4(k/\alpha)^{\frac{1}{2}}\,k\,V,$$
$$b = \tfrac{1}{2}\,V[1-\alpha q^2/(16\,k^3\,V^2)]\quad , \tag{53}$$

and q, V, γ are 3 arbitrary real constants.

6. EXTENSIONS AND APPLICATIONS

(i) If the coefficient B_0 in (38) turns out to vanish, then a different rescaling may be introduced, corresponding to even slower time and space variables and yielding a different nonlinear equation (as in the one-dimensional case[1]). A separate treatment is also required if there occurs a "resonance", for instance if the operator $\beta\,d(2k) - \frac{d}{dy}\,\alpha\,\frac{d}{dy}$ has the eigenvalue $4\,\omega_o^2(k)$ (see (34)).

(ii) Often the setting for water waves is a basin which has a *finite* transversal width. Then the modes are determined by a *regular* Sturm-Liouville problem (which is, at least in principle, simpler than the singular Sturm-Liouville problem resulting from the infinite transversal extension of the domain in the example treated above). One could also deal, however, with more complicated circumstances, yielding a Sturm-Liouville problem with a spectrum featuring both a discrete and a continuum component ; as is for instance the case for water waves in a basin of *finite* depth that is *infinitely* extended in the transversal direction. In such a case one might still focus on a discrete mode, taking care of the continuum by introducing Sommerfeld conditions to account for asymptotically "outgoing" waves.

(iii) Extensions to include the propagation of trapped modes in a rotational context are now being investigated, and related applications.

The study of nonlinear effects in water wave propagation is of interest since such effects play a role in many natural phenomena[6].

REFERENCES

1) Calogero, F. and Eckhaus, W., "Nonlinear evolution equations, rescalings, model PDEs and their integrability. I", Inverse Problems (in press).

2) Eckhaus, W., "The long-time behaviour for perturbed wave equations and related problems", preprint nr.404, Department of Mathematics, University of Utrecht, December 1985. Published in part in Springer Lecture Notes in Physics 246 (1986).

3) Taniuti, T. and Nishihara,K., "Nonlinear Waves", Pitman, Boston London Melbourne (1983).(Translation of Japanese book published in 1977).

4) See the papers by Taniuti and others in Suppl. Progr. Theor. Phys. no.55 (1974) (issue devoted to the "Reductive Perturbation Method for Nonlinear Wave Propagation").

5) Mysak, L.A., "Topographically trapped waves", Ann. Rev. Fluid. Mech. 12, 45-76 (1980).

6) See, for example, Sabatier, P.C., "Formation of waves by ground motion",Chapter 17 of Vol. 6,"Complex flow phenomena and modeling", Encycl. of Fluid Mechanics, Nicholas P. Cheremisinoff Editor, Gulf. Publ. Comp. (1986).

NORMAL FORM AND SOLITONS

Yuji KODAMA

Department of Physics, Nagoya University
Chikusa-ku, Nagoya, 464, JAPAN
Department of Mathematics, The Ohio State University
Columbus, Ohio, 43210, U.S.A.

ABSTRACT

We develop an asymptotic theory of nonlinear dispersive waves where the KdV equation gives the leading order approximation.

First, we define a normal form for unidirectional solution of the wave system. The normal form admits a solitary wave solution in the form of the KdV soliton. In general, the normal form takes an integrable form up to first order, but not to second order.

Then, we study the interaction of two solitary waves of the normal form, and show that the interaction is inelastic and leads to the production of new solitary wave as well as the radiations.

1. INTRODUCTION

The soliton equations or integrable systems, such as the Korteweg-de Vries (KdV) equation are derived as the leading order approximation for a large class of nonlinear dispersive wave equations in an appropriate asymptotic sense [1]. However, as we know, the most of the equations are <u>nonintegrable.</u> In order to see the nonintegrability (or the integrability) of the original systems, we must consider the higher order terms and study the long-time behavior of solitons under these terms as the perturbations. The main purpose of this paper is to discuss this subject for the KdV equation with higher order corrections based on a normal form theory.

The manuscript consists of the followings: In the section 2, we describe the form of the higher order corrections which results naturally from the singular perturbation technique, e.g. the reductive perturbation method [1]. The existence of the integrals for given equation is a key for its integrability. In the section 3, we look for the integrals (adiabatic constants) of the perturbed equation derived in the section 2 in terms of the differential polynomial, and give a condition for the existence of the higher order integrals (<u>Proposition 3.4</u>). With these results, in the section 4 we show that, in general, the perturbed equation is integrable up to first order correction, but not to second order (<u>Collorary 4.6</u>). This is done by using the normal form theorem (<u>Theorem 4.4</u>). Here we first give the definition of the normal form (<u>Definition 4.1</u>) which is slightly different from the previous definition[2]. The normal form defined here is shown to admit a solitary wave in the form of the KdV soliton (<u>Proposition 4.3</u>). Then we construct a map transforming the perturbed equation into the normal form by means of the Lie transformation [3]. Finally, in the section 5, we study the interaction of two solitary waves, and show that the interaction is inelastic and produces a new solitary wave as well as the radiations (<u>Collorary 5.2</u>).

This report is an outline of the forthcoming paper "Nearly Integrable Systems II" where I give the detailed discussion and the proofs of the subjects presented here [4].

2. EVOLUTION EQUATION: PERTURBED INTEGRABLE SYSTEM

Consider the following evolution equation in the expansion form with small parameter ε,

$$u_t + X(u;\varepsilon) = u_t + \sum_{n=0}^{N} \varepsilon^n X^{(n)}(u) + O(\varepsilon^{N+1}) = 0, \tag{2.1}$$

where the vector fields $X^{(n)}(u)$ satisfy the recursion formula,

$$\begin{aligned} X^{(n)}(u) &:= \sum_{i=1}^{M(n)} a_i^{(n)} X_i^{(n)}(u) \\ &= a_1^{(n)} \partial_x^{2n+3} u + \sum_{\substack{n_1+n_2=n-2 \\ 1\le i\le M(n_1) \\ 1\le j\le M(n_2)}} c_{ij}^{n_1 n_2} X_i^{(n_1)}(u)\, D^{-1} X_j^{(n_2)}(u) \qquad \text{for } n \ge -1. \end{aligned} \tag{2.2}$$

with $D^{-1}X(u)=\int_{-\infty}^{x} X(u(x'))dx'$. Here $a_i^{(n)}$ and $c_{ij}^{n_1 n_2}$ are constants determined from the original equation, $X_i^{(n)}(u)$ expresses each monomial in $X^{(n)}(u)$ for $1\le i\le M(n)$ ($M(n)$ is the total number of independent monomials). The first few terms are given by

$$\begin{aligned} X^{(-1)}(u) &= a_1^{(-1)} u_x\,, && M(-1)=1, \\ X^{(0)}(u) &= a_1^{(0)} u_{3x} + a_2^{(0)} u u_x\,, && M(0)=2, \\ X^{(1)}(u) &= a_1^{(1)} u_{5x} + a_2^{(1)} u u_{3x} + a_3^{(1)} u_x u_{2x} + a_4^{(1)} u^2 u_x, && M(1)=4, \end{aligned} \tag{2.3}$$

$$X^{(2)}(u) = a_1^{(2)}u_{7x}+a_2^{(2)}uu_{5x}+a_3^{(2)}u_xu_{4x}+a_4^{(2)}u_{2x}u_{3x}$$
$$+a_5^{(2)}u^2u_{3x}+a_6^{(2)}uu_xu_{2x}+a_7^{(2)}u_x^3+a_8^{(2)}u^3u_x, \quad M(2)=8,$$

$$X^{(3)}(u) = a_1^{(3)}u_{9x}+a_2^{(3)}uu_{7x}+a_3^{(3)}u_xu_{6x}+a_4^{(3)}u_{2x}u_{5x}+a_5^{(3)}u_{3x}u_{4x}$$
$$+a_6^{(3)}u^2u_{5x}+a_7^{(3)}uu_xu_{4x}+a_8^{(3)}uu_{2x}u_{3x}+a_9^{(3)}u_x^2u_{3x}+a_{10}^{(3)}u_xu_{2x}^2$$
$$+a_{11}^{(3)}u^3u_{3x}+a_{12}^{(3)}u^2u_xu_{2x}+a_{13}^{(3)}uu_x^3+a_{14}^{(3)}u^4u_x+a_{15}^{(3)}u_xD^{-1}u_x^3,$$
$$M(3)=15.$$

where $u_{\ell x}=\partial_x^\ell u=\partial^\ell u/\partial x^\ell$. Note that $X^{(0)}(u)$ is the KdV vector field, and hereafter we use a standard form $X_0^{(0)}(u)=u_{3x}+6uu_x$. (This can be done by an appropriate normalization in (x,t,u)). The recursion formula (2.3) is naturally derived from asymptotic perturbation technique for weakly nonlinear-dispersive waves [1,4].

<u>Remark</u> <u>2.1</u> (scaling property): If $u(x,t) = \delta^2v(\delta x,\delta^3t) := \delta^2v(\xi,\tau)$, then $X^{(n)}(u(x,t))=\delta^{2n+5}X^{(n)}(v(\xi,\tau))$. (This is due to the scaling used in the asymptotic perturbation method.)

<u>Definition</u> <u>2.2</u> (Degree of the vector field):

$$\text{Deg } X^{(n)}(u) := 2\cdot(\#\text{ of }u\text{'s in }X^{(n)}) + (\#\text{ of }\partial_x\text{'s in }X^{(n)}) - (\#\text{ of }D^{-1}\text{'s in }X^{(n)}) = 2n+5. \quad (2.4)$$

<u>Definition</u> <u>2.3</u> (Dimension of the vector field):

$$\text{Dim } X^{(n)}(u) := M(n) = (\text{Total number of indep. monomials}). \quad (2.5)$$

3. INTEGRALS OF (2.1)

Recall that the KdV equation $u_t+X_0^{(0)}(u)=0$ has an infinite number of integrals $\{I_\ell^{(0)}[u]\}$ for $\ell\in\Gamma^{(0)}=\{0,1,2,\cdots\}$ which are given by the recursion formula (the bi-Hamiltonian structure) [5],

$$(\partial_x \nabla I_{\ell+1}^{(0)})(u) = (\mathcal{J}\nabla I_\ell^{(0)})(u), \qquad \text{with } I_0^{(0)}(u) = \frac{1}{2}\int_{-\infty}^{\infty} u dx, \tag{3.1}$$

where $(\nabla I_\ell^{(0)})(u)$ is the gradient of the functional $I_\ell^{(0)}[u]$, and the (Hamiltonian) operator $\mathcal{J}$ is given by

$$\mathcal{J} = \partial_x^3 + 2(\partial_x u + u\partial_x) . \tag{3.2}$$

The first few integrals are

$$\begin{aligned} I_1^{(0)}[u] &= \frac{1}{2}\int_{-\infty}^{\infty} u^2 dx , \\ I_2^{(0)}[u] &= \frac{1}{2}\int_{-\infty}^{\infty} (u_x^2 - 2u^3)dx , \\ I_3^{(0)}[u] &= \frac{1}{2}\int_{-\infty}^{\infty} (u_{2x}^2 - 10uu_x^2 + 5u^4)dx . \end{aligned} \tag{3.3}$$

We say that Deg $I_n^{(0)}[u]=2n+2$.

In order to see how the integrals (3.3) are deformed by the higher order terms, we look for the higher order integrals of (2.1) in the formal power series,

$$I_\ell[u;\varepsilon] = \sum_{n=0}^{N} \varepsilon^n I_\ell^{(n)}[u] + O(\varepsilon^{N+1}) \qquad \text{for } \ell \in \Gamma^{(N)}. \tag{3.4}$$

Here $\Gamma^{(N)}$ is a subset of the index set $\Gamma^{(0)}$, $\Gamma^{(N)} \subset \Gamma^{(0)}$. From the condition $dI/dt = -X \cdot \nabla I = O(\varepsilon^{N+1})$, we have the equation for $I_\ell^{(n)}[u]$,

$$(X_0^{(0)} \cdot \nabla I_\ell^{(n)})[u] = -\sum_{m=1}^{n} (X^{(m)} \cdot \nabla I_\ell^{(n-m)})[u], \tag{3.5}$$

where $X \cdot \nabla I$ is the directional derivative of $I[u]$ defined by

$$(X \cdot \nabla I)[u] = \lim_{\delta \to 0} \frac{d}{d\delta} I[u + \delta X]. \tag{3.6}$$

We look for the solution of (3.5) in terms of the functional differential polynomial (FDP).

<u>Definition 3.1</u> (FDP) [2]:

$P_M[u] :=$ (The set of the FDP of degree M)

$$:= \{u^{\ell_0} u_x^{\ell_1} u_{2x}^{\ell_2} \cdots u_{nx}^{\ell_n} \mid 2\ell_0 + 3\ell_1 + \cdots + (n+2)\ell_n = M \quad \text{with } \ell_n \geq 2\}. \tag{3.7}$$

<u>Difinition 3.2</u> (Dimension of FDP) [2]:

$$\text{Dim } P_M[u] : = (\# \text{ of indep. monomials in } P_M[u]) < \infty. \tag{3.8}$$

(i.e. $P_M[u]$ is a finite dimensional vector space).

By direct calculations, we obtain the dimension of the FDP as given in table [6].

Table : The dimension of the FDP

Deg	4	5	6	7	8	9	10	11	12	13	14	15	16	17
Dim	1	0	2	0	3	1	4	2	7	3	10	7	14	11

Note that the operator $X^{(0)}\cdot\nabla$ in (3.5) can be considered as a map on the finite dimensional vector space,

$$X^{(0)}\cdot\nabla : P_{2(n+\ell)+2} \rightarrow P_{2(n+\ell)+5} \quad . \tag{3.9}$$

It should be noted that Dim $P_{2(n+\ell)+2} \neq$ Dim $P_{2(n+\ell)+5}$. For the solvability of (3.5), we first provide;

<u>Lemma</u> <u>3.3</u> (Kernel of the operator $X^{(0)}\cdot\nabla$):

a) $(\ker X^{(0)}\cdot\nabla)\cap P[u]=\{I_\ell^{(0)}[u]\}$ for all $\ell \in \Gamma^{(0)}$ (3.10)

(i.e. $X^{(0)}\cdot\nabla I_\ell^{(0)}=0$, $I_\ell^{(0)}$ is the integral of the KdV equation)

b) Dim $\{(\ker X^{(0)}\cdot\nabla)\cap P_{2\ell}[u]\} = 1$ for each $\ell \geq 1$. (3.11)

Then, the solvability of (3.5) in $P_{2(n+\ell)+2} \ni I_\ell^{(n)}$ can be given by [6]

$$\text{Dim } P_{2(n+\ell)+5} = \text{Dim } P_{2(n+\ell)+2} - 1 \ . \tag{3.12}$$

Note $(X^{(m)}\cdot\nabla I_\ell^{(n-m)}) \in P_{2(n+\ell)+5}$, and -1 is due to the dimension of the ker $X^{(0)}\cdot\nabla$. From the table, the condition (3.12) leads to;

<u>Proposition</u> <u>3.4</u> (Existence of the higher order integrals) [6]:

a) $I_\ell^{(n)}[u]$ for $2 \le \ell+n \le 4$ $(n \ge 1)$ exist.

b) For $I_\ell^{(n)}[u]$ with $\ell+n \ge 5$ $(n \ge 1)$, (3.5) gives an overdetermined system, and the number of the compatibility conditions for the solvability is given by the difference in (3.12), i.e.

$$\mathrm{Dim}\ P_{2(n+\ell)+5} - \mathrm{Dim}\ P_{2(n+\ell)+2} + 1 > 0 . \tag{3.13}$$

(i.e. 1-condition for $I_3^{(2)}$ and for $I_2^{(3)}$, 2-conditions for $I_3^{(3)}$ etc).

It should be noted that the above discussion is general, and that many physical examples have several integrals such as mass, momentum and energy which may be given by the first three integrals $I_\ell[u;\varepsilon]$ for $\ell=0,1,2$. These physical systems, of course, satisfy the conditions for the existence of $I_\ell^{(n)}$ for $\ell=0,1,2$, but not $\ell\ge3$, in general. Here we give the condition for $I_3^{(2)}$ as an example;

<u>Collorary 3.5</u> (Existence of $I_3^{(2)}[u]$)[6]: For the existence of $I_3^{(2)}[u]$, the following relation should hold among the coefficients in $X^{(1)}(u)$ and $X^{(2)}(u)$ given by (2.3);

$$\sum_{\ell=1}^{8} r_\ell a_\ell^{(2)} + f(a^{(1)}) = 0 , \tag{3.14}$$

where the constant vector $r=(r_1,\cdots,r_8)$ and $f(a^{(1)})$ are

$$r = (-560,\ 170,\ -60,\ 24,\ -8,\ -9,\ 24,\ 3),$$

$$f(a^{(1)}) = \frac{1}{3}\{10a_1^{(1)}(240a_1^{(1)}-67a_2^{(1)}+6a_4^{(1)}) + a_2^{(1)}(4a_2^{(1)}+30a_3^{(1)}+a_4^{(1)})-3a_3^{(1)}(2a_3^{(1)}+a_4^{(1)})\}.$$

4. NORMAL FORM

Here we transform (2.1) into a simpler equation (e.g. an integrable form) called "the normal form" based on the integrals of (2.1). The purpose is to determine a prototype equation including higher order terms, and to study the behavior of solitons which will be discussed in the next section.

Definition 4.1 (Normal form): Suppose there exist $I_\ell^{(n)}[u]$ for $\ell \in \Gamma^{(N)}$ and $1 \le n \le N$, such that $I_\ell[u;\varepsilon] = \Sigma\, \varepsilon^n I_\ell^{(n)}[u] + O(\varepsilon^{N+1})$ are the integrals of (2.1). Then the normal form of (2.1) is defined by

$$v_t + X_0(v;\varepsilon) = v_t + \sum_{n=0}^{N} \varepsilon^n X_0^{(n)}(v) + O(\varepsilon^{N+1}) = 0 \tag{4.1}$$

whose integrals, say $J_\ell[v;\varepsilon]$ for $\ell \in \Gamma^{(N)}$, satisfy

$$(X_0^{(0)} \cdot \nabla J_\ell)[v;\varepsilon] = O(\varepsilon^{N+1}). \tag{4.2}$$

From Lemma 3.3, this implies that $J_\ell[v;\varepsilon] = \Sigma\, \varepsilon^n J_\ell^{(n)}[v] + O(\varepsilon^{N+1})$ can be expressed in terms of the integrals of the KdV equation, i.e.

$$J_\ell^{(n)}[v] = C_\ell^{(n)} I_{\ell+n}^{(0)}[v], \tag{4.3}$$

where $C_\ell^{(n)}$ is a constant. From the equation for the integrals, $X_0 \cdot \nabla J_\ell = O(\varepsilon^{N+1})$, the explicit forms of $X_0^{(n)}(v)$ for (2.1) having three integrals I_0, I_1 and I_2 are given by

$$X_0^{(1)}(v) = a_0^{(1)} \partial_x \nabla I_3^{(0)}(v),$$

$$X_0^{(2)}(v) = a_0^{(2)}\partial_x\nabla I_4^{(0)}(v) + \mu_{21}\partial_x R^{(2)}(v), \tag{4.4}$$

$$X_0^{(3)}(v) = a_0^{(3)}\partial_x\nabla I_5^{(0)}(v) - \mu_{21}C_2^{(1)}\ R^{(2)}(v) + \mu_{31}\partial_x R_1^{(3)}(v) + \mu_{32}\partial_x R_2^{(3)}(v).$$

where $\partial_x\nabla I_{\ell+2}^{(0)}:=Y_0^{(\ell)}$ is nothing but the Lax hierarchy of the KdV equation and $v_t+\Sigma\varepsilon^n a_0^{(n)}Y_0^{(\ell)}(v)=0$ is <u>integrable</u>. Here $a_0^{(n)}$ and $\mu_{n\ell}$ are constants, in particular, $\mu_{n\ell}$ vanishes if the additional integrals exist, e.g. $\mu_{21}=0$ if $I_3^{(2)}$ exists (i.e. μ_{21} is proportional to the left hand side of (3.14)). Also the terms $R_i^{(n)}(v)$ are nonlinear functions of v satisfying $(\partial_x R_i^{(n)})\cdot\nabla I_\ell^{(0)}=0$ for $\ell=0,1,2$. Note that the term $\partial_x R^{(2)}$ in $X_0^{(2)}$ becomes $\vartheta R^{(2)} = \{\partial_x{}^3+2(\partial_x u+u\partial_x)\}R^{(2)}=(\vartheta D^{-1})\partial_x R^{(2)}$ in the next order $X^{(3)}$. This is also true for high orders, i.e. this term changes to $(\vartheta D^{-1})(\vartheta D^{-1})\partial_x R^{(2)}$ in $X_0^{(4)}$ and so on. (Note that ϑD^{-1} is the recursion operator for the symmetry of the KdV equation [7]).

The condition (4.2) reduces the dimension of the vector field $X(u;\varepsilon)$, and we give;

<u>Definition 4.2</u> (Dimension of the normal form):

$$\text{Dim } X_0^{(n)}(v): = (\#\text{ of new constants in } X_0^{(n)}), \tag{4.5}$$

(i.e. for the above case, Dim $X_0^{(1)}(v)=1$ with $(a_0^{(1)})$, Dim $X_0^{(2)}(v)=2$ with $(a_0^{(2)},\mu_{21})$, Dim $X_0^{(3)}(v)=3$ with $(a_0^{(3)},\mu_{31},\mu_{32})$).

Note that this dimension can be calculated from the table for the dimension of the FDP. Also (4.2) leads to the following proposition for the solitary wave solution of the normal form;

Proposition 4.3 (Solitary wave solution of the normal form): The normal form admits a solitary wave in the form of the KdV soliton, i.e.

$$v(x,t) = s(x-\lambda t) = 2\kappa^2 \operatorname{sech}^2 \kappa(x-\lambda t), \tag{4.6}$$

where κ and λ are constants (λ depends on κ and $a_0^{(n)}$ in (4.4)).

Note that the KdV soliton is the solution of the variational equation $\nabla I_2^{(0)}+4\kappa^2\nabla I_1^{(0)}=0$, and $\nabla J_2+4\kappa^2\nabla J_1=\{\sum_0 \varepsilon^n \alpha_n (D^{-1}\mathcal{J})^n\}(\nabla I_2^{(0)}+4\kappa^2\nabla I_1^{(0)})=0$. This implies the Proposition 4.3. ($D^{-1}\mathcal{J}$ is the recursion operator for the integrals of the KdV equation[7].)

The main result in this section is the following theorem;

Theorem 4.4 (Normal form theorem): There exists a formal change of coordintes which transforms the perturbed equation (2.1) into the corresponding normal form (4.1).

The change of coordinates may be constructed by the Lie transformation technique [3], i.e.

$$u = T_\varepsilon(v) = \sum_{n=0}^{N} \varepsilon^n T^{(n)}(v) + O(\varepsilon^{N+1}), \tag{4.7}$$

where the transformation operator T_ε satisfies the operator equation,

$$\frac{dT_\varepsilon}{d\varepsilon} = L_\varepsilon T_\varepsilon, \quad T_0 = T^{(0)} = \text{identity}. \tag{4.8}$$

Here the operator L_ε defines the generating function $\varphi(v;\varepsilon)$,

$$L_\varepsilon(v) = \varphi(v;\varepsilon). \tag{4.9}$$

This is a generalization of the Lie canonical transformation where $L_\varepsilon(v) = \mathcal{J}\nabla S(v;\varepsilon)$ with a generator $S[u;\varepsilon]$, and (4.8) gives $\partial u/\partial\varepsilon = \{u,S\}$ with the Poisson bracket $\{\cdot,\cdot\}$ generated by the Hamiltonian operator $\mathcal{J}$[8]. We give an explicit form of the transformation up to ε^3 (the higher order calculations can be carried out straightforwardly, but with care). Expanding $\varphi(v;\varepsilon)$ in a power series of ε,

$$\varphi(v;\varepsilon) = \varphi^{(0)}(v)+2\varepsilon\varphi^{(1)}(v)+\cdots = \frac{\partial}{\partial\varepsilon}\sum_{n=1}\varepsilon^n\varphi^{(n)}, \tag{4.10}$$

the transformation rule for the (contravariant) vector field, i.e.

$$X(T_\varepsilon(v);\varepsilon) = X_0(v;\varepsilon)\cdot\nabla T_\varepsilon(v) \tag{4.11}$$

gives the equations for $\varphi^{(n)}(v)$,

$$\begin{aligned}
[X_0^{(0)}, \varphi^{(1)}] &= X^{(1)} - X_0^{(1)}\ ,\\
[X_0^{(0)}, \varphi^{(2)}] &= X^{(2)} - X_0^{(2)} + \tfrac{1}{2}[\varphi^{(1)}, X^{(1)} + X_0^{(1)}],\\
[X_0^{(0)}, \varphi^{(3)}] &= X^{(3)} - X_0^{(3)} + \tfrac{1}{3}[\varphi^{(1)}, 2X^{(2)} + X_0^{(2)}]\\
&\quad + \tfrac{1}{3}[\varphi^{(2)}, X^{(1)} + 2X_0^{(1)}]\\
&\quad + \tfrac{1}{6}[\varphi^{(1)}, [\varphi^{(1)}, X^{(1)}]],
\end{aligned} \tag{4.12}$$

where $[X,\varphi]:=X\cdot\nabla\varphi-\varphi\cdot\nabla X$ (the Lie bracket). Here the transformation $T^{(n)}(v)$ in (4.7) is expressed in terms of the generating function $\varphi^{(n)}(v)$, i.e.

$$T^{(1)}(v) = \varphi^{(1)}(v),$$

$$T^{(2)}(v) = \varphi^{(2)}(v) + \frac{1}{2}(\varphi^{(1)}\cdot\nabla\varphi^{(1)})(v), \tag{4.13}$$

$$T^{(3)}(v) = \varphi^{(3)}(v) + \frac{1}{3}(\varphi^{(1)}\cdot\nabla\varphi^{(2)} + 2\varphi^{(2)}\cdot\nabla\varphi^{(1)})(v)$$
$$+ \frac{1}{6}(\varphi^{(1)}\cdot\nabla(\varphi^{(1)}\cdot\nabla\varphi^{(1)}))(v).$$

Noting that $[X_0^{(0)},Y_0^{(\ell)}]=0$ $(Y_0^{(\ell)}:=\partial_x\nabla I_{\ell+2}^{(0)})$ for $\ell\geq-1$, the solutions of (4.12) are given by the linear combinations of the terms,

$$\varphi^{(1)}(v) = (v_{2x},\ v^2,\ v_xD^{-1}v),$$

$$\varphi^{(2)}(v) = (v_{4x},\ vv_{2x},\ v_x^2,\ v^3,\ v_xD^{-1}v^2,\ X_0^{(0)}D^{-1}v),$$

$$\varphi^{(3)}(v) = (v_{6x},\ vv_{4x},\ v_xv_{3x},\ v_{2x}^2,\ v^2v_{2x},\ vv_x^2,\ v^4$$
$$v_xD^{-1}v^3,\ v_xD^{-1}v_x^2,\ X_0^{(0)}D^{-1}v^2,\ X_0^{(1)}D^{-1}v\ ;\ X^{(1)}D^{-1}v). \tag{4.14}$$

Theorem 4.4 can be proven purely algebraically by counting the dimensions of $\varphi^{(n)},X^{(n)},X_0^{(n)}$ and the number of constraints for the existence of $I_\ell^{(n)}$ (the dimension of $\varphi^{(n)}$ is defined in the same way as (2.5)). Namely the existence of $\varphi^{(n)}$ derives from

$$\text{Dim}\ \varphi^{(n)} \geq \text{Dim}\ X^{(n)} - \text{Dim}\ X_0^{(n)} - (\#\ \text{of constraints}). \tag{4.15}$$

<u>Remark</u> <u>4.5</u> (Generaling function $\varphi^{(n)}$): Main part of the solution $\varphi^{(n)}$, in which $[X_0^{(0)},\varphi^{(n)}]\in X^{(n)}$, consists of two parts;

a) the <u>differential</u> polynomial of degree 2n+2,

b) the nonlocal terms $Y_0^{(\ell)}(v)D^{-1}P_{2(n-\ell)+2}(v)$ for $\ell\geq -1$ where $Y_0^{(\ell)}=\partial_x \nabla I_{\ell+2}^{(0)}$ and $P_{2(n-\ell)+2}$ is the differential polynomial of degree $2(n-\ell)+2$.

Note that the other part, $X^{(1)}D^{-1}v$ in $\varphi^{(3)}$, does not produce the vector fields $X^{(3)}$ in (2.3), but needs to cancel the term generated by $[\varphi^{(2)},X^{(1)}]$ in the right hand side of $\varphi^{(3)}$-equation. We also note that the nonlocal terms in $\varphi^{(n)}$, e.g. $v_x D^{-1}v$, lead to additional phase shift of the KdV solitons due to the collision with solitary waves and radiations, e.g. $v_x D^{-1}v$ with two solitary waves, $v(x,t) = s_1(x-x_1) + s_2(x-x_2)$ for large $|x_1-x_2|$, contains the term $s_{1x}\int_{-\infty}^{\infty} s_2 dx$ which gives the shift of the position of s_1-solitary wave.

As a consequence of the Theorem 4.4, we obtain;

Collorary 4.6 [6]: In general, the perturbed equation (2.1) can be transformed into an integrable form up to first order, but not to second order. The integrable normal form up to second order results from the existence of the higher order integral $I_3^{(2)}$, or the condition (3.14).

We also note, from the solitary wave of the normal form (4.6) and the transformation (4.13);

Proposition 4.7 (Solitary wave for (2.1))[9]: Eq.(2.1) has the solitary wave in the following form,

$$u(x,t) = \sum_{n=1}^{N+1} \sum_{j=1}^{n} \varepsilon^{n-1}\beta_j^{(n)}\kappa^{2n}\text{sech}^{2j}\kappa(x-\lambda t-x_0) + O(\varepsilon^{N+1}); \tag{4.16}$$

where $\beta_j^{(n)},\kappa,\lambda$ and x_0 are constants.

The form (4.16) has been found by Fenton in the study of the greatest solitary wave of the shallow water waves [10].

5. SOLITARY WAVE INTERACTION

Here we study the interaction of two solitary waves by employing the method of the inverse scattering transform [11]. A key of the method is to write the solution of the normal form in terms of the squared eigenfunctions [5], i.e.

$$v(x,t) = 4\sum_{j=1}^{M}\kappa_j C_j \phi^2(x,t;i\kappa_j) + \frac{2i}{\pi}\int_{-\infty}^{\infty} kr(k,t)\phi^2(x,t;k)dk \tag{5.1}$$

where the set of the functions $\{\kappa_j(t)\geq 0,\ C_j(t)\geq 0\}$ for $j=1,\cdots,M$, and $r(k,t)$ for $k\in R$ is called the scattering data $S(t)$ of the Schrödinger equation for the eigenfunction $\phi(x,t;k)$ [5],

$$\phi_{2x} + (v + k^2)\phi = 0 \tag{5.2}$$

with the boundary condition,

$$\phi(x,t;k) \to e^{-ikx}, \quad \text{as} \quad x \to -\infty .$$

Note that the squared eigenfunction ϕ^2 satisfies

$$\mathcal{J}\phi^2 = \{\partial_x^3 + 2(\partial_x v + v\partial_x)\}\phi^2 = -4k^2\partial_x\phi^2. \tag{5.3}$$

The inverse scattering transform is based on the fact that the mapping between $S(t)$ and $v(x,t)$ is one-to-one [5]. For example, one solitary wave solution corresponds to the case where $M=1$, $k=i\kappa_1$, $C_1>0$ (κ_1, C_1:constants) and $r(k,t)=0$. The time evolution of the scattering data is determined by

$$\frac{dS(t)}{dt} = (v_t \cdot \nabla S)[v] = -(X_0 \cdot \nabla S)[v]. \tag{5.4}$$

(Note that S is a functional of v.) Here the gradient ∇S is evaluated as the squared eigenfunction [9], e.g.

$$\nabla r(k,t) = -\frac{1}{2ika^2(k,t)}\phi^2(x,t;k), \tag{5.5}$$

where a(k,t) is given by

$$a(k,t) = \left(\prod_{j=1}^{M}\frac{k-i\kappa_j}{k+i\kappa_j}\right)\exp\left(\frac{i}{2\pi}\int_{-\infty}^{\infty}\frac{\ln(1-|r(k',t)|^2)}{k'-k-i0}dk'\right).$$

With (5.1), the integrals of the normal form can be expressed in terms of the scattering data;

$$J_\ell[v;\varepsilon] = \sum_{n=0}^{N}\varepsilon^n C_\ell^{(n)} I_{\ell+n}^{(0)}[v] + O(\varepsilon^{N+1}), \tag{5.6}$$

where $I_m^{(0)}[v]$ is given by [5],

$$I_m^{(0)}[v] = \frac{2^{2m+1}}{2m+1}(-1)^{m+1}\left[\sum_{j=1}^{M}\kappa_j^{2m+1} - (-1)^m\Delta_m\right] \tag{5.7}$$

with the radiation part Δ_m

$$\Delta_m(t) = -\frac{2m+1}{2\pi}\int_0^\infty k^{2m}\ln(1-|r(k,t)|^2)dk \geq 0 . \tag{5.8}$$

From the integrals (5.6), we now compute the shift of the soliton parameters κ_j due to the collision. As an initial condition, we consider two solitary waves, i.e. M=2 and r(k,0)=0, as shown in Fig.1.

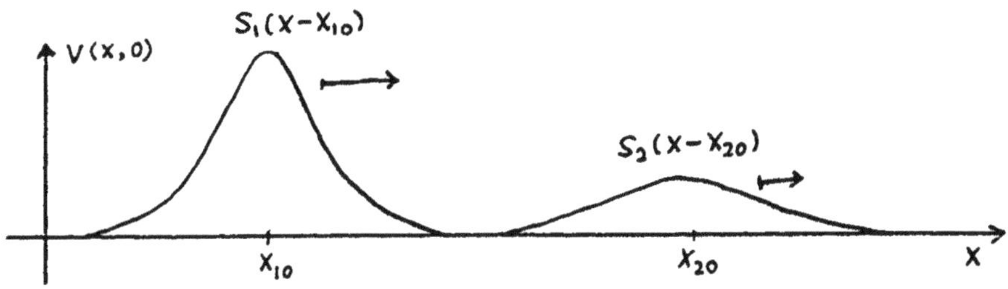

Fig.1. The initial condition: $v(x,0)=s(x-x_{10})+s_2(x-x_{20})$

Here the solitary waves are given by

$$s_j(z) = 2\kappa_j^2\text{sech}^2\kappa_j z \qquad \text{with } \kappa_1 > \kappa_2 \quad , \tag{5.9}$$

where the larger solitary wave propagates faster than the smaller one. Assuming the shift of the soliton parameters $\kappa_j \to \kappa_j + \Delta\kappa_j$ and the production of the radiations $\Delta_m \neq 0$ due to the collision, we obtain, from (5.6), the equations for $\Delta\kappa_j$ in terms of Δ_m;

<u>Proposition</u> <u>5.1</u> (Shift of the soliton parameters):

$$\Delta\kappa_1 = \frac{5\kappa_2{}^2\Delta_1+3\Delta_2}{15\kappa_1{}^2(\kappa_1{}^2-\kappa_2{}^2)} \quad \geq 0 ,$$

$$\Delta\kappa_2 = -\frac{5\kappa_1{}^2\Delta_1+3\Delta_2}{15\kappa_2{}^2(\kappa_1{}^2-\kappa_2{}^2)} \leq 0 \ , \tag{5.10}$$

$$\Delta\kappa_3 = \frac{15\kappa_1{}^2\kappa_2{}^2\Delta_0+5(\kappa_1{}^2+\kappa_2{}^2)\Delta_1+3\Delta_2}{15\kappa_1{}^2\kappa_2{}^2} \geq 0 \ .$$

In the derivation of (5.10) from (5.6), one should include the possibility of the production of new solitary wave with κ_3 which is zero at t=0. From this proposition, we immediately obtain;

Collorary 5.2 (Interaction property):

a) The amplitude of the larger solitary wave is increased, and contrary that of the smaller one is decreased by the interaction. The amount of the shifts has the property, $\Delta\kappa_1 < |\Delta\kappa_2|$ for $\kappa_1 > \kappa_2$, $((\kappa_1/\kappa_2)^2 < |\Delta\kappa_2/\Delta\kappa_1| < (\kappa_1/\kappa_2)^4)$, i.e. the energy expense of the smaller solitary wave is more than the energy gain of the larger one.

b) The interaction produces new solitary wave as well as the radiations.

It should be noted that this property holds commonly for the system which does not have enough integrals in the sense of the section 3. These results except the solitary-wave production are consistent with the numerical observation done by Bona et. al [12] for the regularized long wave (a model of the shallow water waves). It may be difficult to observe new solitary wave from the numerical result, since this solitary wave has long width and small amplitude $(\Delta\kappa_3)^2$.

The inelasticity generating the radiation is due to the non-existence of the higher order integrals, e.g. $I_3^{(2)}[u]$. Let us compute Δ_m for the case where $I_3^{(2)}[u]$ does not exist, i.e. the second order equation of the normal form contains the term $\partial_x R^{(2)}(v)$, i.e. (4.4). From (5.4) and (5.5), we have

$$\frac{dr(k,t)}{dt} = \frac{1}{2ika^2(k,t)}\int_{-\infty}^{\infty}\phi^2(x,t;k)X_0(v(x,t))dx$$

$$= i\omega_k r(k,t)+\varepsilon^2\frac{\mu_{21}}{2ika^2(k,t)}\int_{-\infty}^{\infty}\phi^2(x,t;k)\partial_x R^{(2)}(v(x,t))dx + O(\varepsilon^3). \tag{5.11}$$

where $\omega_k=8k^3(1-\varepsilon 4a_1^{(1)}k^2+\varepsilon^2 16a_1^{(2)}k^4)+O(\varepsilon^3)$. Here we have used the relations (3.1) and (5.3). Noting $r(k,t)=O(\varepsilon^2)$, from (5.11) we obtain the equation of $\Delta_m(t)$ in (5.8),

$$\Delta_m(t) = \frac{2m+1}{2\pi}\int_0^{\infty}D_m(k,t)dk + O(\varepsilon^5) \quad , \tag{5.12}$$

where

$$D_m(k,t) = -k^{2m}\ell n(1-|r(k,t)|^2) = k^{2m}|r(k,t)|^2 + O(\varepsilon^5)$$

$$= \varepsilon^4\ \frac{\mu_{21}^2 k^{2(m-1)}}{4}\left|\int_0^t d\tau\ e^{-i\omega_k\tau}\int_{-\infty}^{\infty}dx\phi^2(x,\tau;k)\partial_x R^{(2)}(v(x,\tau))\right|^2 + O(\varepsilon^5).$$

We integrate (5.12) numerically by means of the perturbation, i.e. at the leading order we assume $v(x,t)$ to be 2-soliton solution of the KdV equation and $\phi(x,t;k)$ to be the corresponding eigenfunction of (5.2). The results are shown in Fig.2 for various values of κ_1 and κ_2 (we fix $\kappa_1+\kappa_2=1$). Here we plot only the behavior of $\Delta_1(t)$, and the other $\Delta_m(t)$ behave similarly.

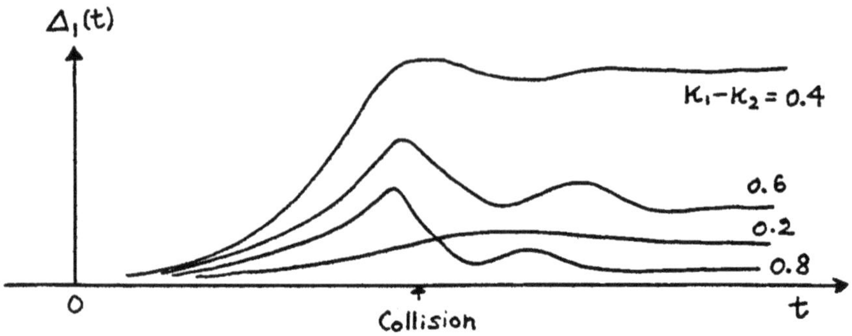

Fig.2 The time evolution of $\Delta_1(t)$ for $\kappa_1-\kappa_2=0.2$, 0.4, 0.6 and 0.8 $(\kappa_1+\kappa_2=1)$.

The graph in Fig.3 shows the asymptotic values of Δ_1 versus various values of $\kappa_1-\kappa_2$ with $\kappa_1+\kappa_2=1$.

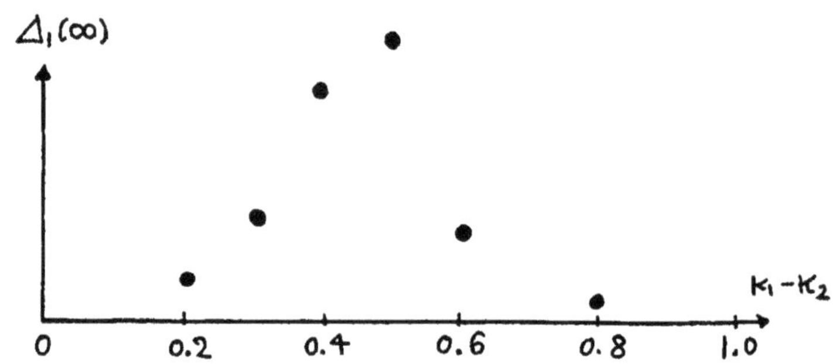

Fig.3 The asymptotic value of $\Delta_1(t)$

One can see from Fig.3 that $\Delta_1(\infty)$ takes the maximum value about $\kappa_1-\kappa_2\simeq 0.5$. This feature was also observed in the numerical calculation in [2]. We will report the explicit calculations for the regularized long wave equation elsewhere [13].

Before ending this section, we notice that recently, Kivshar and Malomed used this perturbed inverse scattering technique to show the inelastic collision of solitons for the nonlinear Schrödinger equation with higher order nonlinear term [14].

Acknowledgements

The author is grateful to H.H. Chen and C.R. Menyuk for useful comments and discussions. He also wishes to thank F. Khorrami and G.C. Santos for numerical evaluations of the radiation part. This work was partially supported by NSF grant, No. DMS 8521055.

References

1. Taniuti, T., Prog. Theor. Phys. 55, 1 (1974);
   Karpman, V.I., Nonlinear-waves in dispersive media (Pergamon, Oxford 1975).
2. Kodama, Y., Physica 16D, 14 (1985).
3. Nayfeh, A., Perturbation methods (Wiley, New York, 1973).
4. Kodama, Y., Nearly Integrable System II (in preparation).
5. Ablowitz, M.J. and Segur, H., Solitons and the inverse scattering transform (SIAM, Philadelphia, 1981).
   Newell, A.C., Solitons in mathematics and physics (SIAM, Phyladelphia, 1985).
6. Kodama, Y., Phys. Lett. 112A, 193 (1985).
7. Fuchssteiner, B. and Focas, A.S., Physica 4D, 47 (1981).
   Olver, P.J., Applications of Lie groups to differential equations (Springer-Verlag, New York 1986).
8. Carry, J.R., Phys. Rep. 79, 129 (1981).
9. Kodama, Y. and Taniuti, T., J. Phys. Soc. Japan 45, 298 (1978).
10. Fenton, J.D., J. Fluid Mech. 53, 257 (1972).
    Longuet-Higgins, M.S. and Fenton, J.D., Proc. R. Soc. Lond. A340, 471 (1974).
11. Kaup, D.J. and Newell, A.C., Proc. R. Soc. Lond. A361, 413 (1978).
    Karpman, V.I., Physica Scripta 20, 462 (1979).
12. Bona, J.L., Prichard, W.G. and Scatt, L.R., Phys. Fluids 23, 438 (1980).
13. Kodama, Y., On the Solitary wave interaction (in preparation).
14. Kivshar, Y.S. and Malomed, B.A., Phys. Lett. 115A, 377 (1986).

List of Participants

M.J. Ablowitz (Clarkson University, Potsdam, N.Y.)

R.W. Beals (Yale University, New Haven)

F. Calogero (Universita Degli Studi, Rome)

H.W.Capel (University of Amsterdam)

H.H. Chen (University of Maryland, College Park)

P. Clarkson (Clarkson University, Potsdam, N.Y.)

A.S. Fokas (Clarkson University, Potsdam, N.Y.)

J.P. Francoise (Universitié de Paris-Sud)

B. Fuchssteiner (University of Paderborn)

D.J. Kaup (Clarkson University, Potsdam, N.Y.)

Y. Kodama (Nagoya University)

Y. Kosmann-Schwarzbach (Universite de Lille, Villeneuve)

M. Kruskal (Princeton University)

Y.S Li (University of Science and Technology of China, Hefei)

F. Magri (Politecnico, Milano)

A.I. Nachman (University of Rochester)

F.W. Nijhoff (University of Utrecht)

W. Oevel (University of Paderborn)

C. Pöppe (Universität Heidelberg)

A. Ramani (Ecole Polytechnique, Palaiseau)

S.N.M. Ruijssenaars (CWI, Amsterdam)

P.C. Sabatier (Universite du Languedoc, Montpellier)

P. Santini (Universita di Roma I, Rome)

J. Satsuma (Tokyo University)

G. Schwake (University of Paderborn)

F. Schwarz (GMD, St.Augustin)

W. Strampp (University of Kassel)

E.R. Tracy (College of William and Mary, Williamsburg)

P. Winternitz (Université de Montreal)

W. Wiwianka (University of Paderborn)

www.ingramcontent.com/pod-product-compliance
Lightning Source LLC
LaVergne TN
LVHW051052210926
848486LV00002B/86
9789971502539